Process Simulation and Control

Process Simulation and Control Using Aspen™

SECOND EDITION

Amiya K. Jana

Assistant Professor
Department of Chemical Engineering
Indian Institute of Technology Kharagpur

PHI Learning Private Limited

Delhi-110092
2014

₹ 350.00

PROCESS SIMULATION AND CONTROL USING ASPEN™, Second Edition
Amiya K. Jana

ISBN-978-81-203-4568-3

The export rights of this book are vested solely with the publisher.

Third Printing (Second edition)   ...   ...   **July, 2014**

Published by Asoke K. Ghosh, PHI Learning Private Limited, Rimjhim House, 111, Patparganj Industrial Estate, Delhi-110092 and Printed by Rajkamal Electric Press, Plot No. 2, Phase IV, HSIDC, Kundli-131028, Sonepat, Haryana.

To my daughter
Lekhoni

Contents

Part II Chemical Plant Simulation using Aspen Plus™

Part III Dynamics and Control using Aspen Dynamics™

Preface

Due to a rapid decline in the cost of computers and improvement in its quality and speed (doubling every 18 months, according to Moore's Law), the use of simulation tool for dynamics prediction and control has become indispensable for industrial plants. This second edition of the book, *Process Simulation and Control Using Aspen*,™ is the outcome of encouraging feedback received from the teachers and students alike. All the features of the first edition have been retained in this second edition, and several new topics and new sections are incorporated to broaden the coverage. The following chapters have been expanded and modified by including several sections and topics:

- Section 1.3 on *Stepwise Aspen Plus Simulation of Flash Drums* is thoroughly updated (Chapter 1)

- Section 3.2 on *Aspen Plus Simulation of the Binary Distillation Columns* is upgraded, a new section on *Simulation of a Reactive Distillation Column* is added (Section 3.6), and a new topic on *Column Sizing* is introduced (Chapter 3)

- A new section on *Aspen Simulation of a Petlyuk Column with Streams Recycling* is included (Chapter 4)

In addition to these above-mentioned inclusions, numerous new problems and illustrations are also incorporated in the book. As the book is meant for undergraduate and postgraduate students of engineering along with the research scientists and practising engineers, considerable modifications have been done to keep them abreast of the latest technologies and trends.

I gratefully acknowledge the useful suggestions provided by my colleague Prof. S. Ray. I wish to express my heartfelt gratitude to Prof. N.C. Pradhan, Head of the Department, and all my esteemed colleagues for gifting so great academic environment in our department. The undergraduate and postgraduate students of IIT Kharagpur have contributed to the book by discussing and exchanging their viewpoints on various topics. Their youthful enthusiasm and genuine interest in the subject encouraged me to complete this project in a satisfactory manner.

I am thankful to my mother and wife for their enormous help and support, without which this project would never have been possible.

Finally, I am greatly indebted to the editorial staff of PHI Learning for their constant encouragement and unstinted support in shaping the book into its present form. All comments and suggestions for improvement of the book would be gratefully acknowledged.

Amiya K. Jana

Preface to the First Edition

The future success of the chemical process industries mostly depends on the ability to design and operate complex, highly interconnected plants that are profitable and that meet quality, safety, environmental and other standards. To achieve this goal, the software tools for process simulation and optimization are increasingly being used in industry.

By developing a computer program, it may be manageable to solve a model structure of a chemical process with a small number of equations. But as the complexity of a plant integrated with several process units increases, the solution becomes a challenge. Under these circumstances, in recent years, the use of process *flowsheet simulator* has been increased considerably as it helps in solving the problems faster and in more reliable manner. In this book, the Aspen^{TM} software package has been used for steady state simulation, process optimization, dynamics and closed-loop control.

To improve the design, operability, safety, and productivity of a chemical process with minimizing capital and operating costs, the engineers concerned must have a solid knowledge of the process behaviour. The process dynamics can be predicted by solving the mathematical model equations. Within a short time period, this can be achieved quite accurately and efficiently by using Aspen flowsheet simulator. This software tool is not only useful for plant simulation but can also automatically generate several control structures, suitable for the used process flow diagram. In addition, the control parameters, including the constraints imposed on the controlled as well as manipulated variables, are also provided by Aspen to start the simulation run. However, we have the option to modify or even replace them.

This well organized book is divided into three parts. Part I (Steady State Simulation and Optimization using Aspen Plus^{TM}) includes three chapters. Chapter 1 presents the introductory concepts with solving the flash chambers. The computation of bubble point and dew point temperatures is also focused. Chapters 2 and 3 are devoted to simulation of several reactor models and separating column models, respectively.

Part II (Chemical Plant Simulation using Aspen Plus^{TM}) consists of only one chapter (Chapter 4). It addresses the steady state simulation of large chemical plants. Several individual processes are interconnected to form the chemical plants. The Aspen Plus^{TM} simulator is used in both Part I and Part II.

The Aspen Dynamics™ package is employed in Part III (Dynamics and Control using Aspen Dynamics™) that comprises Chapters 5 and 6. Chapter 5 is concerned with the dynamics and control of flow-driven chemical processes. In the closed-loop control study, the servo as well as regulatory tests have been conducted. Dynamics and control of pressure-driven processes have been discussed in Chapter 6.

The target readers for this book are undergraduate and postgraduate students of chemical engineering. It will be also helpful to research scientists and practising engineers.

Amiya K. Jana

Acknowledgements

It is a great pleasure to acknowledge the valuable contributions provided by many of my well-wishers. I wish to express my heartfelt gratitude and indebtedness to Prof. A.N. Samanta, Prof. S. Ganguly and Prof. S. Ray, Department of Chemical Engineering, IIT Kharagpur. I am also grateful to Prof. D. Mukherjee, Head, Department of Chemical Engineering, IIT Kharagpur. My special thanks go to all of my colleagues for having created a stimulating atmosphere of academic excellence. The chemical engineering students at IIT Kharagpur also provided valuable suggestions that helped to improve the presentations of this material.

I am greatly indebted to the editorial staff of PHI Learning Private Limited, for their constant encouragement and unstinted efforts in bringing the book in its present form.

No list would be complete without expressing my thanks to two most important people in my life—my mother and my wife. I have received their consistent encouragement and support throughout the development of this manuscript.

Any further comments and suggestions for improvement of the book would be gratefully acknowledged.

Part I

Steady State Simulation and Optimization using Aspen Plus™

CHAPTER 1

Introduction and Stepwise Aspen Plus™ Simulation
Simple Examples

1.1 ASPEN: AN INTRODUCTION

By developing a computer program, it may be manageable to solve a model structure of a chemical process with a small number of equations. However, as the complexity of a plant integrated with several process units increases, solving a large equation set becomes a challenge. In this situation, we usually use the process flowsheet simulator, such as Aspen Plus™ (AspenTech), ChemCad™ (Chemstations), HYSYS™ (Hyprotech) and PRO/II™ (SimSci-Esscor). In 2002, Hyprotech was acquired by AspenTech and at present, most widely used commercial process simulation software is the Aspen software.

During the 1970s, the researchers have developed a novel technology at the Massachusetts Institute of Technology (MIT) with United States Department of Energy funding. The undertaking, known as the *Advanced System for Process Engineering (ASPEN) Project*, was originally intended to design nonlinear simulation software that could aid in the development of synthetic fuels. In 1981, AspenTech, a publicly traded company, was founded to commercialize the simulation software package. AspenTech went public in October 1994 and has acquired 19 industry-leading companies as part of its mission to offer a complete, integrated solution to the process industries (http://www.aspentech.com/corporate/careers/faqs.cfm#whenAT).

The sophisticated Aspen software tool can simulate large processes with a high degree of accuracy. It has a model library that includes mixers, splitters, phase separators, heat exchangers, distillation columns, reactors, pressure changers, manipulators, etc. By interconnecting several unit operations, we are able to develop a process flow diagram (PFD) for a complete plant. To solve the model structure of either a single unit or a chemical plant, required Fortran codes are built-in in the Aspen simulator. Additionally, we can also use our own subroutine in the Aspen package.

The Aspen simulation package has a large experimental databank for thermodynamic and physical parameters, and this is the most important advantage of Aspen compared to the other simulation packages. Therefore, we need to give limited input data for solving even a process plant having a large number of units with avoiding human errors and spending a minimum time.

Aspen simulator has been developed for the simulation of a wide variety of processes, such as chemical and petrochemical, petroleum refining, polymer, and coal-based processes. Previously, this flowsheet simulator was used with limited applications. Nowadays, different Aspen packages are available for simulations with promising performance. Briefly, some of them are presented below.

Aspen Plus—This process simulation tool is mainly used for steady state simulation of chemicals, petrochemicals and petroleum industries. It is also used for performance monitoring, design, optimization and business planning.

Aspen Dynamics—This powerful tool is extensively used for dynamics study and closed-loop control of several process industries. Remember that Aspen Dynamics is integrated with Aspen Plus.

Aspen BatchCAD—This simulator is typically used for batch processing, reactions and distillations. It allows us to derive reaction and kinetic information from experimental data to create a process simulation.

Aspen Chromatography—This is a dynamic simulation software package used for both batch chromatography and chromatographic simulated moving bed processes.

Aspen Properties—It is useful for thermophysical properties calculation.

Aspen Polymers Plus—It is a modeling tool for steady state and dynamic simulation, and optimization of polymer processes. This package is available within Aspen Plus or Aspen Properties rather than via an external menu.

Aspen HYSYS—This process modeling package is typically used for steady state simulation, performance monitoring, design, optimization and business planning for petroleum refining, and oil and gas industries.

It is clear that Aspen simulates the performance of the designed process. A solid understanding of the underlying chemical engineering principles is needed to supply reasonable values of input parameters and to analyze the results obtained. For example, a user must have good idea of the distillation column behavior before attempting to use Aspen for simulating that column. In addition to the process flow diagram, required input information to simulate a process are: setup, components, properties, streams and blocks.

1.2 GETTING STARTED WITH ASPEN PLUS SIMULATION

Aspen Plus is a user-friendly steady state process flowsheet simulator. It is extensively used both in the educational arena and industry to predict the behavior of a process by using material balance equations, equilibrium relationships, reaction kinetics, etc. Using Aspen Plus, which is a part of Aspen software package, we will mainly perform in this book the steady state simulation and optimization. For process dynamics and closed-loop control, Aspen Dynamics (formerly DynaPLUS) will be used in subsequent chapters. The standard Aspen notation is used throughout this book. For example, distillation column stages are counted from the top of the column: the condenser is Stage 1 and the reboiler is the last stage.

As we start Aspen Plus from the *Start* menu or by double-clicking the Aspen Plus icon on our desktop, the *Aspen Plus Startup* dialog appears. There are three choices and we can create our work from scratch using a *Blank Simulation*, start from a *Template* or *Open an Existing Simulation*. Let us select the *Blank Simulation* option and click *OK* (see Figure 1.1).

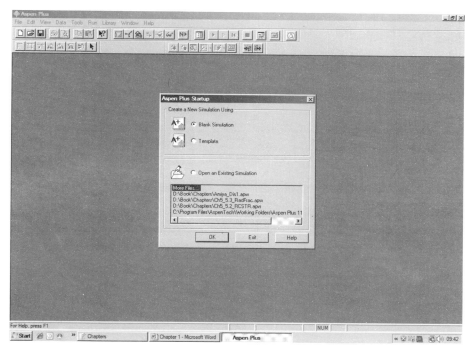

FIGURE 1.1

The simulation engine of Aspen Plus is independent from its Graphical User Interface (GUI). We can create our simulations using the GUI at one computer and run them connecting to the simulation engine at another computer. Here, we will use the simulation engine at 'Local PC'. Default information are *OK*.

Hit *OK* in the *Connect to Engine* dialog (Figure 1.2). Notice that this step is specific to the installation.

The next screen shows a blank *Process Flowsheet Window*. The first step in developing a simulation is to create the process flowsheet. *Process flowsheet* is simply defined as a blueprint of a plant or part of it. It includes all input streams, unit operations, streams that interconnect the unit operations and the output streams. Several process units are listed by category at the bottom of the main window in a toolbar known as the *Model Library*. If we want to know about a model, we can use the Help menu from the menu bar. In Figure 1.3, different useful items are highlighted briefly.

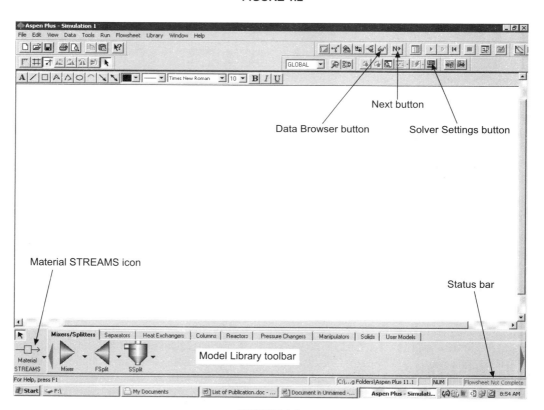

FIGURE 1.2

FIGURE 1.3

To develop a flowsheet, first choose a unit operation available in the *Model Library*. Proprietary models can also be included in the flowsheet window using *User Models* option. Excel workbook or Fortran subroutine is required to define the user model. In the subsequent step, using *Material STREAMS* icon, connect the inlet and outlet streams with the process. A process is called as a *block* in Aspen terminology. Notice that clicking on *Material STREAMS*, when we move the cursor into the flowsheet area red and blue arrows appear around the model block. These arrows indicate places to attach streams to the block. Red arrows indicate required streams and blue arrows are optional.

When the flowsheet is completed, the status message changes from *Flowsheet Not Complete* to *Required Input Incomplete*. After providing all required input data using input forms, the status bar shows *Required Input Complete* and then only the simulation results are obtained. In the *Data Browser*, we have to enter information at locations where there are red semicircles. When one has finished a section, a blue checkmark appears.

In this chapter, simple flash chamber problems are solved in Subsection 1.3. presenting a detailed stepwise simulation procedure in Aspen Plus. In addition, the problems related to azeotrope formation and data regression have also been discussed with their solution approaches subsequently.

1.3 STEPWISE ASPEN PLUS SIMULATION OF FLASH DRUMS

1.3.1 Built-in Flash Drum Models

In the *Model Library*, there are five built-in separators. A brief description of these models is as follows:

Flash 2: It is used for equilibrium calculations of two-phase (vapor-liquid) and three-phase (vapor-liquid-liquid) systems. In addition to inlet stream(s), this separator can include three product streams: one liquid stream, one vapor stream and an optional water decant stream. It can be used to model evaporators, flash chambers and other single-stage separation columns.

Flash 3: It is used for equilibrium calculations of a three-phase (vapor-liquid-liquid) system. This separator can handle maximum three outlet streams: two liquid streams and one vapor stream. It can be used to model single-stage separation columns.

Decanter: It is typically used for liquid-liquid distribution coefficient calculations of a two-phase (liquid-liquid) system. This separator includes two outlet liquid streams along with inlet stream(s). It can be used as the separation columns. If there is any tendency of vapor formation with two liquid phases, it is recommended to use Flash3 instead of Decanter.

Sep: It is a multi-outlet component separator since two or more outlet streams can be produced from this process unit. It can be used as the component separation columns.

Sep 2: It is a two-outlet component separator since two outlet streams can be withdrawn from this process unit. It is also used as the component separation columns.

At this point it is important to mention that for additional information regarding a built-in model, select that model icon in the *Model Library* toolbar and then press F1 on the keyboard.

1.3.2 Simulation of a Flash Drum

Problem statement

A 100 kmol/h feed consisting of 10, 20, 30, and 40 mole% of propane, *n*-butane, *n*-pentane, and *n*-hexane, respectively, enters a flash chamber at 105 psia and 200°F. The flash drum (Flash2) is shown in Figure 1.4 and it operates at 100 psia and 200°F. Applying the SYSOP0 property method, compute the composition of the exit streams.

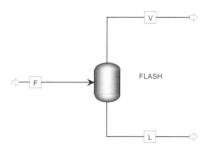

FIGURE 1.4 A flowsheet of a flash drum.

Simulation approach

From the desktop, select *Start* button followed by *Programs*, *AspenTech*, *Aspen Engineering Suite*, *Aspen Plus Version* and *Aspen Plus User Interface*. Then choose *Template* option in the *Aspen Plus Startup* dialog (Figure 1.5).

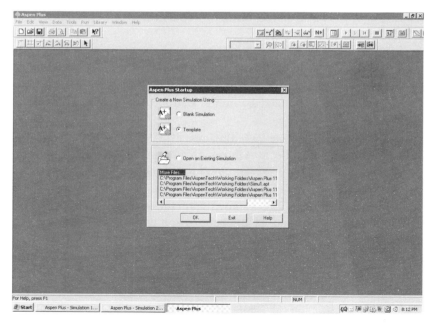

FIGURE 1.5

As the next window appears after hitting *OK* in the above screen, select *General with English Units* (Figure 1.6).

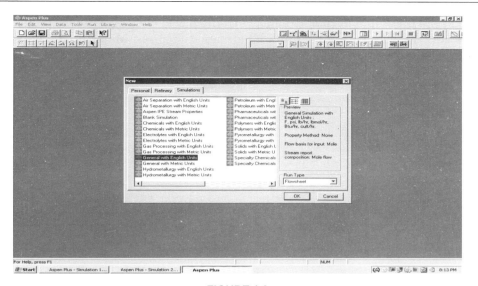

FIGURE 1.6

Then click *OK*. Again, hit *OK* when the Aspen Plus engine window pops up and subsequently, proceed to create the flowsheet.

Creating flowsheet

Select the *Separators* tab from the *Model Library* toolbar. As discussed earlier, there are five built-in models. Among them, select Flash2 and place this model in the window. Now the *Process Flowsheet Window* includes the flash drum as shown in Figure 1.7. By default, the separator is named as B1.

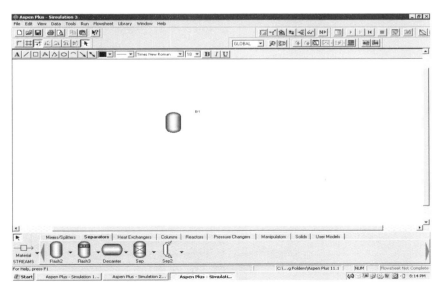

FIGURE 1.7

To add the input and output streams with the block, click on *Streams* section (lower left-hand corner). There are three different stream categories (Material, Heat and Work), as shown in Figure 1.8.

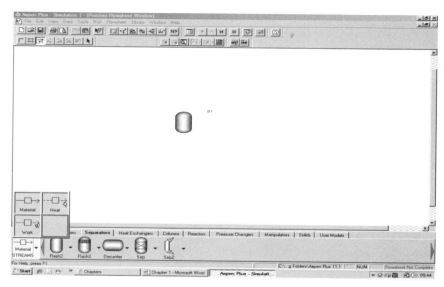

FIGURE 1.8

Block B1 includes three red arrows and one blue arrow as we approach the block after selecting the *Material STREAMS* icon (Figure 1.9). Now we need to connect the streams with the flash chamber using red arrows and the blue arrow is optional.

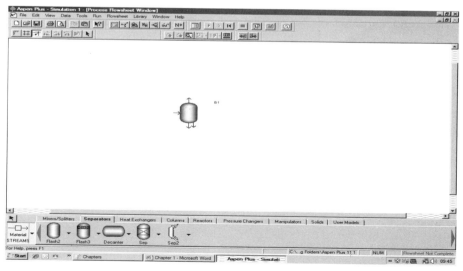

FIGURE 1.9

The connection procedure is as follows.

Clicking on *Material STREAMS*, move the mouse pointer over the red arrow at the inlet of the flash chamber. Click once when the arrow is highlighted and move the cursor so that the stream is in the position we want. Then click once more. We should see a stream labeled 1 entering the drum as a feed stream. Next, click the red arrow coming out at the bottom of the unit and drag the stream away and click. This stream is marked as 2. The same approach has been followed to add the product stream at the top as Stream 3. Now the flowsheet looks like Figure 1.10. Note that in the present case, only the red arrows have been utilized. In case any line representing a stream (e.g., feed or product streams) is not straight in the schematic, right click highlighting that line and then click "Align Blocks" (or Ctrl+B).

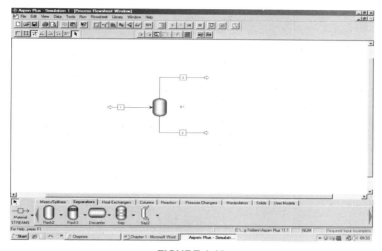

FIGURE 1.10

We can rename the stream(s) and block(s). To do that highlight the object (say, a block) we want to rename and click the right mouse button. Select *Rename Block* and then give a new name, as shown in Figure 1.11 for Block B1.

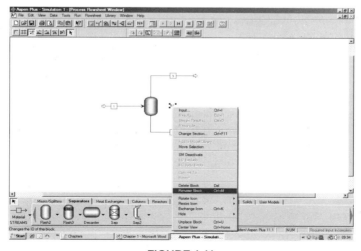

FIGURE 1.11

Alternatively, highlight the object, press Ctrl + M on the keyboard, change the name, and finally hit *Enter* or *OK*. After renaming Stream 1 to F, Stream 2 to L, Stream 3 to V and Block B1 to FLASH, the flowsheet finally resembles Figure 1.12.

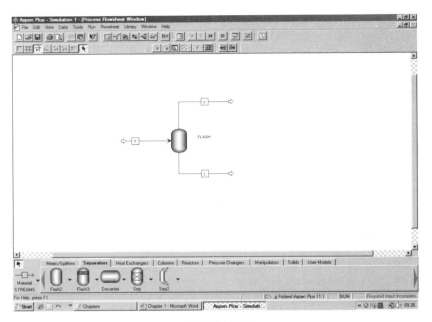

FIGURE 1.12

In order to inspect completeness for the entire process flowsheet, look at the status indicator. If the message includes *Flowsheet Not Complete*, click on *Material STREAMS*. If any red arrow(s) still exists in the flowsheet window, it indicates that the process is not precisely connected with the stream(s). Then we need to try again for proper connection. To find out why the connectivity is not complete, hit the *Next* button on the *Data Browser* toolbar. However, if we made a mistake and want to remove a stream (or block) from the flowsheet, highlight it, right click on it, hit *Delete Stream* (or *Delete Block*), and finally click *OK*.

Anyway, suppose that the flowsheet connectivity is complete. Accordingly, as mentioned earlier, the status message changes from *Flowsheet Not Complete* to *Required Input Incomplete*.

We have defined the unit operation to be simulated and set up the streams into and out of the process. Next we need to enter the rest of the information using several input forms required to complete the simulation. Within Aspen Plus, the easiest way to find the next step is to use one of the followings:

1. click the *Next* button ![N>]
2. find *Next* in the *Tools* menu
3. use shortcut key F4

As a consequence, Figure 1.13 appears.

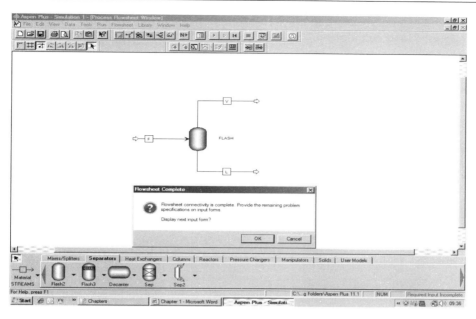

FIGURE 1.13

Configuring settings

As we click *OK* on the message, Aspen Plus opens the *Data Browser* window containing the *Data Browser* menu tree and *Setup/Specifications/Global* sheet.

Alternatively, clicking on *Solver Settings* and then choosing *Setup/Specifications* in the left pane of the *Data Browser* window, we can also obtain this screen (Figure 1.14).

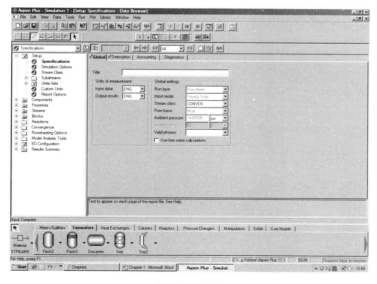

FIGURE 1.14

Although optional, it is a good practice to fill up the above form for our project giving the *Title* (e.g., Flash calculations) and keeping the other items unchanged (Figure 1.15).

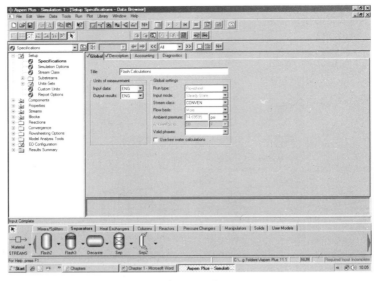

FIGURE 1.15

In the next step, we may provide the Aspen Plus *accounting* information (required at some installations). In this regard, a sample copy is displayed in Figure 1.16 with the following information:

User name: AKJANA
Account number: 1
Project ID: ANYTHING
Project name: AS YOU WISH

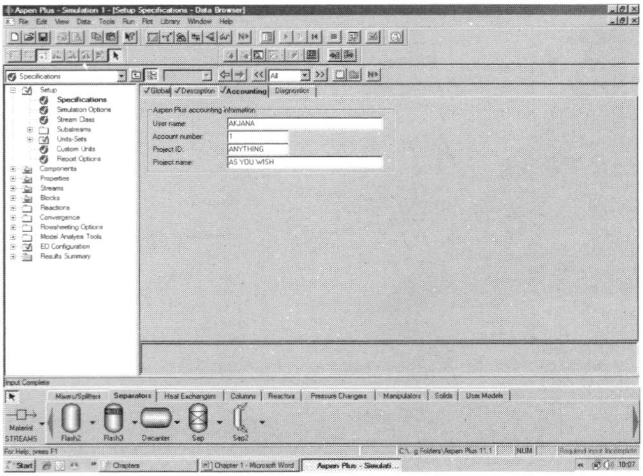

FIGURE 1.16

We may wish to have streams results summarized with mole fractions or some other basis that is not set by default. For this, we can use the *Report Options* under *Setup* folder. In the subsequent step, select *Stream* sheet and then choose *Mole* fraction basis.

As filled out, the form shown in Figure 1.17, final results related to all inlet and product streams will be shown additionally in terms of mole fraction. Remember that all values in the final results sheet should be given in the British unit as chosen it previously.

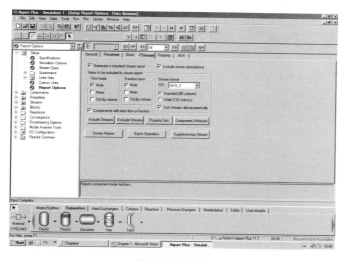

FIGURE 1.17

Specifying components

Clicking on *Next* button or double-clicking on *Components* in the column at the left side and then selecting *Specifications*, we get the following opening screen (Figure 1.18).

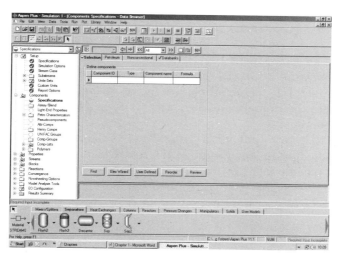

FIGURE 1.18

Next, we need to fill up the table in Figure 1.18. A *Component ID* is essentially an alias for a component. It is enough to enter the formulas or names of the components as their IDs. Based on these component IDs, Aspen Plus fills out the *Type, Component name* and *Formula* columns. But sometimes Aspen Plus does not find an exact match in its library. Like, in the present simulation, we have the following screen (Figure 1.19) after inserting chemical formulas of the components in the *Component ID* column.

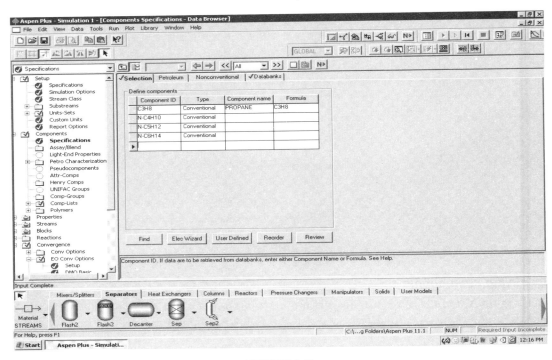

FIGURE 1.19

Obviously, only for Component ID C3H8, Aspen Plus provided the *Component name* (PROPANE) and *Formula* (C3H8). This simulator does not recognize the other three components by their IDs. Therefore, we have to search in the following way (Figure 1.20) to obtain their names and formulas. Click on a component ID (say, N-C4H10), then hit *Find* button.

Now, we have to give a hint with *Component name or formula* (butane) and then hit *Enter* or *Find now* button (see Figure 1.21). Apart from component name or formula, we can also search a component by giving component class or molecular weight (range) or boiling point (range) or CAS (Chemical Abstracts Service) number. Click on *Advanced* button on Figure 1.21 to get these options.

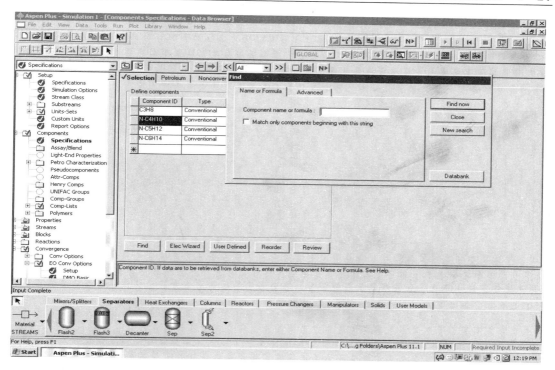

FIGURE 1.20

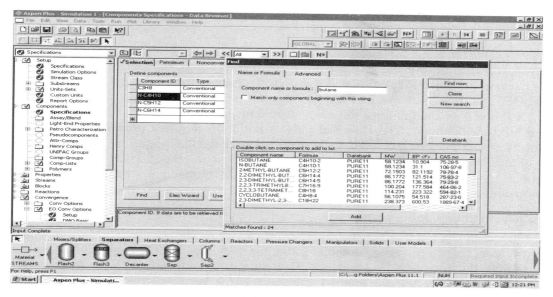

FIGURE 1.21

Aspen Plus suggests a number of possibilities. Among them, select a suitable component name (N-BUTANE) and then click on *Add*. Automatically, the *Component name* and *Formula* for *Component ID* N-C4H10 enter into their respective columns. For last two components, we follow the same approach. When all the components are completely defined, the filled component input form looks like Figure 1.22.

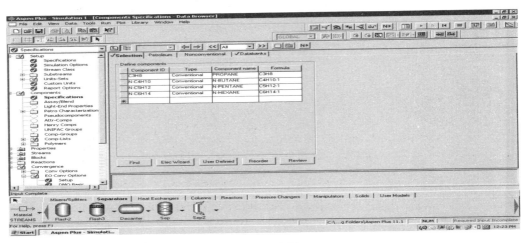

FIGURE 1.22

The *Type* is a specification of how Aspen calculates the thermodynamic properties. For fluid processing of organic chemicals, it is usually suitable to use 'Conventional' option. Notice that if we make a mistake adding a component, right click on the row and then hit *Delete Row* or *Clear*.

Specifying property method

Press *Next* button or choose *Properties/Specifications* from the Data Browser. Then if we click on the down arrow under *Base method* option, a list of choices appears. Set the SYSOP0[1] method as shown in Figure 1.23.

A *Property method* defines the methods and models used to describe pure component and mixture behavior. The chemical plant simulation requires property data. A wide variety of methods are available in Aspen Plus package for computing the properties.

Each *Process type* has a list of recommended property methods. Therefore, the selected *Process type* can guide us to choose a suitable base property method. If there is any confusion, we may select 'All' option as *Process type*.

Specifying stream information

In the list on the left, double click on *Streams* folder or simply use *Next* button. Inside that folder, there are three subfolders, one for each stream. Click on inlet stream F, and enter the temperature, pressure, flow rate and composition (mole fraction). No need to provide any data for product streams L and V because these streams are asked to simulate in the present problem. Finally, we have Figure 1.24.

---

[1] This property method assumes ideal behavior for vapor as well as liquid phase.

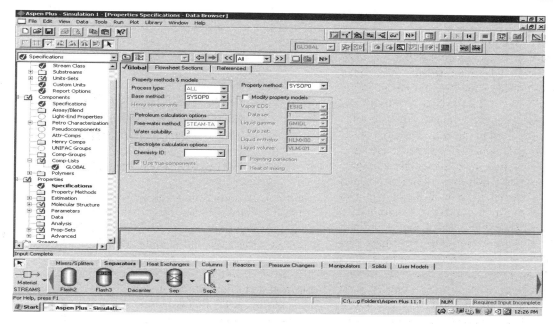

FIGURE 1.23

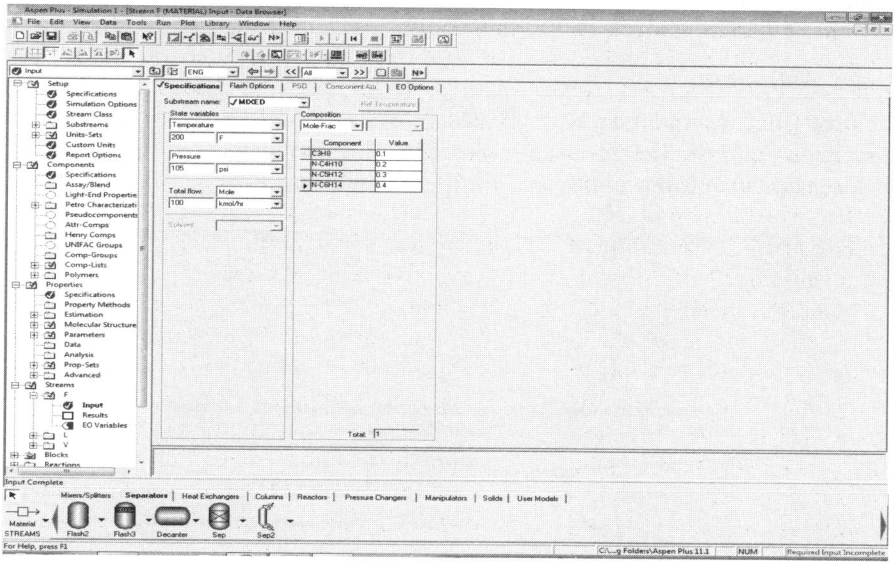

FIGURE 1.24

Specifying block information

Hitting *Next* button or selecting *Blocks/FLASH* in the column at the left side, we get the block input form. After inserting the operating temperature and pressure, one obtains Figure 1.25.

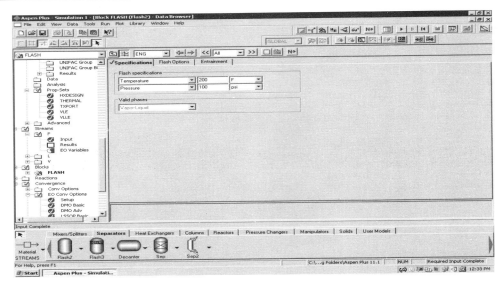

FIGURE 1.25

Now the *Status message* (i.e., *Required Input Complete*) implies that all necessary information have been inserted adequately. Moreover, all the icons on the left are blue. It reveals that all the menus are completely filled out. If any menu is still red, carefully enter the required information to make it blue.

Running the simulation

Click on *Next* button and get the following screen (see Figure 1.26). To run the simulation, press *OK* on the message. We can also perform the simulation selecting *Run* from the *Run* pulldown menu or using shortcut key F5.

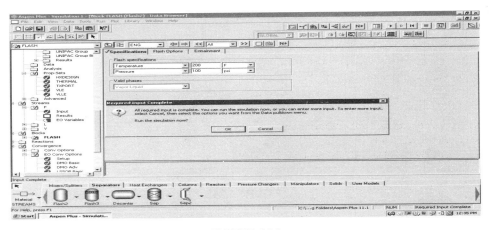

FIGURE 1.26

The *Control Panel*, as shown in Figure 1.27, shows the progress of the simulation. It presents all warnings, errors, and status messages.

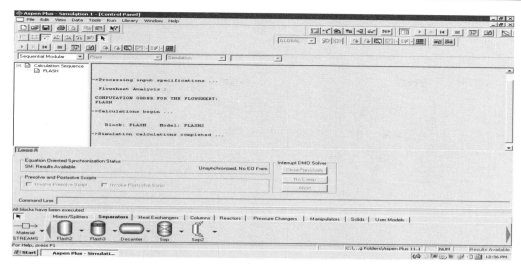

FIGURE 1.27

Viewing results

Hitting *Next* button and then clicking *OK*, the *Run Status* screen appears first (see Figure 1.28).

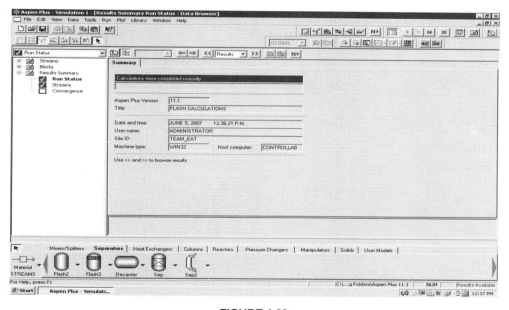

FIGURE 1.28

From the Data Browser, choose *Results Summary / Streams* and get the following screen that includes the final results of the given problem (see Figure 1.29).

Save the work by choosing *File / Save As / ...*from the menu list on the top. We can name the file whatever we want. Note that an Aspen Plus Backup file (*.bkp) takes much less space than a normal Aspen Plus Documents file (*.apw).

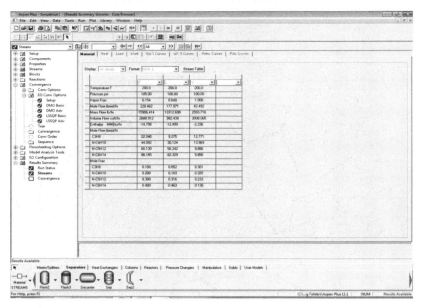

FIGURE 1.29

If we click on *Stream Table* button, the results table takes a place in the *Process Flowsheet Window*, as shown in Figure 1.30.

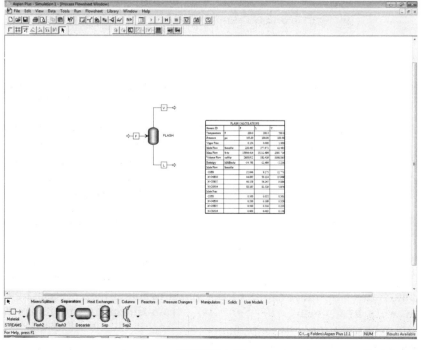

FIGURE 1.30

Viewing input summary

To obtain the input information, press Ctrl + Alt + I on the keyboard or select *Input Summary* from the *View* pulldown menu. The supervisor may ask to include the results, shown in Figure 1.30, along with the input summary (Figure 1.31) in the final report on the present project.

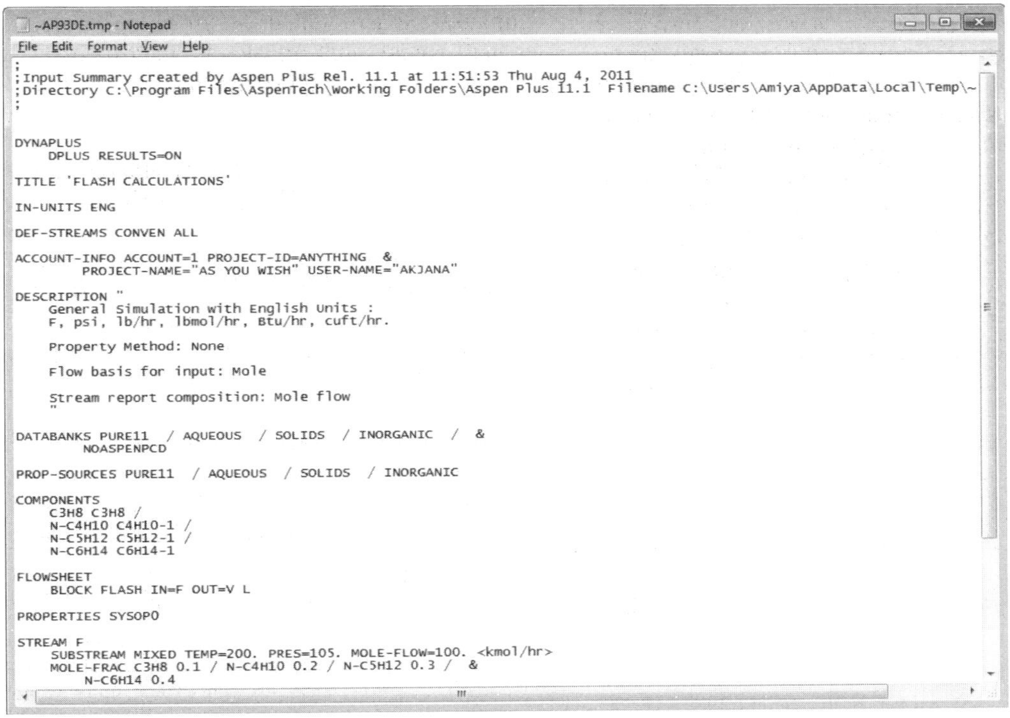

```
~AP93DE.tmp - Notepad

File  Edit  Format  View  Help

;Input Summary created by Aspen Plus Rel. 11.1 at 11:51:53 Thu Aug 4, 2011
;Directory C:\Program Files\AspenTech\working Folders\Aspen Plus 11.1  Filename C:\Users\Amiya\AppData\Local\Temp\~


DYNAPLUS
    DPLUS RESULTS=ON

TITLE 'FLASH CALCULATIONS'

IN-UNITS ENG

DEF-STREAMS CONVEN ALL

ACCOUNT-INFO ACCOUNT=1 PROJECT-ID=ANYTHING  &
        PROJECT-NAME="AS YOU WISH" USER-NAME="AKJANA"

DESCRIPTION "
    General Simulation with English Units :
    F, psi, lb/hr, lbmol/hr, Btu/hr, cuft/hr.

    Property Method: None

    Flow basis for input: Mole

    Stream report composition: Mole flow
    "

DATABANKS PURE11  / AQUEOUS  / SOLIDS  / INORGANIC  /  &
        NOASPENPCD

PROP-SOURCES PURE11  / AQUEOUS  / SOLIDS  / INORGANIC

COMPONENTS
    C3H8 C3H8 /
    N-C4H10 C4H10-1 /
    N-C5H12 C5H12-1 /
    N-C6H14 C6H14-1

FLOWSHEET
    BLOCK FLASH IN=F OUT=V L

PROPERTIES SYSOP0

STREAM F
    SUBSTREAM MIXED TEMP=200. PRES=105. MOLE-FLOW=100. <kmol/hr>
    MOLE-FRAC C3H8 0.1 / N-C4H10 0.2 / N-C5H12 0.3 /  &
        N-C6H14 0.4
```

FIGURE 1.31

Creating report file

To create a detailed report of the work we have done, including input summary, stream information, etc., select *Export* (Ctrl + E) from the *File* dropdown menu. Then save the work as a report file (e.g., C/Program Files/AspenTech/Working Folders/Aspen Plus Version/ Flash.rep). Subsequently, we may open the saved report file (Flash.rep) going through *My Computer* with using a program, such as the *Microsoft Office Word* or *WordPad* or *Notepad*. A report file for the present problem is opened and shown here.

ASPEN PLUS IS A TRADEMARK OF
ASPEN TECHNOLOGY, INC.
TEN CANAL PARK
CAMBRIDGE, MASSACHUSETTS 02141
617/949-1000

HOTLINE:
U.S.A. 888/996-7001
EUROPE (32) 2/724-0100

PLATFORM: WIN32   AUGUST 4, 2011
VERSION: 11.1 Build 192   THURSDAY
INSTALLATION: TEAM_EAT   11:22:01 A.M.
ASPEN PLUS PLAT: WIN32 VER: 11.1   08/04/2011 PAGE I
          FLASH CALCULATIONS

TABLE OF CONTENTS

ASPEN PLUS PLAT: WIN32   VER: 11.1   08/04/2011   PAGE 1
                         FLASH CALCULATIONS
                         RUN CONTROL SECTION
RUN CONTROL INFORMATION

THIS COPY OF ASPEN PLUS LICENSED TO

TYPE OF RUN: NEW

INPUT FILE NAME: _1229hak.inm

OUTPUT PROBLEM DATA FILE NAME: _1229hak VERSION NO. 1

LOCATED IN:
PDF SIZE USED FOR INPUT TRANSLATION:
NUMBER OF FILE RECORDS (PSIZE) = 0
NUMBER OF IN-CORE RECORDS = 256
PSIZE NEEDED FOR SIMULATION = 256

CALLING PROGRAM NAME: apmain
LOCATED IN: C:\PROGRA~1\ASPENT~1\ASPENP~3.1\Engine\xeq

SIMULATION REQUESTED FOR ENTIRE FLOWSHEET

DESCRIPTION

GENERAL SIMULATION WITH ENGLISH UNITS : F, PSI, LB/HR, LBMOL/HR,
BTU/HR, CUFT/HR. PROPERTY METHOD: NONE FLOW BASIS FOR INPUT: MOLE
STREAM REPORT COMPOSITION: MOLE FLOW

ASPEN PLUS PLAT: WIN32   VER: 11.1   08/04/2011   PAGE 2
FLASH CALCULATIONS
FLOWSHEET SECTION

FLOWSHEET CONNECTIVITY BY STREAMS

STREAM	SOURCE	DEST	STREAM	SOURCE	DEST
F	_____	FLASH	V	FLASH	_____
L	FLASH	_____			

FLOWSHEET CONNECTIVITY BY BLOCKS

BLOCK	INLETS	OUTLETS
FLASH	F	V L

COMPUTATIONAL SEQUENCE

SEQUENCE USED WAS:

FLASH

OVERALL FLOWSHEET BALANCE

*** MASS AND ENERGY BALANCE ***

CONVENTIONAL COMPONENTS	IN	OUT (LBMOL/HR)	RELATIVE DIFF.
C3H8	22.0462	22.0462	0.101867E-09
N-C4H10	44.0925	44.0925	0.326964E-10
N-C5H12	66.1387	66.1387	−0.113614E-10
N-C6H14	88.1849	88.1849	−0.332941E-10

TOTAL BALANCE
  MOLE(LBMOL/HR)   220.462   220.462   0.000000E+00
  MASS(LB/HR)   15906.4   15906.4   −0.782159E-11
  ENTHALPY(BTU/HR)   −0.147978E+08   −0.147349E+08   −0.425384E-02

ASPEN PLUS PLAT: WIN32   VER: 11.1   08/04/2011   PAGE 3
                          FLASH CALCULATIONS
                          PHYSICAL PROPERTIES SECTION

COMPONENTS

ID	TYPE	FORMULA	NAME OR ALIAS	REPORT NAME
C3H8	C	C3H8	C3H8	C3H8
N-C4H10	C	C4H10-1	C4H10-1	N-C4H10
N-C5H12	C	C5H12-1	C5H12-1	N-C5H12
N-C6H14	C	C6H14-1	C6H14-1	N-C6H14

ASPEN PLUS PLAT: WIN32   VER: 11.1   08/04/2011   PAGE 4
                          FLASH CALCULATIONS
                          U-O-S BLOCK SECTION

BLOCK: FLASH   MODEL: FLASH2
    INLET STREAM: F
    OUTLET VAPOR STREAM: V
    OUTLET LIQUID STREAM: L
    PROPERTY OPTION SET: SYSOP0   IDEAL LIQUID / IDEAL GAS

                *** MASS AND ENERGY BALANCE ***

                          IN   OUT   RELATIVE DIFF.
TOTAL BALANCE
MOLE(LBMOL/HR)   220.462   220.462   0.000000E+00
MASS(LB/HR)   15906.4   15906.4   −0.782136E-11
ENTHALPY(BTU/HR)   −0.147978E+08   −0.147349E+08 −0.425384E-02

                    *** INPUT DATA ***

TWO PHASE TP FLASH
SPECIFIED TEMPERATURE   F   200.000
SPECIFIED PRESSURE   PSI   100.000
MAXIMUM NO. ITERATIONS   30
CONVERGENCE TOLERANCE   0.000100000

                    *** RESULTS ***

OUTLET TEMPERATURE   F   200.00

OUTLET PRESSURE   PSI   100.00
HEAT DUTY   BTU/HR   62948
VAPOR FRACTION   0.19274

V-L PHASE EQUILIBRIUM:

COMP	F(I)	X(I)	Y(I)	K(I)
C3H8	0.10000	0.52117E-01	0.30055	5.7668
N-C4H10	0.20000	0.16926	0.32874	1.9422
N-C5H12	0.30000	0.31602	0.23290	0.73697
N-C6H14	0.40000	0.46260	0.13781	0.29790

ASPEN PLUS PLAT: WIN32   VER: 11.1   08/04/2011   PAGE 5
FLASH CALCULATIONS
STREAM SECTION

F L V

STREAM ID	F	L	V
FROM :	_____	FLASH	FLASH
TO :	FLASH	_____	_____

SUBSTREAM: MIXED

PHASE:	MIXED	LIQUID	VAPOR
COMPONENTS: LBMOL/HR			
C3H8	22.0462	9.2754	12.7709
N-C4H10	44.0925	30.1237	13.9688
N-C5H12	66.1387	56.2424	9.8963
N-C6H14	88.1849	82.3291	5.8558
COMPONENTS: MOLE FRAC			
C3H8	0.1000	5.2117-02	0.3005
N-C4H10	0.2000	0.1693	0.3287
N-C5H12	0.3000	0.3160	0.2329
N-C6H14	0.4000	0.4626	0.1378
TOTAL FLOW:			
LBMOL/HR	220.4623	177.9706	42.4917
LB/HR	1.5906+04	1.3313+04	2593.7158
CUFT/HR	2688.9117	382.4385	3008.0650
STATE VARIABLES:			
TEMP F	200.0000	200.0000	200.0000
PRES PSI	105.0000	100.0000	100.0000
VFRAC	0.1540	0.0	1.0000
LFRAC	0.8460	1.0000	0.0
SFRAC	0.0	0.0	0.0

```
ENTHALPY:
   BTU/LBMOL          -6.7122+04   -7.0232+04   -5.2612+04
   BTU/LB             -930.3052    -938.9019    -861.9118
   BTU/HR             -1.4798+07   -1.2499+07   -2.2356+06
ENTROPY:
   BTU/LBMOL-R        -116.9519    -123.3349     -87.8846
   BTU/LB-R             -1.6209      -1.6488      -1.4398
DENSITY:
   LBMOL/CUFT         8.1989-02       0.4654    1.4126-02
   LB/CUFT              5.9156       34.8100       0.8623
AVG MW                72.1503       74.8028       61.0406
```

ASPEN PLUS PLAT: WIN32 VER: 11.1   08/04/2011   PAGE 6

FLASH CALCULATIONS
PROBLEM STATUS SECTION

BLOCK STATUS

```
**********************************************************************
*                                                                   *
* Calculations were completed normally                              *
*                                                                   *
* All Unit Operation blocks were completed normally                 *
*                                                                   *
* All streams were flashed normally                                 *
*                                                                   *
**********************************************************************
```

1.3.3 Computation of Bubble Point Temperature

Problem statement

Compute the bubble point temperature at 18 bar of the following hydrocarbon mixture (see Table 1.1) using the RK-Soave property method.

TABLE 1.1

Component	Mole fraction
C_1	0.05
C_2	0.1
C_3	0.15
$i\text{-}C_4$	0.1
$n\text{-}C_4$	0.2
$i\text{-}C_5$	0.25
$n\text{-}C_5$	0.15

Assume the mixture inlet temperature of 25°C, pressure of 19 bar and flow rate of 120 kmol/h.

Simulation approach

After starting the Aspen Plus simulator, the *Aspen Plus Startup* dialog appears first. Among the three choices, select *Template* option and then click *OK* (see Figure 1.32).

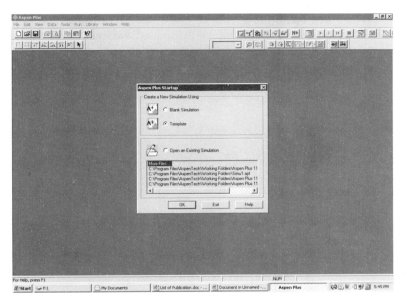

FIGURE 1.32

When the next window pops up (see Figure 1.33), select *General with Metric Units* and then hit *OK*.

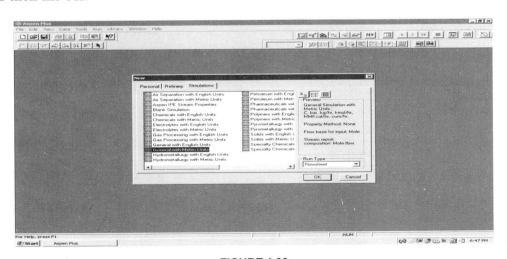

FIGURE 1.33

In the next, press *OK* in the *Connect to Engine* dialog. Once Aspen Plus connects to the simulation engine, we are ready to begin entering the process system.

Creating flowsheet

Using the Flash2 separator available in the equipment *Model Library*, develop the following process flow diagram (see Figure 1.34) in the *Flowsheet Window* by connecting the input and output streams with the flash drum. Recall that red arrows are required ports and blue arrows are optional ports. To continue the simulation, we need to click either on *Next* button or *Solver Settings* as discussed earlier. Note that whenever we have doubts on what to do next, the simplest way is to click the *Next* button.

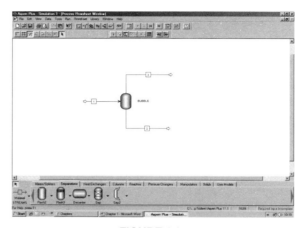

FIGURE 1.34

Configuring settings

From the Data Browser, choose *Setup/Specifications*. The *Title* of the present problem is given as 'Bubble Point Calculations'. Other items in Figure 1.35 remain untouched. However, we can also change those items (e.g., *Units of measurement, Input mode*, etc).

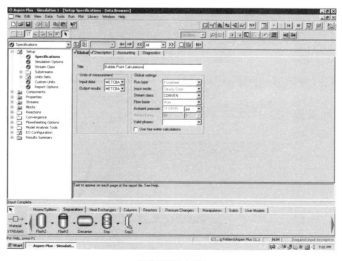

FIGURE 1.35

In the next, the Aspen Plus *accounting* information is given in Figure 1.36.

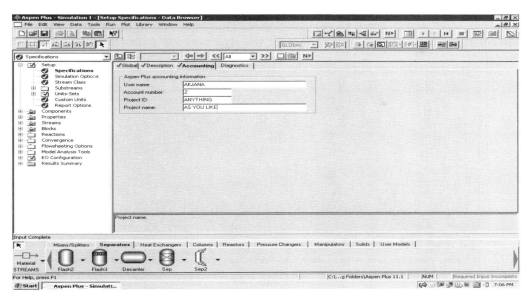

FIGURE 1.36

Specifying components

Click on *Next* button or choose *Components/Specifications* in the list on the left. Then define all components and obtain the following window (see Figure 1.37).

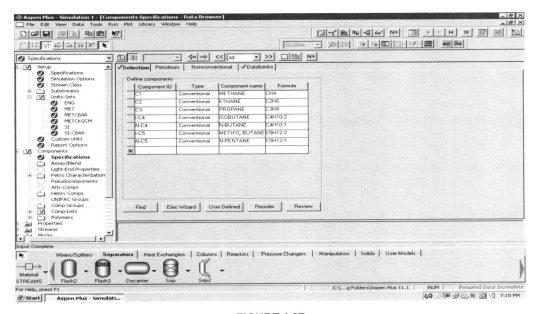

FIGURE 1.37

Specifying property method

Hit *Next* button or select *Properties / Specifications* in the column at the left side. In *Property method*, scroll down to get RK-Soave. This equation of state model is chosen for thermodynamic property predictions for the hydrocarbon mixture (see Figure 1.38).

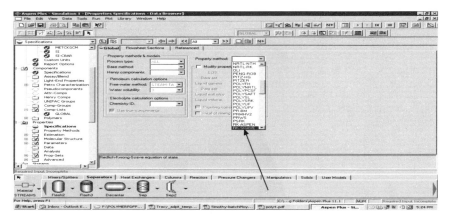

FIGURE 1.38

Hitting *Next* button twice, we have the following picture (see Figure 1.39). The binary parameters are tabulated below. When we close this window or click *OK* on the message, it implies that we approve the parameter values. However, we have the opportunity to edit or enter the parameter values in the table. In blank spaces of the table, zeros are there. It does not reveal that the ideal mixture assumption is used because many thermodynamic models predict non-ideal behavior with parameter values of zero.

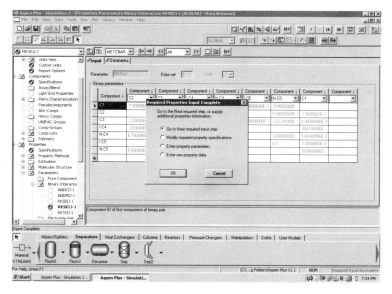

FIGURE 1.39

Specifying stream information

Click *OK*. Alternatively, use the Data Browser menu tree to navigate to the *Streams/1/ Input/Specifications* sheet. Then insert all specifications for Stream 1 as shown in Figure 1.40.

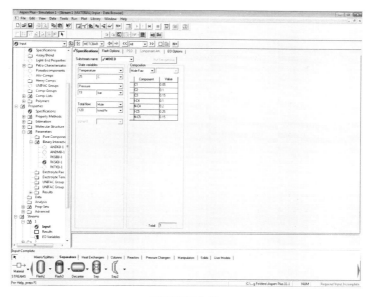

FIGURE 1.40

Specifying block information

Hit *Next* or select *Blocks/BUBBLE* from the Data Browser. After getting the blank input form, enter the required inputs (Pressure = 18 bar and Vapor fraction = 0) for block *BUBBLE* (see Figure 1.41).

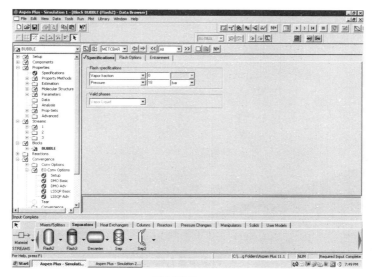

FIGURE 1.41

Running the simulation

Press *Next* button and then hit *OK* to run the simulation. The following *Control Panel* demonstrates the status of our simulation work (see Figure 1.42).

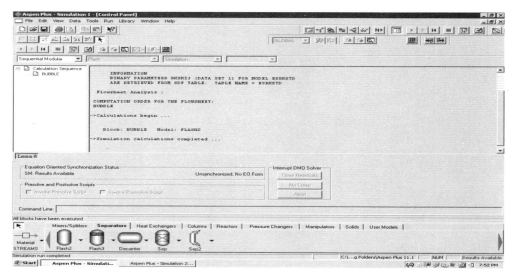

FIGURE 1.42

Viewing results

Clearly, Figure 1.42 includes the *Status message: Results Available.* As the simulation calculations completed, click on *Solver Settings* and then double-click on *Blocks* to obtain the following screen (see Figure 1.43).

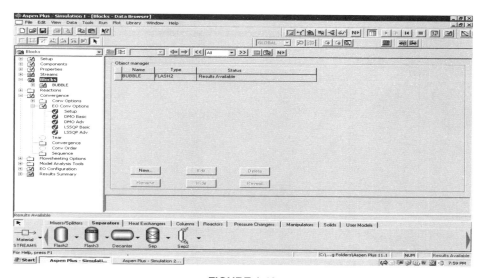

FIGURE 1.43

Choosing *Blocks*/*BUBBLE*/*Results* in the column at the left side, we get the following results summary for the present problem (see Figure 1.44).

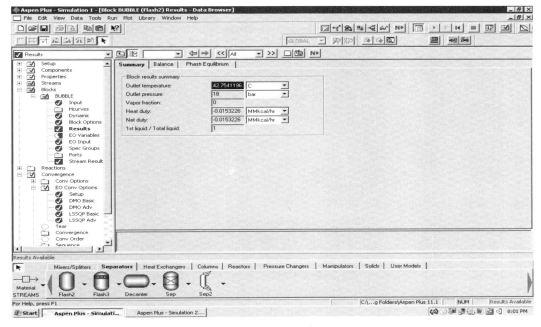

FIGURE 1.44

From the results sheet, we obtain the bubble point temperature = 42.7541196°C.

1.3.4 Computation of Dew Point Temperature

Problem statement

Compute the dew point temperature at 1.5 bar of the hydrocarbon mixture, shown in Table 1.2, using the RK-Soave property method.

TABLE 1.2

Component	Mole fraction
C_1	0.05
C_2	0.1
C_3	0.15
$i\text{-}C_4$	0.1
$n\text{-}C_4$	0.2
$i\text{-}C_5$	0.25
$n\text{-}C_5$	0.15

Assume the mixture inlet temperature of 25°C, pressure of 5 bar and flow rate of 120 kmol/h.

Simulation approach

As we start Aspen Plus from the *Start* menu or by double-clicking the Aspen Plus icon on our desktop, the *Aspen Plus Startup* dialog appears (see Figure 1.45). Select *Template* option.

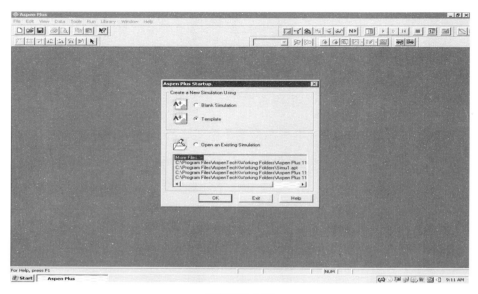

FIGURE 1.45

As Aspen Plus presents the window after clicking *OK* on Figure 1.45, choose *General with Metric Units*. Then press *OK* (see Figure 1.46).

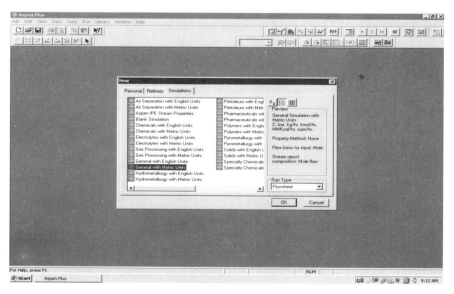

FIGURE 1.46

Subsequently, click *OK* when the Aspen Plus engine window pops up.

Creating flowsheet

In the next, we obtain a blank *Process Flowsheet Window*. Then we start to develop the process flowsheet by adding the Flash2 separator from the *Model Library* toolbar and joining the inlet and product streams by the help of *Material STREAMS* (Figure 1.47).

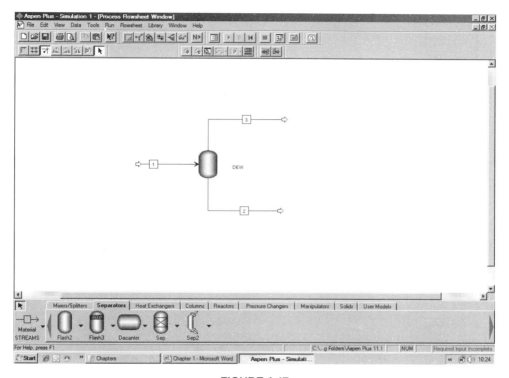

FIGURE 1.47

Now the process flow diagram is complete. The *Status bar* in the bottom right of the above window (see Figure 1.47) reveals *Required Input Incomplete* indicating that input data are required to continue the simulation.

Configuring settings

Hitting *Next* button and then clicking *OK*, we get the setup input form. The present problem is titled as 'Dew Point Calculations' (see Figure 1.48).

In Figure 1.49, the Aspen Plus *accounting* information are provided.

Specifying components

Here we have to enter all the components we are using in the simulation. In the list on the left, choose *Components/Specifications* and fill up the table following the procedure explained earlier (see Figure 1.50).

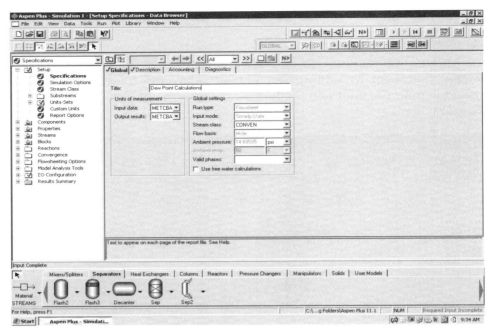

FIGURE 1.48

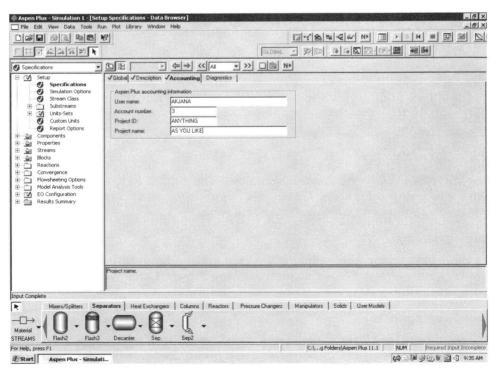

FIGURE 1.49

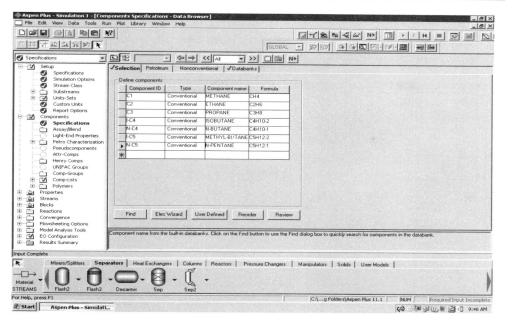

FIGURE 1.50

Specifying property method

From the Data Browser, select *Properties/Specifications* to obtain a blank property input form. From the *Base method* pulldown menu, select RK-Soave (see Figure 1.51).

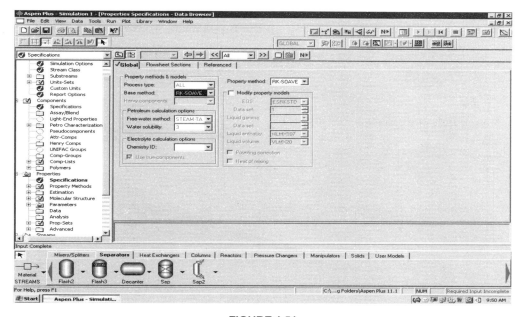

FIGURE 1.51

Specifying stream information

In the column at the left side, choose *Streams*/1. As a result, a stream input form opens. Entering all required information, one obtains the screen as shown in Figure 1.52.

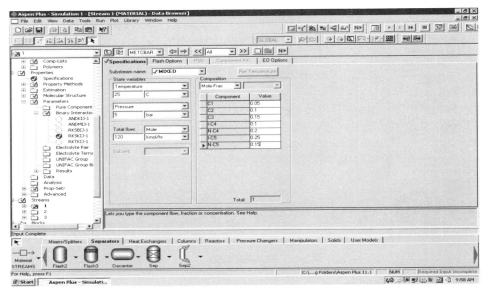

FIGURE 1.52

Specifying block information

The final area that requires input is the *Blocks* tab. In the list on the left, double-click on *Blocks* and then select *DEW*. Filling up the input form, we have Figure 1.53.

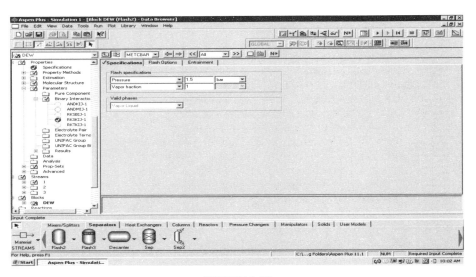

FIGURE 1.53

Running the simulation

Running the simulation, the following progress report is obtained (see Figure 1.54).

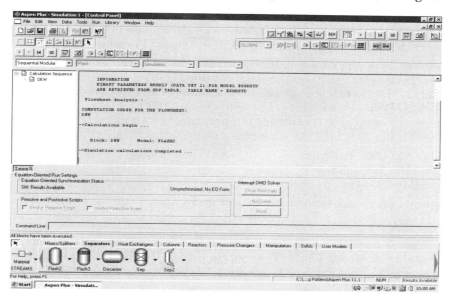

FIGURE 1.54

Viewing results

First click on *Solver Settings*. From the Data Browser, choose *Blocks/DEW/Results* (see Figure 1.55) to get the dew point temperature equal to 22.1945384°C.

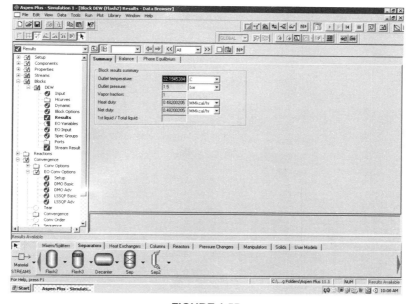

FIGURE 1.55

1.3.5 T-xy and P-xy Diagrams of a Binary Mixture

Problem statement

A binary mixture, consisting of 60 mole% ethanol and 40 mole% water, is introduced into a flash chamber (Flash2) with a flow rate of 120 kmol/h at 1.013 bar and 70°C. The chamber operates at its input conditions. Using the NRTL property method:

(a) Produce T-xy plot at a constant pressure (1.013 bar)
(b) Produce xy plot based on the data obtained in part (a)
(c) Analyze the azeotropic behaviour of ethanol-water system
(d) Produce P-xy plot at a constant temperature (90°C)
(e) Produce P-xy plot at a constant temperature (90°C) using the NRTL parameters given in the DECHEMA[2] Chemistry Data Series (Vol. I, Part 1A, p. 129) as:

a_{ij} = −517.9603 cal/mol
a_{ji} = 1459.309 cal/mol
α_{ij} = 0.0878

where, i and j correspond to ethanol and water, respectively.

Simulation approach

As usual, start Aspen Plus and select *Template*. Click *OK* to get the next screen and choose *General with Metric Units*. Then again hit *OK*. In the subsequent step, click *OK* in the *Connect to Engine* window to obtain a blank *Process Flowsheet Window*.

Creating flowsheet

From the equipment *Model Library* at the bottom of the Aspen Plus process flowsheet window, select the *Separators* tab and insert the Flash2 separator. Then connect the separation unit with the incoming and outgoing streams. The complete process is shown in Figure 1.56.

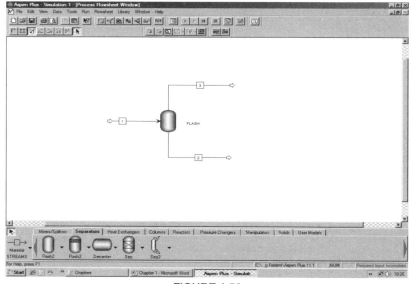

FIGURE 1.56

---

[2]DECHEMA stands for *Gesellschaft für Chemische Technik und Biotechnologie e.V.* It is a society for chemical engineering and biotechnology based in Frankfurt am Main, Germany and founded in 1926.

Configuring settings

After clicking on *Solver Settings*, select *Setup/Specifications* in the list on the left. The *Title* of the present problem is given as 'TXY and PXY Diagrams'. Subsequently, the Aspen Plus *accounting* information is also provided [see Figures 1.57(a) and (b)].

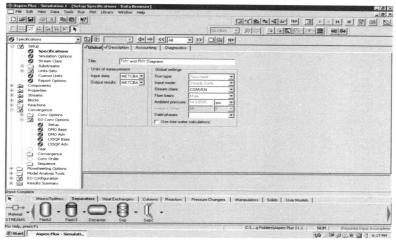

FIGURE 1.57(a)

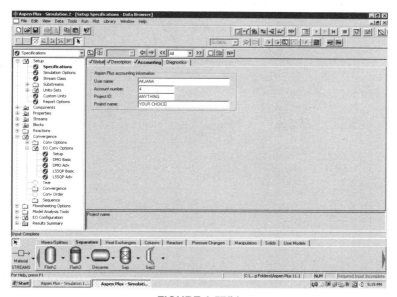

FIGURE 1.57(b)

Specifying components

Hitting *Next* button and defining the components (ethanol and water) in the input form, one obtains Figure 1.58.

Specifying property method

The user input under the *Properties* tab is probably the most critical input required to run a successful simulation. Clicking *Next* button, we obtain the property input form. For this problem, choose the NRTL model by scrolling down (see Figure 1.59).

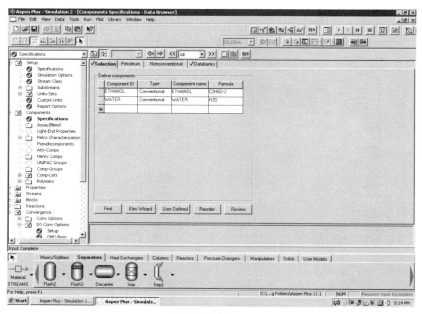

FIGURE 1.58

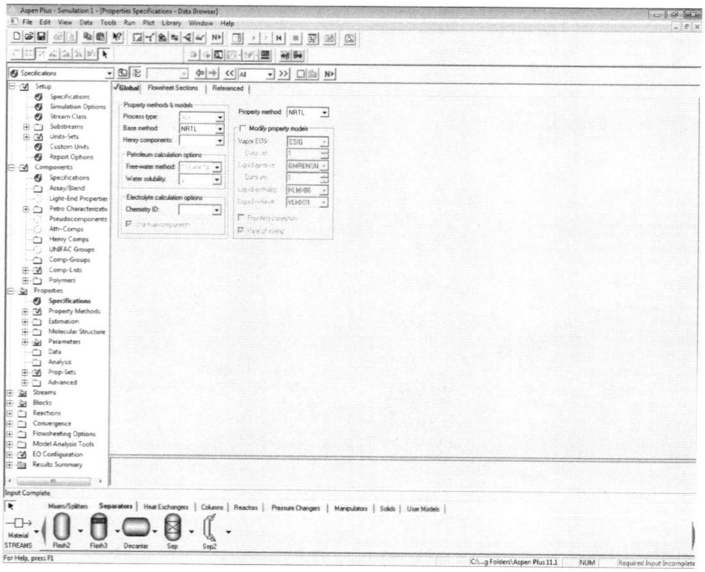

FIGURE 1.59

Once the base property method has been selected and we click the *Next* button, a window pops up asking whether to continue to the next step or to modify the properties (see Figure 1.60).

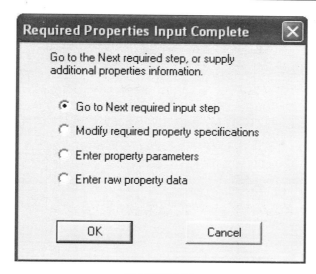

FIGURE 1.60

Specifying stream information

The next window includes a stream input form. Specifying temperature, pressure, flow rate and components mole fraction, one obtains Figure 1.61 as shown.

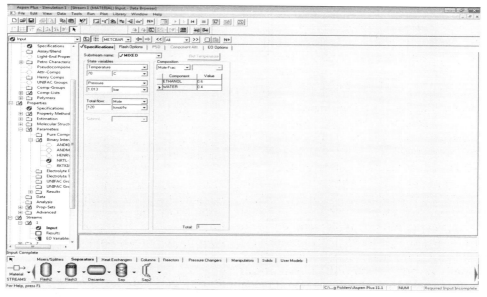

FIGURE 1.61

Specifying block information

Providing the operating temperature (70°C) as well as pressure (1.013 bar), one obtains Figure 1.62.

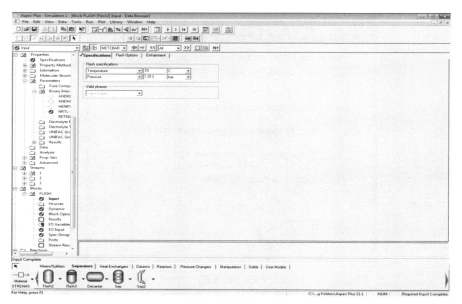

FIGURE 1.62

(a) Creating T-xy plot: Selecting *Tools / Analysis / Property / Binary*, we have Figure 1.63.

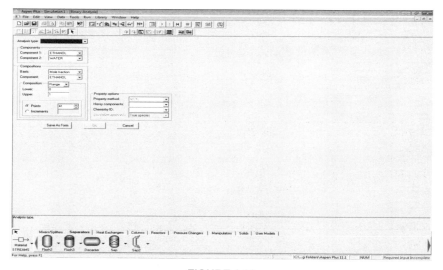

FIGURE 1.63

We must note that this form has to be filled up to generate any one of T-xy, P-xy or Gibbs energy of mixing diagrams. Select 'Txy' for the present problem. We aim to do an analysis on the mixture of ethanol and water; so select these components accordingly. The user has the option of specifying, which component will be used for the x-axis (which component's mole fraction will be diagrammed). The default is whichever component is indicated as component 1. Make sure that we are creating the diagram for the mole fraction of ethanol. Entering required information, Figure 1.63 takes the following form (see Figure 1.64).

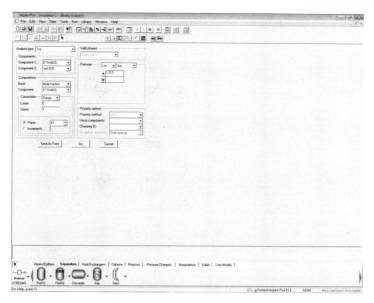

FIGURE 1.64

Click on *Go* and get the T-xy plot at a constant pressure (1.013 bar) as shown in Figure 1.65. Although the *Status bar* shows *Required Input Incomplete*, but there is no problem to get the plot based on the given information.

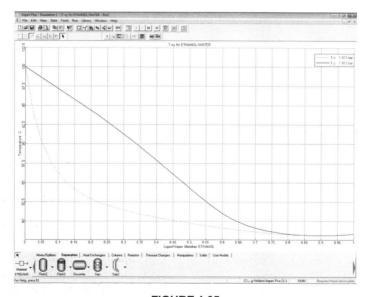

FIGURE 1.65

It should be noted that if we move the T-xy plot slightly or close it, we find Figure 1.66 having a databank. Some of these values have been used to make the plot in Figure 1.65.

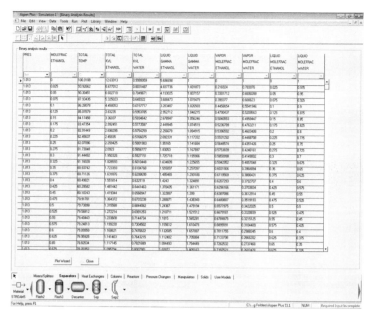

FIGURE 1.66

(b) Creating xy plot: Click on *Plot Wizard* shown in Figure 1.66 and then hit *Next* button. In the subsequent step, select YX plot type and click on *Next*. Filling up the form, again hit *Next* button. Providing (correcting) the input information, click on *Finish* and finally get the following vapor-liquid equilibrium diagram (see Figure 1.67).

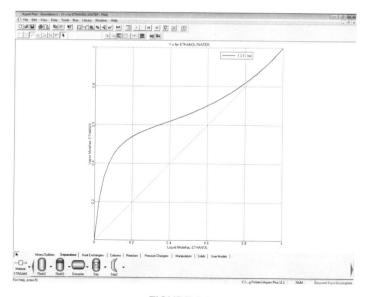

FIGURE 1.67

For ternary systems, the vapor-liquid equilibrium information can be seen by clicking *Tools / Analysis / Properties / Residue*. Consequently, an equilateral ternary diagram is obtained and it shows the residue curves which can be used for further analysis, including the azeotropic behavior.

(c) We know that the vapor and liquid compositions are identical at the azeotropic point. Therefore, all K^3-values are unity and no separation can take place. It is obvious from Figures 1.65 and 1.67 that the ethanol and water form an azeotrope.

There are several nice analysis tools available in the Aspen flowsheet simulator. Here, *Aspen Split* can be used to have additional insight. Clicking *Tools / Aspen Split / Azeotrope Search*, we obtain Figure 1.68. Under *Component List*, select ethanol and water. In the list on the left, click *Azeotropes* and then hit *OK* button to get Figure 1.69.

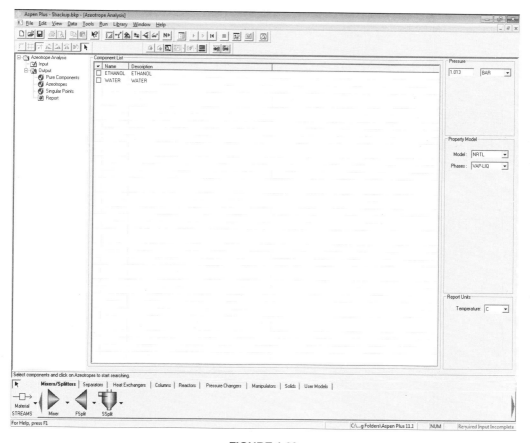

FIGURE 1.68

---

[3]Vapor-liquid equilibrium coefficient.

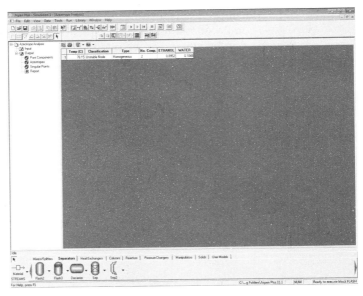

FIGURE 1.69

Tabulated results in Figure 1.69 indicate that for a pressure of 1.013 bar, the azeotropic mixture occurs at 78.15°C and 89.52 mol% ethanol. Alternatively, clicking *Report*, instead of *Azeotropes*, in Figure 1.68, we have Figure 1.70 that also reports the same azeotropic information.

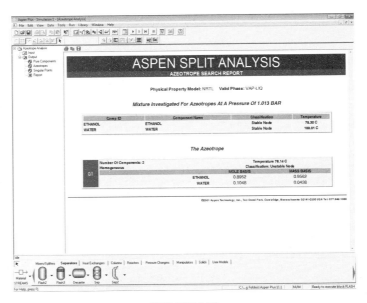

FIGURE 1.70

In order to examine the shifting of azeotropic composition with pressure, we further run the simulator at 3 bar pressure. It is clear from Figure 1.71 that for a pressure of 3 bar, the azeotrope forms at 108.41°C and 86.38 mol% ethanol. Clearly, the azeotrope shifts with pressure.

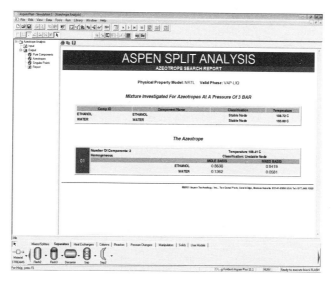

FIGURE 1.71

(d) Creating P-xy Plot: To avoid any possibility of error in the simulation, it is always suggested to click *Run / Reinitialize* (or press *Shift+F5*) and then hit *OK* button. In order to produce P-xy plot, the required data are entered in the sheet shown in Figure 1.72. The same sheet we used previously for producing T-xy diagram in Part (a).

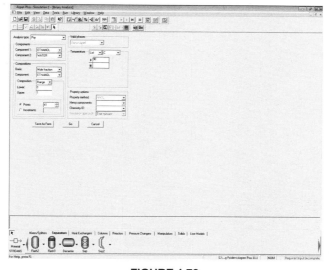

FIGURE 1.72

Clicking on *Go* button, we have the following P-xy plot [see Figure 1.73(a)] at a constant temperature (90°C) and respective databank produced [see Figure 1.73(b)]. Figure 1.73(a) confirms the formation of azeotrope by ethanol-water mixture.

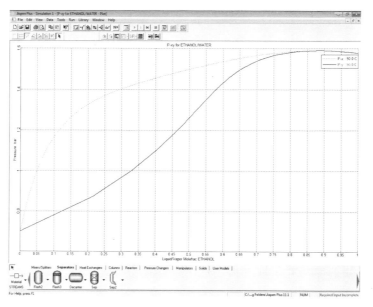

FIGURE 1.73(a)

FIGURE 1.73(b)

Notice that the plot window can be edited by right clicking on that window and selecting *Properties*. In the properties window, the user can modify the title, axis scale, font, color of the plot, etc. Alternatively, double-click on the different elements of the plot and modify them as we like to improve the presentation and clarity.

(e) It should be noted that the DECHEMA Chemistry Data Series include a large number of data sets for binary parameters of the NRTL, Wilson, and UNIQUAC models. Interestingly, these binary parameters are not compatible with the form of the model equations used in the Aspen flowsheet simulator. However, a suitable provision is there to directly use the DECHEMA data in the Aspen. For this, follow the sequential steps given below for the example binary system.

Step 1: Click *Run/Reinitialize* and then *OK* button.

Step 2: In the left pane of the Data Browser, double-click the *Properties* followed by the *Parameters* folder and then select the *Binary Interaction* folder to open the *Object manager*, as shown in Figure 1.74.

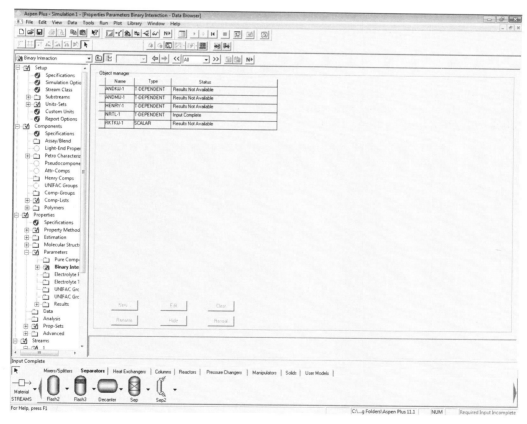

FIGURE 1.74

Step 3: Under *Object manager*, click *NRTL-1* and then hit *Edit* button to get Figure 1.75[4].

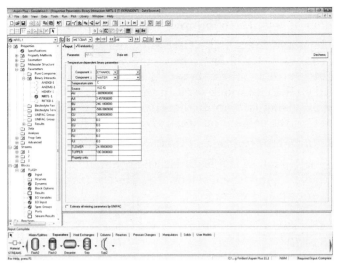

FIGURE 1.75

Step 4: On the *Input* sheet, Component *i* is appeared by default as ethanol and Component *j* as water. Select the column that contains the binary parameter values. As we click the *Dechema* button, Figure 1.76 appears.

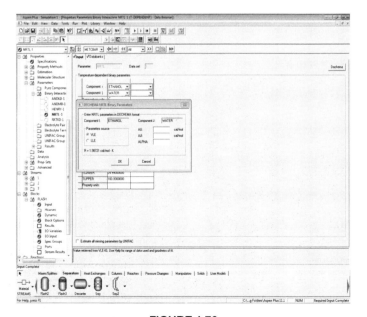

FIGURE 1.76

---

[4]If one continues to use the NRTL model, this input sheet allows to modify the binary parameter values.

Step 5: Enter the given binary parameter values in the suitable boxes (see Figure 1.77). Notice that we can also specify whether the parameters came from the VLE (vapour-liquid equilibrium) or LLE (liquid-liquid equilibrium) collection.

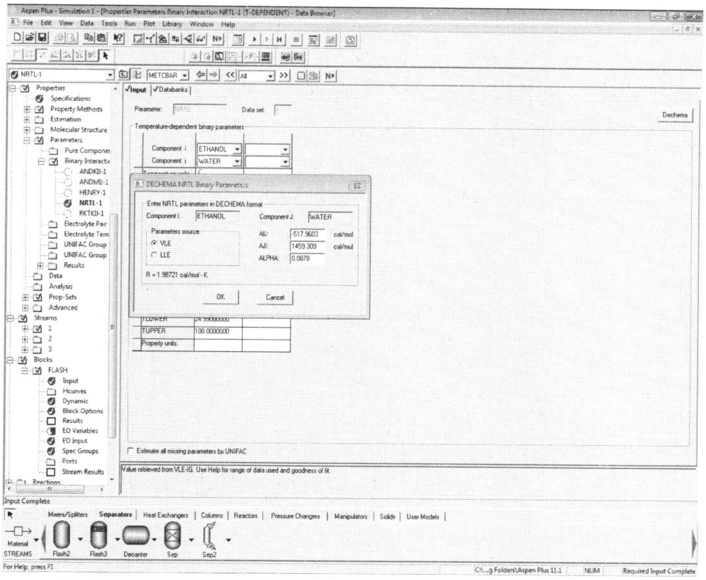

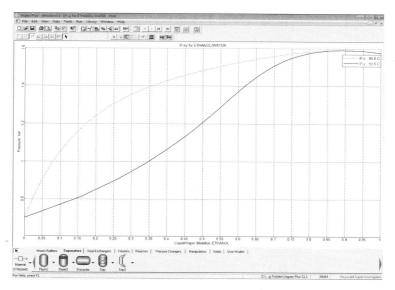

FIGURE 1.77

Step 6: Click *OK* and then produce the P-xy plot as shown in Figure 1.78 following the steps discussed earlier [part (d)].

FIGURE 1.78

Very close to this plot is obtained earlier in Figure 1.73(a) using the NRTL model in-built in the Aspen simulator.

1.4 DATA REGRESSION

Problem statement

For a binary benzene/ethanol system, the experimental vapor-liquid equilibrium data are given at 1 atm in Table 1.3.

TABLE 1.3

Temperature (T) (°C)	Benzene in liquid (x) (mol fract.)	Benzene in vapor (y) (mol fract.)
78.35	0.0	0.0
77.5	0.0148	0.074
75.1	0.0501	0.278
72.4	0.121	0.422
70.0	0.215	0.538
68.6	0.305	0.602
67.5	0.682	0.678
68.5	0.805	0.726
72.3	0.912	0.817
75.0	0.951	0.877
77.4	0.978	0.952
80.0	1.0	1.0

(i) Perform the regression of the Wilson model parameters on the basis of the experimental data summarized in Table 1.3.

(ii) Produce a T-xy plot to compare the estimated results with the experimental data provided.

Simulation approach

(i) Start Aspen Plus for a new simulation using *Template*. On the *New* dialog box (Figure 1.79), select *General with Metric Units* and then *Data Regression* in the *Run Type* list box.

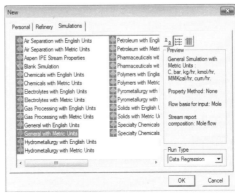

FIGURE 1.79

Configuring settings

Under *Setup / Specifications*, fill up *Global* and *Accounting* sheets.

Specifying components

In Figure 1.80, we define two components, namely benzene and ethanol.

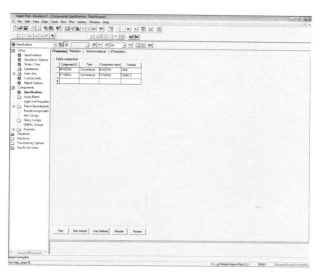

FIGURE 1.80

Specifying property method

In this simulation, our purpose is to use the experimental VLE data for finding the Wilson parameters. Selecting the *Wilson* as a *Base method*, we have Figure 1.81.

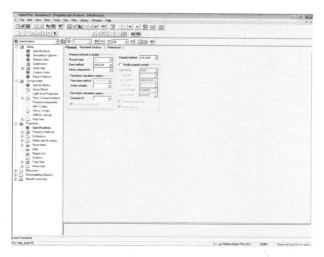

FIGURE 1.81

Entering phase equilibrium and mixture data

To input experimental data for phase equilibrium and mixture properties as a function of temperature (T), pressure (P), and composition (x for liquid and y for vapour), use the Data Browser menu tree to navigate to the *Properties/Data* sheet. Clicking *New* on the *Object manager*, enter an *ID* (i.e., VLE) or accept the default in the *Create new ID* dialog box. Before pressing *OK* button, additionally choose MIXTURE[5] in Figure 1.82.

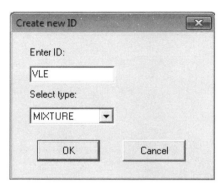

FIGURE 1.82

As the *Setup* form appears, we fill it as shown in Figure 1.83. Our problem deals with the vapour-liquid equilibrium under which the following options are available:

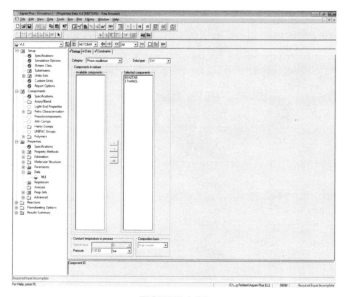

FIGURE 1.83

---

[5]Choosing PURE-COMP in the *Select type* list box, we can go for fitting pure component property data, such as vapor pressure (temperature dependent), heat capacity (temperature dependent), etc.

Option	Data type
TXY	VLE at constant pressure (Isobaric)
PXY	VLE at constant temperature (Isothermal)
TPXY	VLE at varying T and P
ALPHA	Relative volatility (with respect to the first component listed on the *Component/Specifications/Selection* form)

Next open the *Data* sheet to enter the experimental VLE data. Here the user needs to either enter the standard deviation[6] (Std-Dev) values, if known, for the property data or accept the system defaults. Note that we are free to assign a set of standard deviation values to:

- A single data point
- A multiple data points
- All data points in a data group

To include a standard deviation row in the *Data* sheet, use the *Usage* field, and select *Std-Dev*. It is interesting to mention that the standard deviation values we enter in a row will apply to all subsequent data points until another *Std-Dev* row is encountered. On the other hand, if we wish Aspen Plus to ignore some data, including standard deviations that have already been entered, go to the *Usage* field, click on the row, and then select *Ignore*.

A variable that has a standard deviation value of zero is considered as error-free. Remember that only variables with little or no random error can have standard deviations of zero. There are several properties (e.g., vapor pressure, density) that will never be treated as error-free. We should ensure that all standard deviation values in *Data* sheet will not be zero.

For the case of phase equilibrium, the number of nonzero standard deviations should be greater than or equal to the number of phase equilibrium constraints (or equivalently, the number of components in the mixture that participate in phase equilibrium)[7]. To illustrate it, let us take the example of TPXY data of two components. We can assign a standard deviation of zero to only two variables [i.e., either T or P, and either $x(1)$ or $y(1)$]. At this moment we should note that an exception is TPX data. We can set the standard deviation of x and either T or P to zero.

For the example binary system, we adopt the default standard deviation values. Entering the TXY data, we obtain Figure 1.84.

Specifying the regression case

Select *Regression* under *Properties* folder and then press *New* button on the *Object manager*. Accepting the default ID (i.e., R-1), hit *OK*. In the *Setup* sheet, specify property method (the default is the Wilson since it was selected earlier in *Global* sheet under *Properties/Specifications*), Henry components ID, Chemistry ID, and calculation type.

---

[6]The standard deviation of a measurement variable measures the spread of the data about the mean value.
[7]Aspen Plus 11.1 User Guide (September 2001).

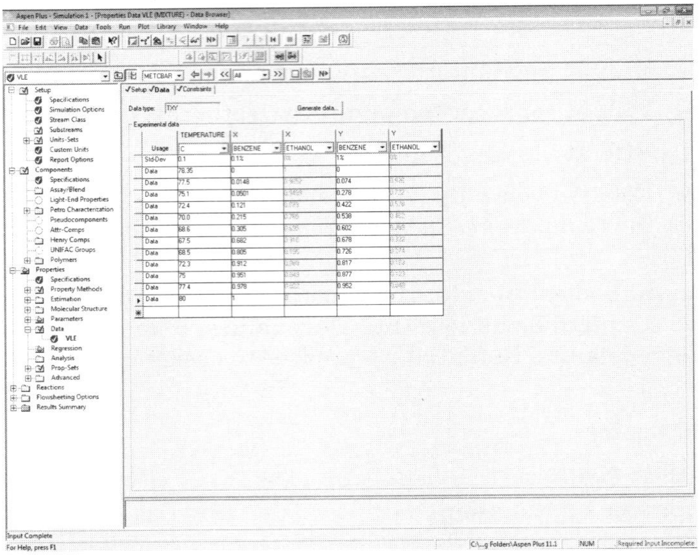

FIGURE 1.84

The table at the bottom of the *Setup* sheet is filled by default. Note that to assign more (or less) weight to data sets, enter a value greater (or smaller) than 1 in the *Weight* field. Here we wish to include thermodynamic consistency test and therefore, a check mark is inserted in the *Perform test* box (see Figure 1.85).

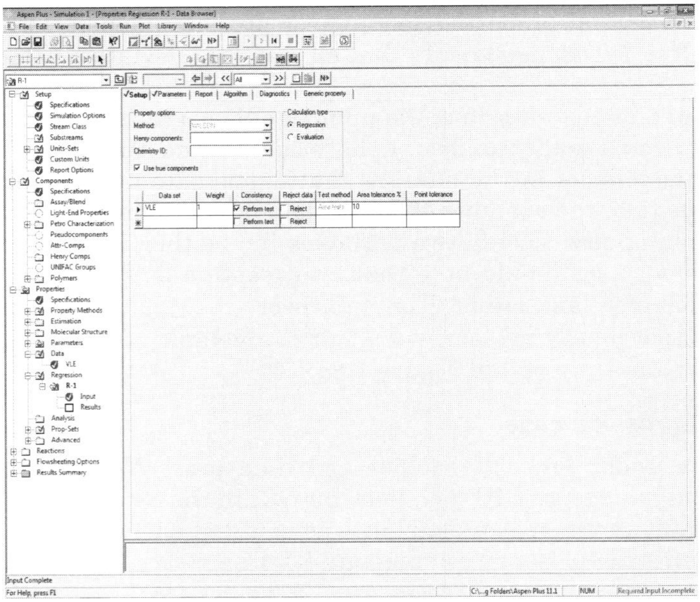

FIGURE 1.85

There are two methods available in the Aspen Plus for testing consistency and they are:

- The area test of Redlich-Kister (default option)
- The point test of van Ness-Fredenslund

We can use either of them or both. For details of these tests, go through the Help menu from the menu bar.

Aspen Plus selects species 1 and 2 of the Wilson binary parameters for regression (Figure 1.86) and the simulator uses the databank values for the binary parameters as initial guesses in the regression.

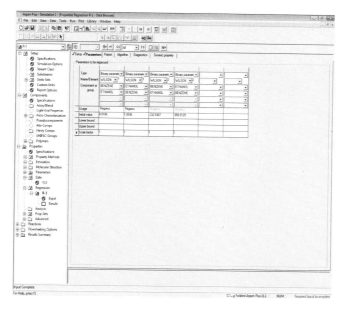

FIGURE 1.86

Running the regression

Select *Run* from the *Run* menu (or press F5), and then click *OK* on Figure 1.87.

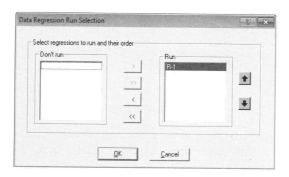

FIGURE 1.87

Analysis of regression results

For the benzene/ethanol system, the VLE data are regressed without any convergence problem. We should note that if the Data Regression run fails to converge, check the data contained in the *Data* form.

Another possibility is that the regression run may converge without error but the results may be unsatisfactory. This can be identified by some of the following indicators:

- For a regressed parameter, the standard deviation of zero implies that the parameter is at a bound. To verify this, open the *Parameters* sheet of *R-1 / Results* under *Regression* subfolder. Figure 1.88 shows the results for the representative binary system and obviously, the standard deviation value of zero is there in first two rows.

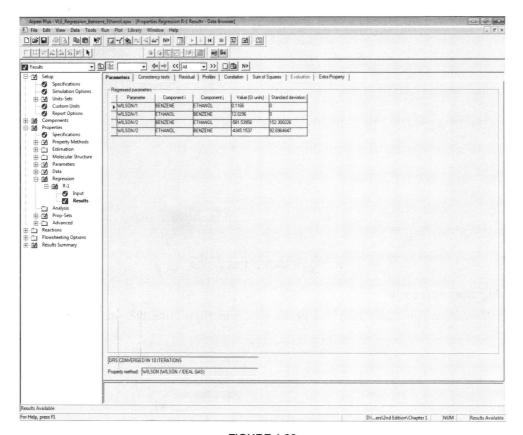

FIGURE 1.88

- Usually, the residual root mean square error value should be less than 10 for VLE data and less than 100 for LLE data. For our case, we have Figure 1.89.

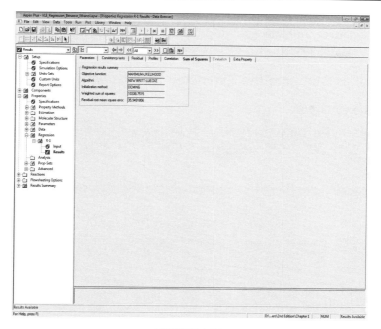

FIGURE 1.89

- The VLE data fail the thermodynamic consistency test. Our binary VLE data fail to clear the consistency test (i.e., area test) and it is obvious in Figure 1.90.

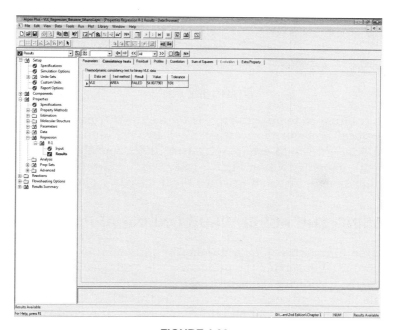

FIGURE 1.90

At this point, we must note that if any of the above three conditions exists, check the original data source and the data and units on the *Data* form for errors. And this is also applicable to the representative benzene/ethanol system.

1.4.1 T-xy Plot to Compare the Estimated Results with the Experimental Data

Plotting regression results

Let us take a look on the regressed values at the side of the corresponding experimental data. For this, open the *Residual* form under *Regression / R-1 / Results* and obtain Figure 1.91.

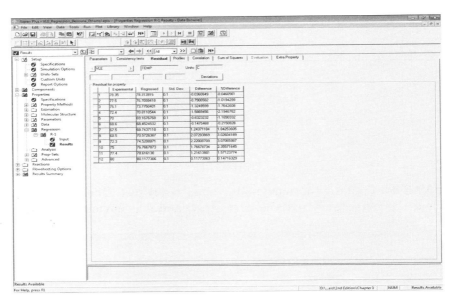

FIGURE 1.91

To plot the residual of temperature, use *Plot Wizard* from the *Plot* menu. Pressing *Next*, we see several types of plot (e.g., T-xy, P-xy, T-x, etc). Choosing *Residual* plot, click on *Next*. Under *VLE* data group, we select *TEMP* variable (i.e., temperature) and hit *Finish* and subsequently obtain Figure 1.92.

By the similar fashion, we can produce the T-xy plot as shown in Figure 1.93.

1.5 ACCESSING THE ASPEN PLUS ONLINE APPLICATIONS LIBRARY

Aspen Plus online library provides a number of simulated examples to illustrate how Aspen Plus is employed in solving the problems. The examples include a range of process industries, namely chemicals, gas processing, petroleum refining, petrochemicals, pharmaceuticals, metals processing, etc. After learning the Aspen Plus simulation of a particular process available in online library, we can modify and run it to simulate as per our requirements.

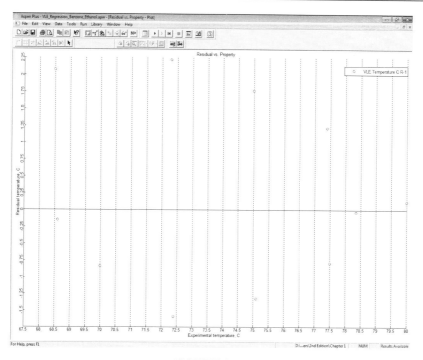

FIGURE 1.92

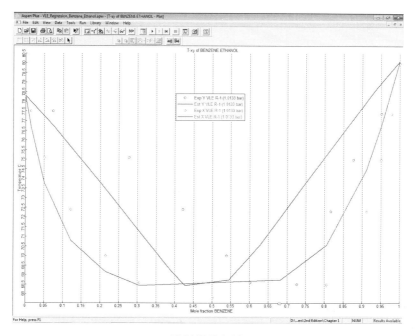

FIGURE 1.93

The following steps can be followed to access the Aspen Plus online library.

1. Select Aspen Plus User Interface. When the Aspen Plus window pops up, choose *Open an Existing Simulation* followed by *More Files* (see Figure 1.94).

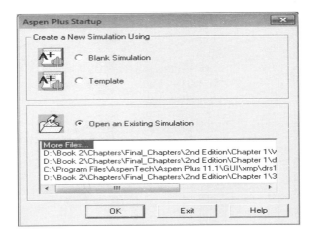

FIGURE 1.94

2. Click *OK* and as the *Open* dialog box appears, click the Favorites button as shown in Figure 1.95.

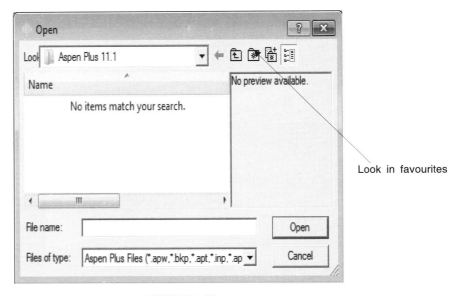

FIGURE 1.95

3. Select the *Applications* or *Examples* directory in Figure 1.96.

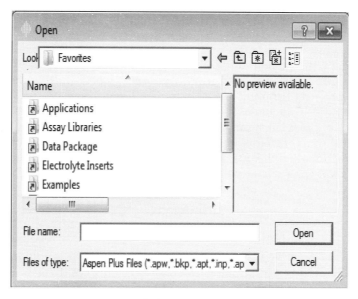

FIGURE 1.96

4. Before opening a backup file (*.bkp), we can have a little description of a file, if available, by highlighting it as shown in Figure 1.97.

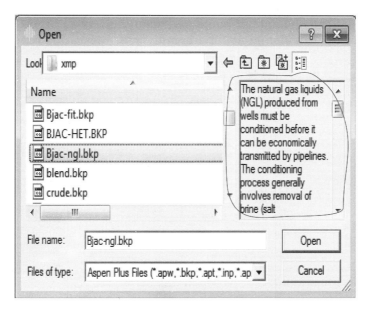

FIGURE 1.97

5. Hit *Open* button selecting a Aspen file of our interest.

SUMMARY AND CONCLUSIONS

In this chapter, a brief introduction of the Aspen simulator is presented first. It is well recognized that the Aspen software is an extremely powerful simulation tool, in which, a large number of parameter values are stored in the databank and the calculations are pre-programmed. At the preliminary stage of this software course, this chapter may help to accustom with several items and stepwise simulation procedures. Here, a range of simple problems, dealing with flash calculation, bubble point calculation, dew point calculation and T-xy as well as P-xy plot generation, has been solved showing all simulation steps. In addition, we also learn here how to regress a set of experimental data.

PROBLEMS

1.1 A liquid mixture, consisting of 60 mole% benzene and 40 mole% toluene, is fed with a flow rate of 100 kmol/h at 3 bar and 25°C to a flash chamber (Flash2) operated at 1.2 atm and 100°C. Applying the SYSOP0 method, compute the amounts of liquid and vapor products and their compositions.

1.2 A liquid mixture, consisting of 60 mole% benzene, 30 mole% toluene and 10 mole% o-xylene, is flashed at 1 atm and 110°C. The feed mixture with a flow rate of 100 kmol/h enters the flash drum (Flash2) at 1 atm and 80°C. Using the SYSOP0 property method,

(a) compute the amounts of liquid and vapor outlets and their compositions
(b) Repeat the calculation at 1.5 atm and 120°C (operating conditions)

1.3 A hydrocarbon mixture with the composition, shown in Table 1.4, is fed to a flash drum at 50°F and 20 psia.

TABLE 1.4

Component	Flow rate (lbmol/h)
i-C_4	12
n-C_4 (LK)	448
i-C_5 (HK)	36
n-C_5	15
C_6	23
C_7	39.1
C_8	272.2
C_9	31
	876.3

The flash chamber (Flash2) operates at 180°F and 80 psia. Applying the SYSOP0 thermodynamic model, determine the amounts of liquid and vapor products and their compositions.

1.4 Find the bubble point and dew point temperatures of a mixture of 0.4 mole fraction toluene and 0.6 mole fraction *iso*-butanol at 101.3 kPa. Assume ideal mixture and inlet temperature of 50°C, pressure of 1.5 atm, and flow rate of 100 kmol/h.

1.5 Find the bubble point and dew point temperatures and corresponding vapor and liquid compositions for a mixture of 33 mole % *n*-hexane, 33 mole% *n*-heptane and 34 mole % *n*-octane at 1 atm pressure. The feed mixture with a flow rate of 100 kmol/h enters at 50°C and 1 atm. Consider ideality in both liquid and vapor phases.

1.6 Compute the bubble point and dew point temperatures of a solution of hydrocarbons with the following composition at 345 kN/m² (see Table 1.5).

TABLE 1.5

Component	Mole fraction
C_3	0.05
n-C_4	0.25
n-C_5	0.4
C_6	0.3

The ideal solution with a flow rate of 100 kmol/h enters at 50°C and 1 atm.

1.7 Calculate the bubble point pressure at 40°C of the following hydrocarbon stream (see Table 1.6).

TABLE 1.6

Component	Mole fraction
C_1	0.05
C_2	0.1
C_3	0.15
i-C_4	0.1
n-C_4	0.2
i-C_5	0.15
n-C_5	0.15
C_6	0.1

Use the SRK thermodynamic model and consider the inlet temperature of 30°C, pressure of 4.5 bar and flow rate of 100 kmol/h.

1.8 A binary mixture, consisting of 50 mole% ethanol and 50 mole% 1-propanol, is fed to a flash drum (Flash 2) with a flow rate of 120 kmol/h at 3.5 bar and 30°C.

(a) Produce T-xy plot at a constant pressure (1.013 bar)
(b) Produce P-xy plot at a constant temperature (75°C)
(c) Produce xy plot based on the data obtained in part (b)

Consider the RK-Soave thermodynamic model as a base property method.

1.9 A ternary mixture with the following component-wise flow rates is introduced into a decanter model run at 341.1 K and 308.9 kPa. To identify the second liquid phase, select *n*-pentane as a key component (see Table 1.7).

TABLE 1.7

Component	Flow rate (kmol/h)
n-pentane	10
ethanol	3
water	7.5

Applying the NRTL property method, simulate the decanter block to compute the flow rates of two product streams.

1.10 A ternary mixture having the following flow rates is fed to a separator (Sep2) at 50°C and 5 bar (see Table 1.8).

TABLE 1.8

Component	Flow rate (kmol/h)
n-pentane	33.623
ethanol	0.473
water	3.705

To solve the present problem using the Aspen Plus, the following specifications are provided along with a T/F ratio of 0.905478 (see Table 1.9).

TABLE 1.9

Component	Split fraction in Stream T
n-pentane	0.999
ethanol	0.9
water	(calculated by Aspen)

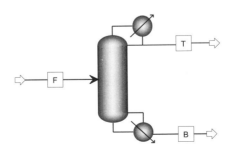

FIGURE 1.98   A flowsheet of a separator.

Applying the SRK property method, simulate the flowsheet, shown in Figure 1.98, and determine the product compositions.

1.11 Repeat the above problem with replacing the separator Sep2 by Sep and using split fraction of water 0.4 in Stream T.

1.12 A dryer, as specified in Figure 1.99, operates at 200°F and 1 atm. Apply the SOLIDS base property method and simulate the dryer model (Flash2) to compute the recovery of water in the top product.

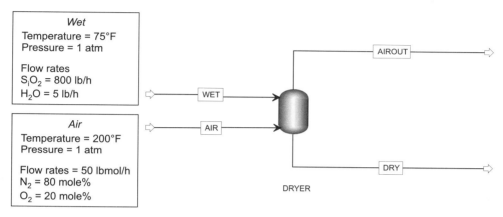

FIGURE 1.99   A flowsheet of a dryer.

1.13 The vapor pressure data for toluene are listed in Table 1.10.

TABLE 1.10

Temperature (°C)	Vapor pressure (torr)
18.5	20
31.9	40
40.2	60
52.0	100
69.4	200
89.6	400
110.5	760
135.8	1500

(i)   Regress the vapor pressure data using the NRTL model.

(ii)   Produce a plot to compare the estimated results with the experimental data provided (Est vs. Exp).

REFERENCE

AspenTech Official Site "When was the Company Founded?, http://www.aspentech.com/corporate/careers/faqs.cfm#whenAT.

CHAPTER 2

Aspen Plus™ Simulation
of Reactor Models

2.1 BUILT-IN REACTOR MODELS

In the Aspen Plus™ model library, seven built-in reactor models are available. They are RStoic, RYield, REquil, RGibbs, RCSTR, RPlug and RBatch. The stoichiometric reactor, RStoic, is used when the stoichiometry is known but the reaction kinetics is either unknown or unimportant. The yield reactor, RYield, is employed in those cases where both the reactions kinetics and stoichiometry are unknown but the product yields are known to us. For single-phase chemical equilibrium or simultaneous phase and chemical equilibrium calculations, we choose either REquil or RGibbs. REquil model solves stoichiometric chemical and phase equilibrium equations. On the other hand, RGibbs solves its model by minimizing Gibbs free energy, subject to atom balance constraints. RCSTR, RPlug and RBatch are rigorous models of continuous stirred tank reactor (CSTR), plug flow reactor (PFR) and batch (or semi-batch) reactor, respectively. For these three reactor models, kinetics is known. RPlug and RBatch handle rate-based kinetic reactions, whereas RCSTR simultaneously handles equilibrium and rate-based reactions. It should be noted that the rigorous models in Aspen Plus can use built-in Power law or Langmuir-Hinshelwood-Hougen-Watson (LHHW) or user defined kinetics. The user can define the reaction kinetics in Fortran subroutine or in excel worksheet.

One of the most important things to remember when using a computer simulation program, in any application, is that incorrect input data or programming can lead to solutions that are "correct" based on the program's specifications, but unrealistic with regard to real-life applications. For this reason, a good knowledge is must on the reaction engineering. In the following, we will simulate several reactor models using the Aspen Plus software package. Apart from these solved examples, interested reader may simulate the reactor models given in the exercise at the end of this chapter.

2.2 ASPEN PLUS SIMULATION OF A RStoic MODEL

Problem statement

Styrene is produced by dehydrogenation of ethylbenzene. Here we consider an irreversible reaction given as:

$$C_6H_5\text{–}C_2H_5 \rightarrow C_6H_5\text{–}CH = CH_2 + H_2$$

ethylbenzene   styrene   hydrogen

Pure ethylbenzene enters the RStoic reactor with a flow rate of 100 kmol/h at 260°C and 1.5 bar. The reactor operates at 250°C and 1.2 bar. We can use the fractional conversion of ethylbenzene equals 0.8. Using the Peng-Robinson thermodynamic method, simulate the reactor model.

Simulation approach

As we start Aspen Plus from the *Start* menu or by double-clicking the Aspen Plus icon on our desktop, first the *Aspen Plus Startup* dialog appears Choose *Template* option and then click *OK* in Figure 2.1.

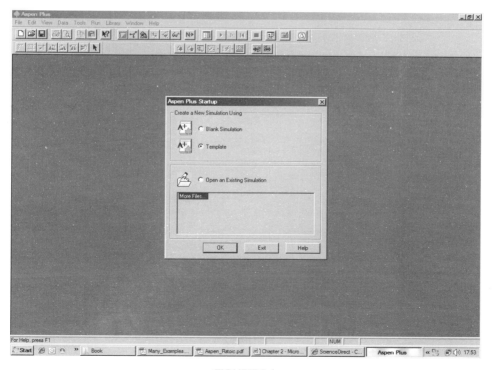

FIGURE 2.1

As the next window pops up (see Figure 2.2), select *General with Metric Units* and hit *OK* button.

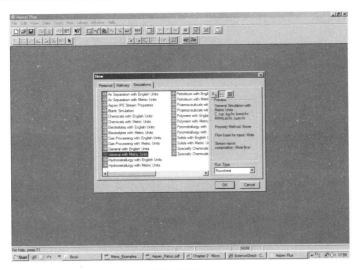

FIGURE 2.2

Here we use the simulation engine at 'Local PC'. Click *OK* when the *Connect to Engine* dialog is displayed (see Figure 2.3). Note that this step is specific to the installation.

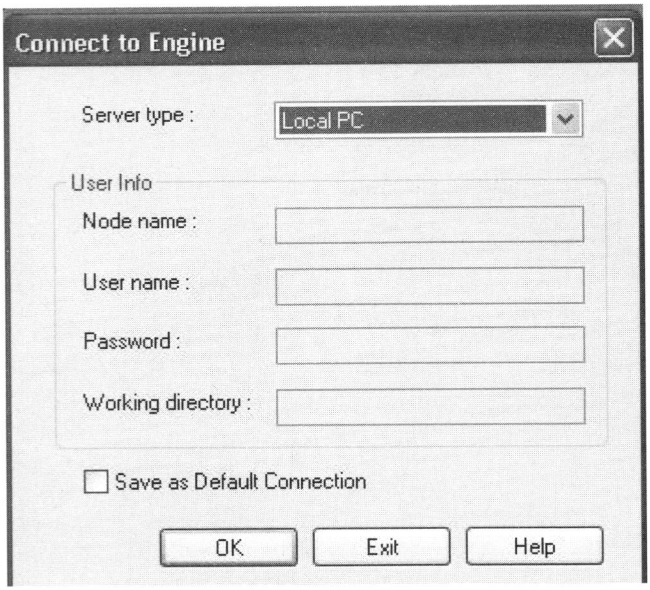

FIGURE 2.3

Creating flowsheet

We are now ready to develop the process flow diagram. Select the *Reactors* tab from the *Model Library* toolbar, then choose *RStoic* icon and finally place this unit in the blank *Process Flowsheet Window*. In order to connect the feed and effluent streams with the reactor block, click on *Material STREAMS* tab in the lower left-hand corner. As we move the cursor, now a crosshair, onto the process flowsheet, block B1 includes two red arrows and one blue arrow. Remember that red arrows are required ports and blue arrows are optional ports.

Click once on the starting point, expand the feed line and click again. This feed stream is labeled as 1. Adding the outlet stream to the reactor in a similar way, finally we make the image as shown in Figure 2.4.

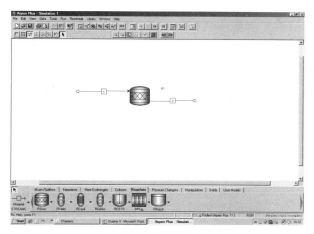

FIGURE 2.4

After renaming Stream 1 to F, Stream 2 to P and Block B1 to REACTOR, the flowsheet looks like Figure 2.5.

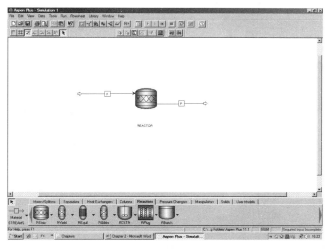

FIGURE 2.5

Obviously, the *Status indicator* in the bottom right of the main window has changed the message from *Flowsheet Not Complete* to *Required Input Incomplete*. Next, we need to enter the remaining data using several input forms required to complete the simulation.

Configuring settings

Hitting *Next* icon and clicking *OK* on the message sheet displayed, we get the setup input form. First the title of the present problem is given as 'Simulation of the RStoic Reactor' [see Figure 2.6(a)]. In the next, the Aspen Plus *accounting* information (required at some installations) is provided in Figure 2.6(b) with:

User name: AKJANA
Account number: 5
Project ID: ANYTHING
Project name: YOUR CHOICE

Finally, select *Report Options* under *Setup* folder, choose 'Mole' as well as 'Mass' fraction item under *Stream* tab [see Figure 2.6(c)].

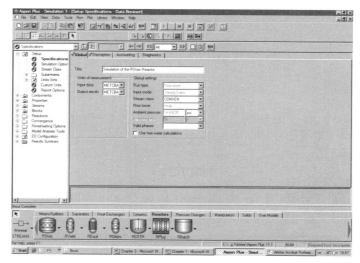

FIGURE 2.6(a)

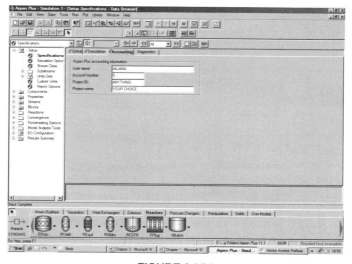

FIGURE 2.6(b)

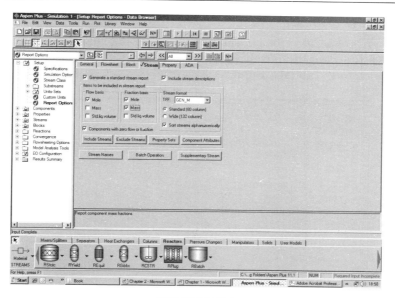

FIGURE 2.6(c)

Specifying components

In the Data Browser window, choose *Components / Specifications* to obtain the component input form. Now fill out the table for three components, ethylbenzene, styrene and hydrogen (see Figure 2.7). If Aspen Plus does not recognize the components by their IDs as defined by the user, use the *Find* button to search them. Select the components from the lists and then *Add* them. A detailed procedure is presented in Chapter 1.

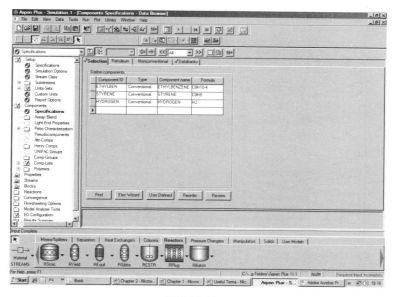

FIGURE 2.7

Specifying property method

Choosing *Properties/Specifications* in the column at the left side, one obtains the property input form. Use the Peng-Robinson thermodynamic package by selecting PENG-ROB under the *Base method* tab (see Figure 2.8).

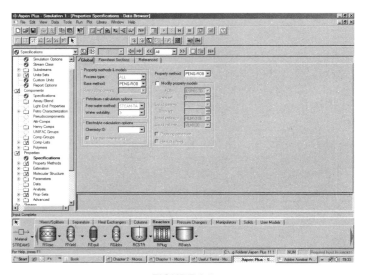

FIGURE 2.8

Specifying stream information

The *Streams/F/Input/Specifications* sheet appears with the Data Browser menu tree in the left pane. Entering the values for state variables (temperature, pressure and total flow) and composition (mole fraction), we, finally, have Figure 2.9.

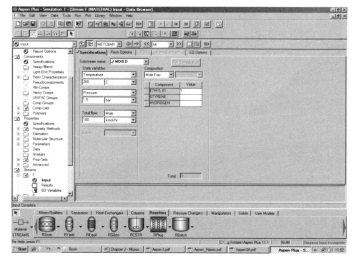

FIGURE 2.9

Specifying block information

From the Data Browser, select *Blocks/REACTOR*. Specifying operating conditions for the reactor model, the form looks like Figure 2.10.

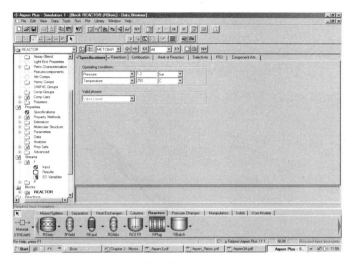

FIGURE 2.10

Specifying reaction information

In the next, either hit *Next* button or *Reactions* tab under *Blocks/REACTOR*. Click *New*, to choose the reactants and products using the dropdown list, input the stoichiometric coefficients and specify the fractional conversion. In the Aspen Plus simulator, coefficients should be negative for reactants and positive for products (see Figure 2.11).

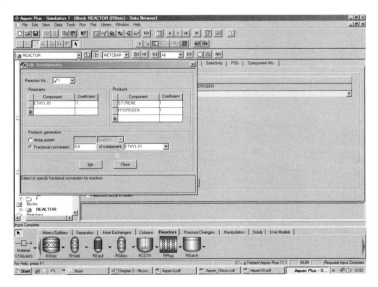

FIGURE 2.11

Running the simulation

In Figure 2.11, *Status message* includes *Required Input Complete*. It implies that all required input information have been inserted by the user. There are a few ways to run the simulation. We could select either the *Next* button in the toolbar which will tell us that all of the required inputs are complete and ask if we would like to run the simulation. We can also run the simulation by selecting the *Run* button in the toolbar (this is the button with a block arrow pointing to the right). Alternatively, we can go to *Run* on the menu bar and select 'Run' (F5). Finally, we have Figure 2.12.

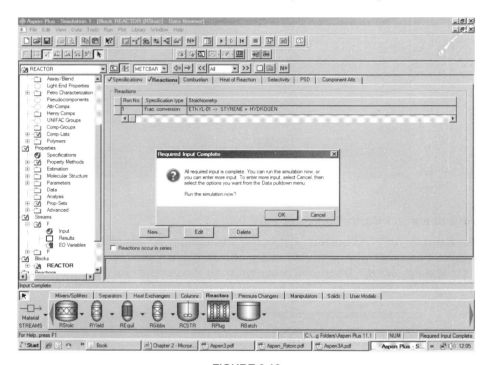

FIGURE 2.12

Viewing results

As we click *OK* on the above message, the *Control Panel* appears showing the progress of the simulation (see Figure 2.13). After the simulation is run and converged, we notice that the *Results Summary* tab on the Data Browser window has a blue checkmark. If the simulation has converged, also we observe no error or warning statement included in the control panel.

Pressing *Next* button and then *OK*, we get the *Run Status* screen. In the subsequent step, select *Results Summary/Streams* in the list on the left and obtain the final results (see Figure 2.14). Save the work done by choosing *File/Save As/...*in the menu list on the top.

If we click on *Stream Table* knob located just above the results table, the results are recorded in the *Process Flowsheet Window*, as shown in Figure 2.15.

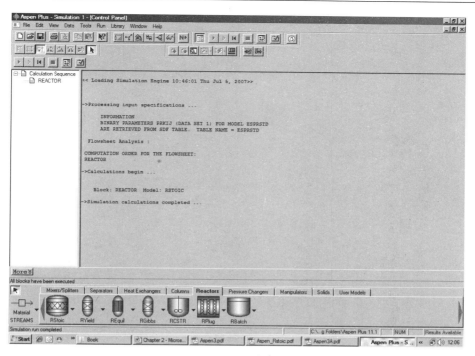

FIGURE 2.13

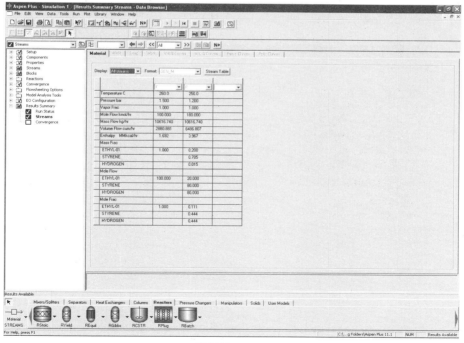

FIGURE 2.14

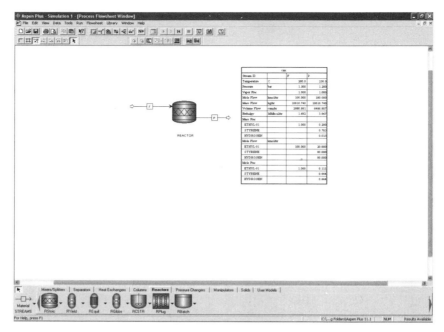

FIGURE 2.15

Viewing input summary

For input information, press Ctrl + Alt + I on the keyboard or select *Input Summary* from the *View* pulldown menu (see Figure 2.16).

FIGURE 2.16

If one may wish to generate a report file (*.rep) for the present problem, follow the instructions as presented in Chapter 1.

2.3 ASPEN PLUS SIMULATION OF A RCSTR MODEL

Problem statement

The hydrogenation of aniline produces cyclohexylamine in a CSTR, according to the following reaction:

$$C_6H_5NH_2 + 3H_2 \rightarrow C_6H_{11}NH_2$$

aniline   hydrogen   cyclohexylamine

The reactor operates at 40 bar and 120°C, and its volume is 1200 ft³ (75% liquid). For the liquid-phase reaction, the inlet streams have the specifications, shown in Table 2.1.

TABLE 2.1

Reactant	Temperature (°C)	Pressure (bar)	Flow rate (kmol/h)
Pure aniline	43	41	45
Pure hydrogen	230	41	160

Fake reaction kinetics data for the Arrhenius law are given as:

Pre-exponential factor = 5×10^5 m³/kmol · s

Activation energy = 20,000 Btu/lbmol

[C_i] basis = Molarity

Use the SYSOP0 base property method in the simulation. The reaction is first-order in aniline and hydrogen. The reaction rate constant is defined with respect to aniline. Simulate the CSTR model and compute the component mole fractions in both the liquid as well as vapour product.

Simulation approach

Start with the *General with Metric Units* Template, as shown in Figures 2.17(a) and (b).

Click *OK* in the above screen. When the *Connect to Engine* dialog appears, again hit *OK* knob to obtain a blank *Process Flowsheet Window*.

Creating flowsheet

Select the *Reactors* tab from the *Model Library* toolbar. As stated earlier, there are total seven built-in models available. Among them, choose RCSTR and place it in the flowsheet window. Adding inlet and product streams and renaming them, the process flow diagram looks like Figure 2.18.

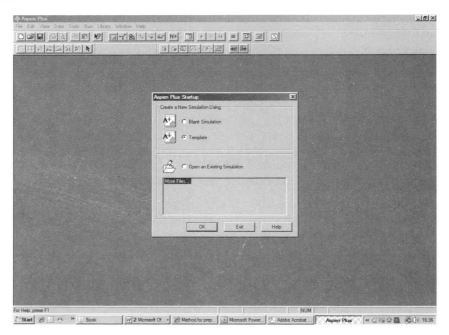

FIGURE 2.17(a)

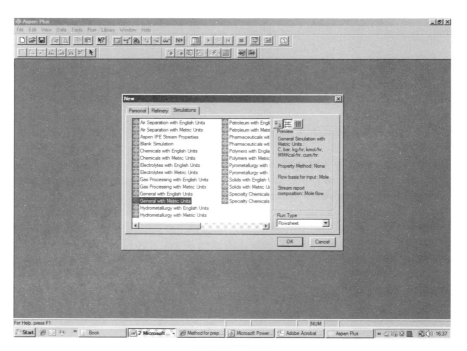

FIGURE 2.17(b)

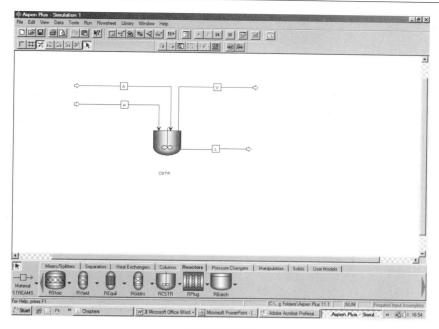

FIGURE 2.18

Configuring settings

Hit *Next* button and then *OK* and get the setup input form. The present project is titled as 'Simulation of the RCSTR Reactor' and the *accounting* information are given as 'AKJANA/ 6/ANYTHING/YOUR CHOICE' [see Figures 2.19(a) and (b)].

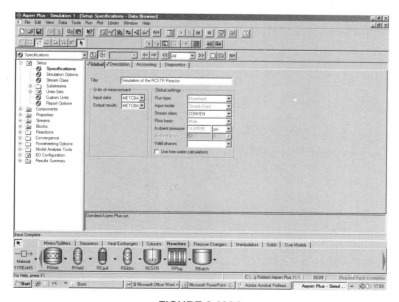

FIGURE 2.19(a)

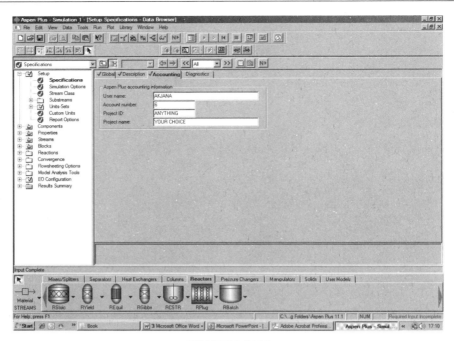

FIGURE 2.19(b)

In the subsequent step, choose *Setup / Report Options / Stream* from the Data Browser window and select 'Mole' as well as 'Mass' fraction basis (see Figure 2.20).

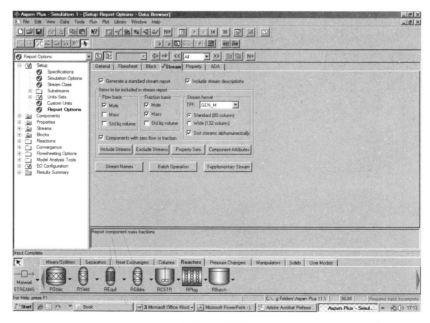

FIGURE 2.20

Specifying components

The example reaction system includes three components. They are aniline, hydrogen and cyclohexylamine. Defining all these species in the component input form, one obtains Figure 2.21.

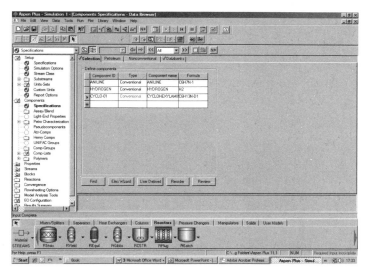

FIGURE 2.21

Specifying property method

We know that a property method is a bank of methods and models used to compute physical properties. For the sample reactor model, select SYSOP0 base property method (see Figure 2.22) after clicking on *Next* icon in Figure 2.21.

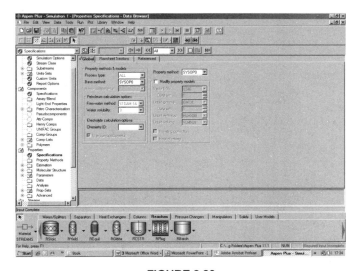

FIGURE 2.22

Specifying stream information

As we hit *Next* followed by *OK*, a stream input form appears. For Stream A (pure aniline) and Stream H (pure hydrogen), values of state variables and composition are inserted in the following two forms, shown in Figures 2.23(a) and (b).

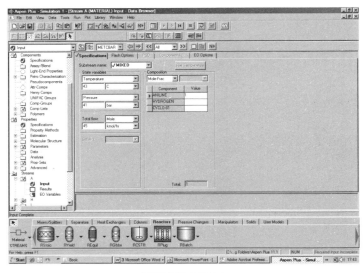

FIGURE 2.23(a)

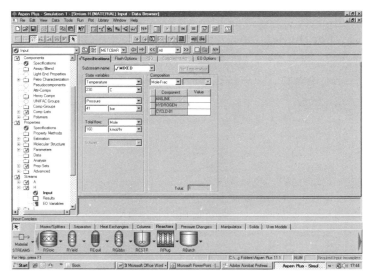

FIGURE 2.23(b)

Specifying block information

In the next, there is a block input form. Providing required information for the CSTR block, we have the screen as shown in Figure 2.24.

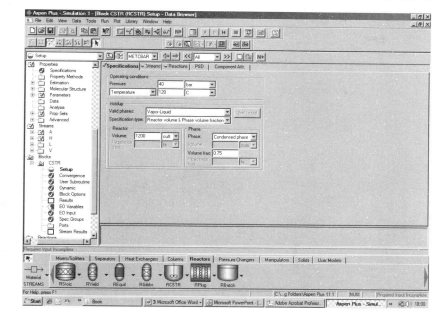

FIGURE 2.24

Product streams have been defined in Figure 2.25 with their phases.

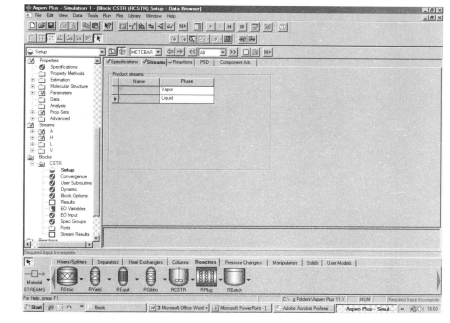

FIGURE 2.25

Press *Next* button or click on *Reactions* and get the window as shown in Figure 2.26.

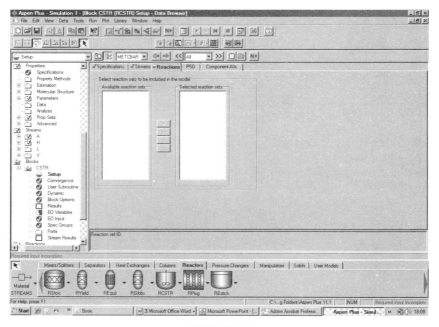

FIGURE 2.26

Right click on *Available reaction sets*, hit *New* button, then either accept default name R-1 or give a name as we want for the reaction set and finally click on *OK*. Subsequently, select POWERLAW in the *Enter Type* list and hit *OK* to get the screen as shown in Figure 2.27.

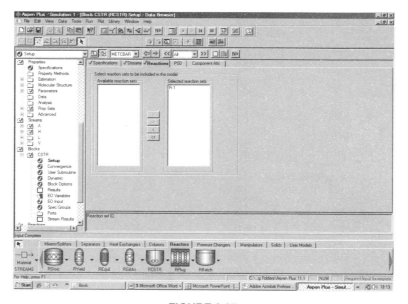

FIGURE 2.27

Specifying reaction information

Hitting *Next* knob, we obtain the screen, shown in Figure 2.28.

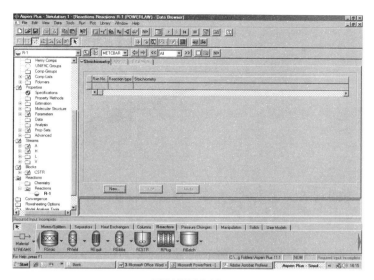

FIGURE 2.28

As we click on *New* button, a form is displayed as shown in Figure 2.29. In this form, we need to enter the stoichiometric coefficient as well as exponent for all components. The exponents represent the *order* of the reaction with respect to each component. Note that there are two types of reactions [kinetic (rate-controlled reactions) and equilibrium] permitted under Power law reaction ID.

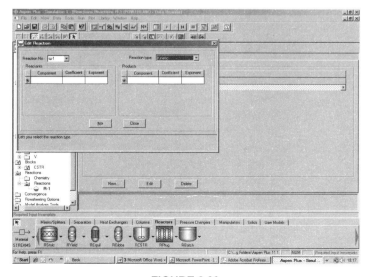

FIGURE 2.29

As stated, the reaction

$$C_6H_5NH_2 + 3H_2 \rightarrow C_6H_{11}NH_2$$

is first-order in aniline and hydrogen. Also, the reaction rate constant is defined with respect to aniline. Accordingly, we may use the following information to specify the reaction (see Table 2.2).

TABLE 2.2

Component	Coefficient	Exponent
aniline	−1	1
hydrogen	−3	1
cyclohexylamine	1	0

Recall that in Aspen Plus terminology, coefficients must be negative for reactants and positive for products. As we fill up the form, it looks like Figure 2.30.

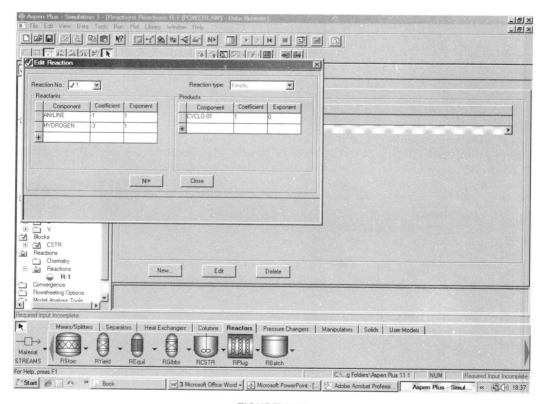

FIGURE 2.30

If we do not specify the exponent for a species, Aspen Plus takes a default value of zero. In Figure 2.31, the resulting reaction is displayed in the stoichiometry sheet.

In the subsequent step (see Figure 2.32), we move on to *Kinetic* tab.

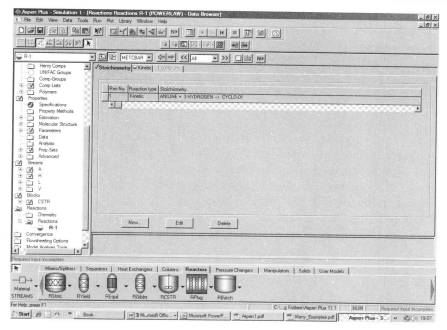

FIGURE 2.31

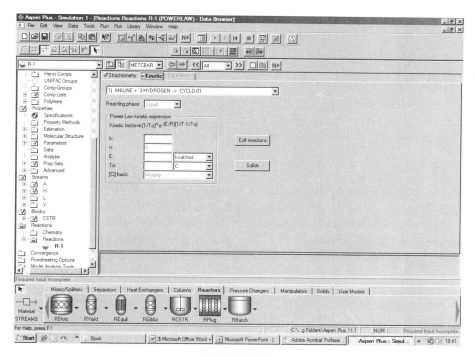

FIGURE 2.32

As directed in the problem statement, we use 'Molarity' basis. Accordingly, the Power law is expressed as:

$$r = k \left(\frac{T}{T_0} \right)^n \exp \left[-\frac{E}{R} \left(\frac{1}{T} - \frac{1}{T_0} \right) \right] \Pi (C_i)^{\alpha_i} \tag{2.1}$$

$$= K \Pi (C_i)^{\alpha_i}$$

where, r is the rate of reaction, K the reaction rate constant (kinetic factor in Aspen Plus terminology), k the pre-exponential or frequency factor, T the temperature in K, T_0 the datum temperature in K, n the temperature exponent, E the activation energy, R the universal gas constant, C the molarity in kmol/m^3, α the concentration exponent, i the component index, and Π the product operator.

If T_0 is ignored, the Power law expression has the following form:

$$r = k T^n \exp \left(-\frac{E}{RT} \right) \Pi (C_i)^{\alpha_i} \tag{2.2}$$

where

$$K = k T^n \exp \left(-\frac{E}{RT} \right) \tag{2.3}$$

In most of our simple cases, the reaction rate constant is represented by the Arrhenius law, that is

$$K = k \exp \left(-\frac{E}{RT} \right) \tag{2.4}$$

Notice that when the Arrhenius formula is used, we put zero for n and nothing for T_0 in the Aspen Plus window. Also, the units of the pre-exponential factor are identical to those of the rate constant and vary depending on the order of the reaction. As we know, the dimensions of the rate constant for an nth order reaction are:

$$(\text{time})^{-1} (\text{concentration})^{1-n}$$

Next come back to the problem. The kinetic data are required to provide in Figure 2.32. Here we use the Arrhenius law to represent the reaction rate constant. It is important to mention that the pre-exponential factor must be specified in SI unit. For the example CSTR problem, the pre-exponential factor and activation energy are given as 5×10^5 m^3/kmol·s and 20,000 Btu/lbmol respectively (see Figure 2.33).

Running the simulation

In the window shown in Figure 2.33, the *Status bar* clearly indicates that all required inputs are now complete. Hitting *Next* knob and clicking on *OK*, we have the following *Control Panel* (see Figure 2.34).

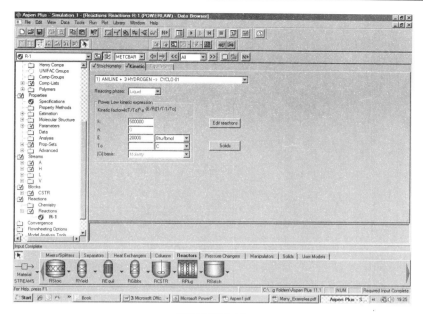

FIGURE 2.33

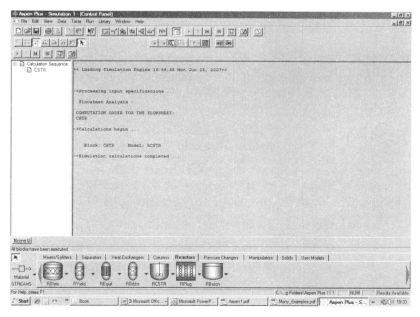

FIGURE 2.34

Viewing results

In the next, select *Solver Settings*, choose *Results Summary/Streams* in the list on the left and finally get the results shown in Figure 2.35 in a tabulated form.

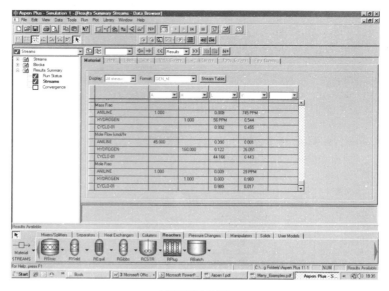

FIGURE 2.35

Save the simulation work in a folder giving a suitable file name.

2.4 ASPEN PLUS SIMULATION OF A RPlug MODEL

Problem statement

The combination of two benzene molecules forms one molecule of diphenyl and one of hydrogen (Fogler, 2005). The elementary reversible vapor-phase reaction occurs in a plug flow reactor (PFR).

$$2C_6H_6 \leftrightarrow C_{12}H_{10} + H_2$$
benzene diphenyl hydrogen

The forward and reverse reaction rate constants are defined with respect to benzene. The vaporized benzene (pure) with a flow rate of 0.02 lbmol/h enters the reactor at 1250°F and 15 Psia. The data for the Arrhenius law are given below.

$$\text{Forward reaction: } k = 3.2 \times 10^{-6} \text{ kmol/s} \cdot \text{m}^3 \cdot (\text{N/m}^2)^2$$

$$E = 30200 \text{ cal/mol}$$

$$\text{Reverse reaction: } k = 1.0 \times 10^{-5} \text{ kmol/s} \cdot \text{m}^3 \cdot (\text{N/m}^2)^2$$

$$E = 30200 \text{ cal/mol}$$

$$[C_i] \text{ basis} = \text{Partial pressure}$$

The reactor length is 36 in and diameter is 0.6 in. It operates at inlet temperature. Applying the SYSOP0 thermodynamic model,

(a) compute the component mole fraction in the product stream, and
(b) produce a plot of 'reactor molar composition' (mole fraction) vs 'reactor length' (in).

Simulation approach

Select Aspen Plus User Interface. When the Aspen Plus window pops up, choose *Template* and click on *OK* (see Figure 2.36).

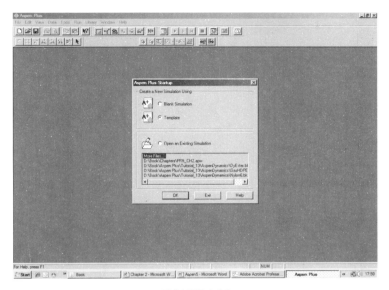

FIGURE 2.36

In the next step (see Figure 2.37), select *General with English Units* and hit *OK* button.

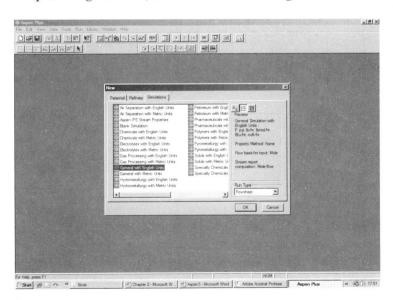

FIGURE 2.37

Click *OK* when the Aspen Plus engine window appears.

Creating flowsheet

In the *Model Library*, select the *Reactors* tab. Expanding the *RPlug* icon, the following screen is obtained (see Figure 2.38).

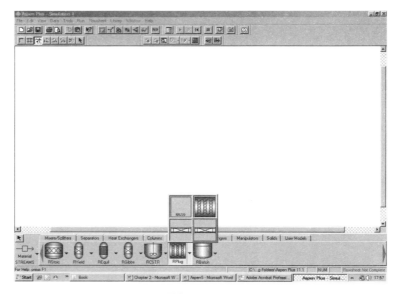

FIGURE 2.38

Inserting the left bottom symbol in the *Process Flowsheet Window*, adding the feed and product streams, and renaming the block as well as streams, finally, we have Figure 2.39.

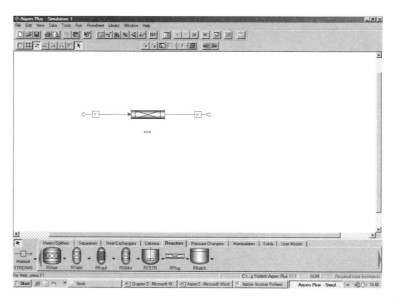

FIGURE 2.39

Configuring settings

At this moment, we are sure that the process flow diagram is drawn correctly. The *Status message* directs us to provide the input information. Hitting *Next* knob and clicking on *OK*, we obtain a form for setup specifications. First we input the *Title* of the present project (Simulation of the RPlug Model), followed by the *accounting* information (AKJANA/7/ANYTHING/AS YOU WANT) and *Report Options* [see Figures 2.40(a) to (c)].

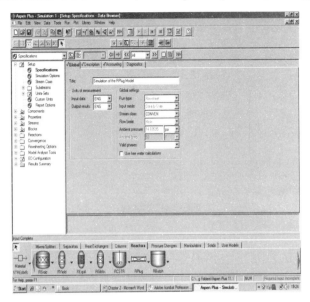

FIGURE 2.40(a)

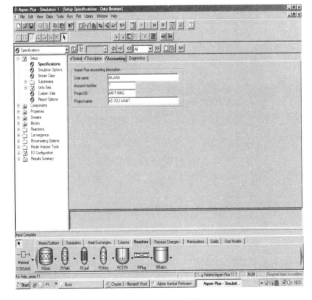

FIGURE 2.40(b)

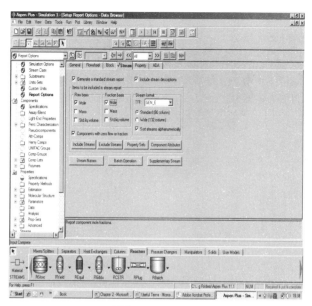

FIGURE 2.40(c)

Specifying components

From the Data Browser, select *Specifications* under the *Components* folder. As we provide the chemical formula of the components in the *Component ID* column, the other columns of the table are automatically filled up (see Figure 2.41).

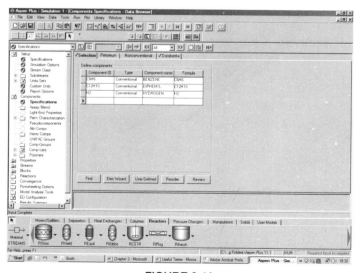

FIGURE 2.41

Specifying property method

In the list on the left, choose *Properties/Specifications* to obtain the property input form. Then choose SYSOP0 by scrolling down (see Figure 2.42).

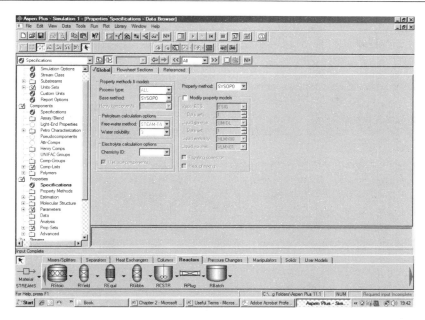

FIGURE 2.42

Specifying stream information

In the left pane of the Data Browser window, select *Streams/F* and enter the values for all state variables and composition as shown in Figure 2.43.

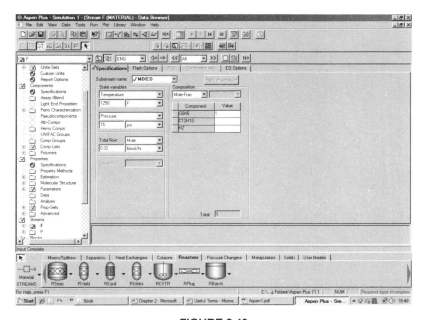

FIGURE 2.43

Specifying block information

In the next, select *PFR* by opening the *Blocks* folder. The reactor is specified in the window, shown in Figure 2.44.

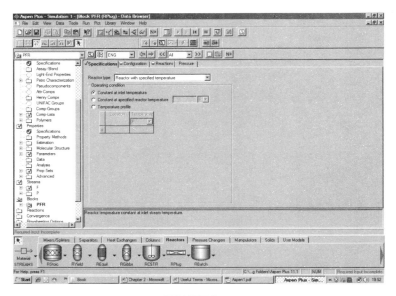

FIGURE 2.44

Open the *Configuration* sheet and enter the reactor dimensions (see Figure 2.45).

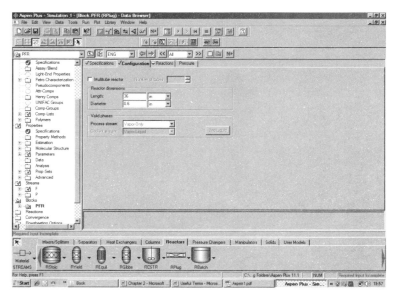

FIGURE 2.45

In the subsequent step, we define a reaction set for the simulation. The default name R-1 has been accepted. Then select Power law kinetics and obtain the picture, shown in Figure 2.46.

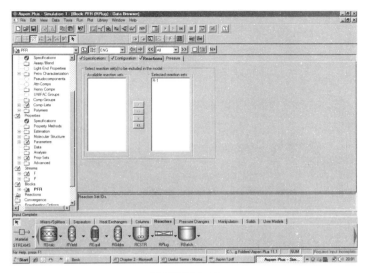

FIGURE 2.46

Specifying reaction information

Hitting *Next* button and clicking on *New*, we have the following forms [see Figures 2.47(a) and (b)] for reaction number 1 ($2C_6H_6 \rightarrow C_{12}H_{10} + H_2$). Since the reaction rate constants are defined with respect to benzene, we convert the stoichiometric coefficient of benzene to unity for both the reactions. Obviously, the reactions are second-order.

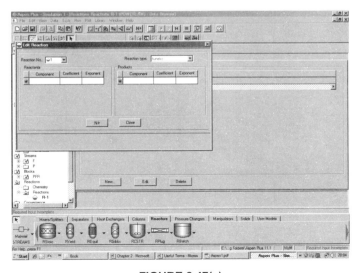

FIGURE 2.47(a)

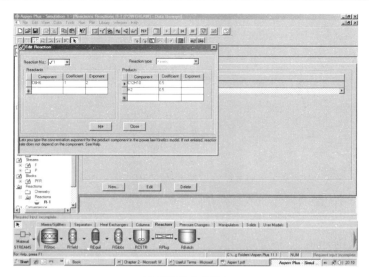

FIGURE 2.47(b)

As mentioned previously, when we do not specify the exponent for a component, Aspen Plus uses a default value of zero. As the message on the screen, shown in Figure 2.47(b) reveals, it is true that the forward reaction rate does not depend on the product components. After completing the first reaction, select 'New' from the *Reaction No.* list. Enter '2' for the reverse reaction ($2C_6H_6 \rightarrow C_{12}H_{10} + H_2$) and click *OK* (see Figure 2.48).

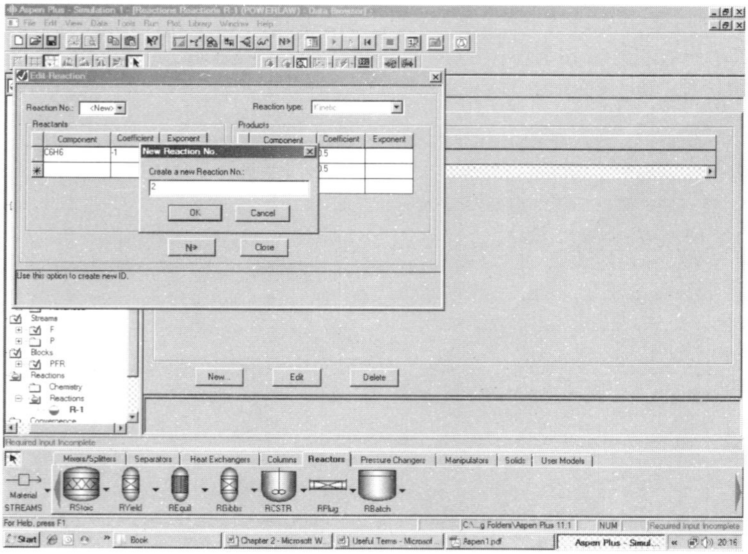

FIGURE 2.48

Subsequently provide the stoichiometric coefficients along with exponents, and get the screen, shown in Figure 2.49.

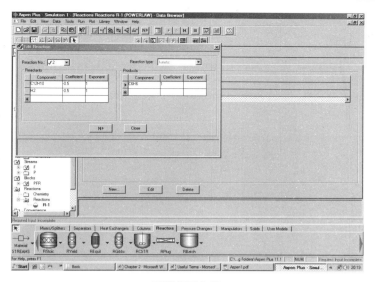

FIGURE 2.49

Hit *Next* knob and obtain two stoichiometric relations as shown in Figure 2.50.

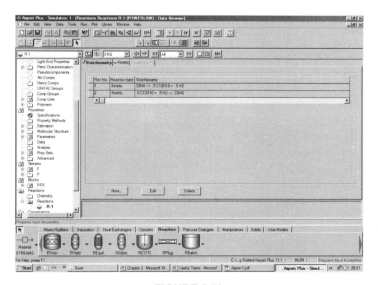

FIGURE 2.50

In the simulation of the present problem, we use partial pressure basis (applicable for vapor only) and therefore, the Power law expression has the following form:

$$r = k \left(\frac{T}{T_0} \right)^n \exp \left[-\frac{E}{R} \left(\frac{1}{T} - \frac{1}{T_0} \right) \right] \Pi (P_i)^{\alpha_i} \qquad (2.5)$$

where, P represents the partial pressure (N/m^2). If T_0 is not specified, the above equation is replaced by:

$$r = kT^n \exp\left(-\frac{E}{RT}\right)\Pi(P_i)^{\alpha_i} \tag{2.6}$$

For the prescribed reactions, values of the pre-exponential factor and activation energy are provided in the two forms, shown in Figures 2.51(a) and (b). To apply the Arrhenius law, we put zero for temperature exponent n and left the box, allotted for datum temperature T_0, empty.

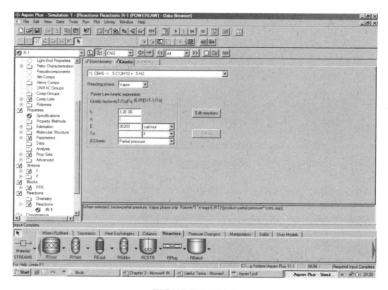

FIGURE 2.51(a)

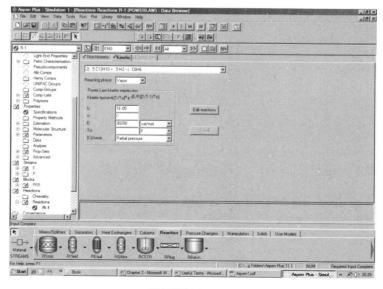

FIGURE 2.51(b)

Running the simulation

Hitting *Next* button and running the simulation, we obtain the *Control Panel* (Figure 2.52) showing the progress of the present simulation.

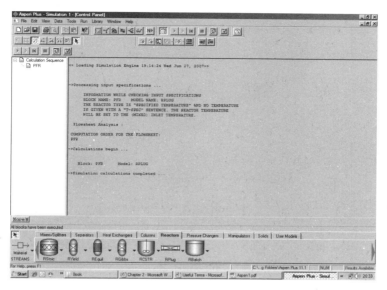

FIGURE 2.52

(a) **Viewing results:** Click on *Solver Settings* knob, choose *Results Summary/Streams* in the column at the left side, and finally, obtain the results for all streams, in Figure 2.53.

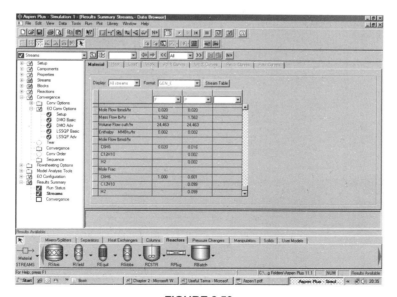

FIGURE 2.53

(b) Producing a plot of mole fraction vs length: Use the Data Browser menu tree to navigate to the *Blocks/PFR/Profiles* sheet (see Figure 2.54).

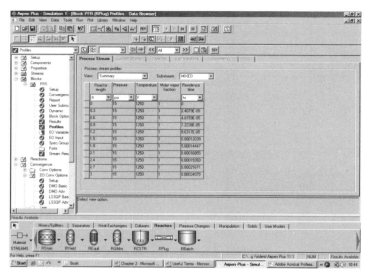

FIGURE 2.54

In the next, select *Plot Wizard* from the *Plot* pulldown menu. Alternatively, press Ctrl+Alt+W on the keyboard and obtain Figure 2.55.

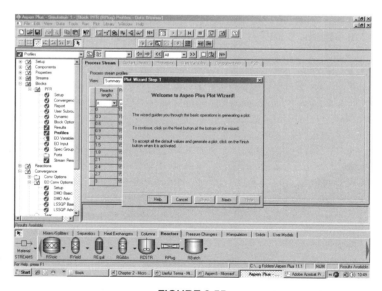

FIGURE 2.55

Click on *Next* icon and get a variety of plot options (see Figure 2.56).

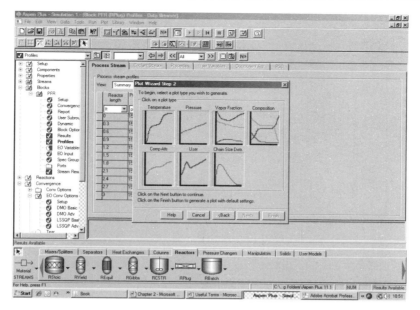

FIGURE 2.56

Among the available options, select one plot type that is titled as 'Composition' and press *Next* button (see Figure 2.57).

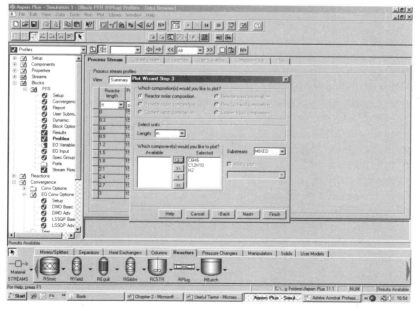

FIGURE 2.57

Again click on *Next* and get the form, shown in Figure 2.58.

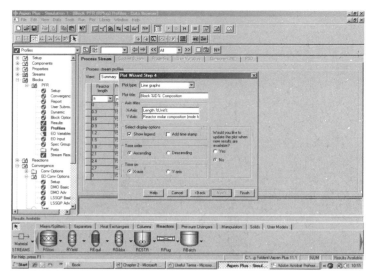

FIGURE 2.58

Check whether the information displayed in Figure 2.58, is *OK* or not. Hitting *Finish* knob, Figure 2.59 is obtained by plotting 'reactor molar composition' (mole fraction) as ordinate against 'reactor length' (in) as abscissa.

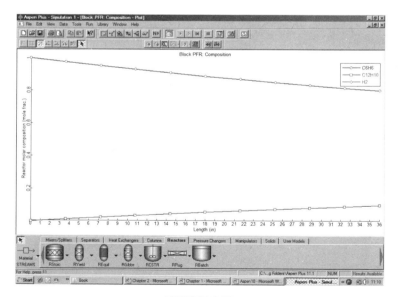

FIGURE 2.59

Note that the plot window can be edited by right clicking on that window and selecting *Properties*. In the properties window, the user can modify the title, axis scale, font and color of the plot. Alternatively, double-click on the different elements of the plot and modify them as we like to improve the presentation and clarity.

2.5 ASPEN PLUS SIMULATION OF A RPlug MODEL USING LHHW KINETICS

Problem statement

In acetic anhydride manufacturing, the cracking of acetone produces ketene and methane according to the following irreversible vapor-phase reaction:

$$CH_3COCH_3 \rightarrow CH_2CO + CH_4$$
$$\text{acetone} \qquad \text{ketene} \quad \text{methane}$$

This reaction is first-order with respect to acetone. Pure acetone feed with a flow rate of 130 kmol/h enters a PFR at 725°C and 1.5 atm. The kinetic data for the Aspen Plus simulation are given below.

$$k = 1.1 \text{ s}^{-1}$$
$$E = 28.5 \times 10^7 \text{ J/kmol}$$
$$n = 0$$
$$T_0 = 980 \text{ K}$$

The unit of pre-exponential factor clearly indicates the $[C_i]$ basis. To use the Langmuir-Hinshelwood-Hougen-Watson (LHHW) kinetic model, set zero for all coefficients under Term 1 and that for all coefficients except A under Term 2. Take a very large negative value for coefficient A. The sample adiabatic PFR is 3 m in length and 0.6 m in diameter. Applying the SYSOP0 base method, compute the component mole fraction in the product stream.

Simulation approach

As we select Aspen Plus User Interface, first the *Aspen Plus Startup* window appears, as shown in Figure 2.60. Choose *Template* option and press *OK*.

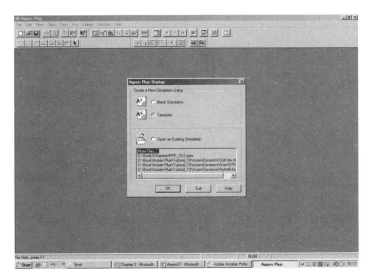

FIGURE 2.60

In the next, select *General with Metric Units* and again hit *OK* button in Figure 2.61.

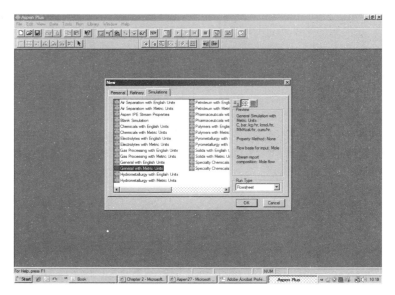

FIGURE 2.61

As the *Connect to Engine* dialog pops up, click *OK*.

Creating flowsheet

From the *Model Library* toolbar, we have selected RPlug reactor and developed the process flow diagram as displayed in Figure 2.62.

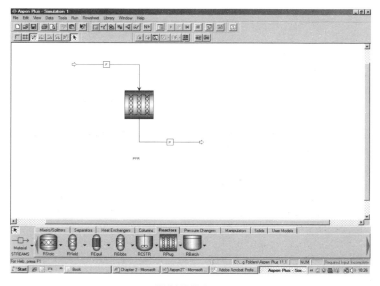

FIGURE 2.62

Configuring settings

In the list on the left, choose *Setup / Specifications*. For the present problem, we wish to give the *Title* as 'Simulation of the PFR', and *accounting* information as 'AKJANA/8/ ANYTHING/AS WE LIKE'. In addition, choose 'Mole' and 'Mass' fraction basis for the streams under *Report Options* [see Figures 2.63(a), to (c)].

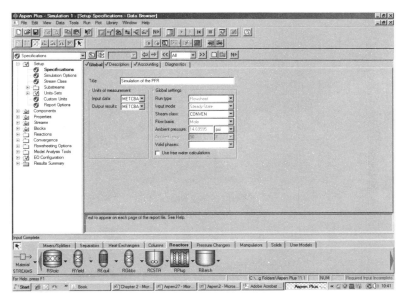

FIGURE 2.63(a)

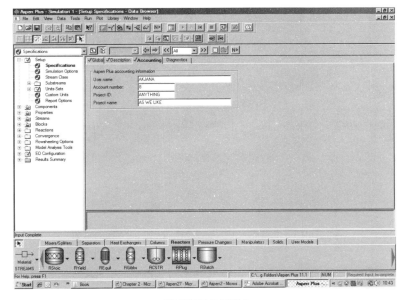

FIGURE 2.63(b)

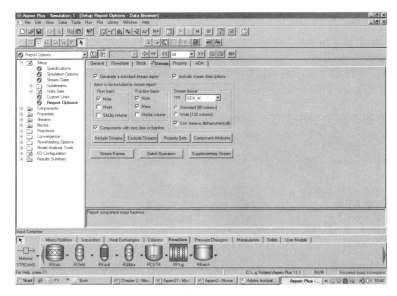

FIGURE 2.63(c)

Specifying components

Select *Specifications* under *Components* folder in the Data Browser window. As we fill out the *Component ID* column, Aspen Plus provides the rest of the information in the component input form, shown in Figure 2.64.

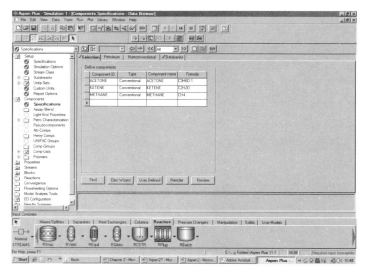

FIGURE 2.64

Specifying property method

Hit *Next* button and under base method (see Figure 2.65), scroll down to get SYSOP0.

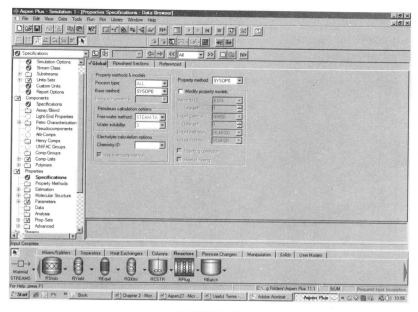

FIGURE 2.65

Specifying stream information

In the left pane of the Data Browser window, select *Streams/F*. Inputting the values for temperature, pressure, total flow and mole fraction, we have the picture as displayed in Figure 2.66.

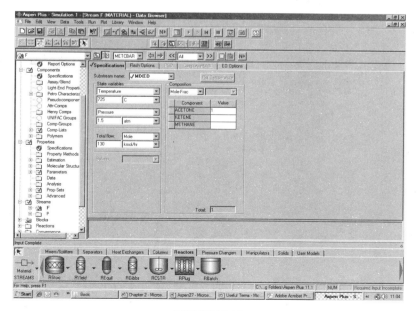

FIGURE 2.66

Specifying block information

In the subsequent step (see Figure 2.67), select *PFR* under *Blocks* folder. Specify the reactor as an adiabatic one.

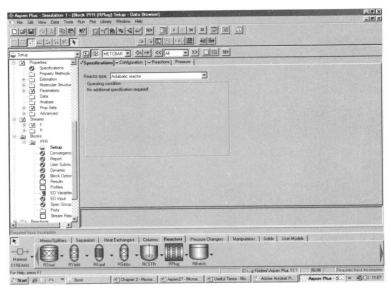

FIGURE 2.67

In the *Configuration* sheet, reactor length as well as diameter is given (see Figure 2.68).

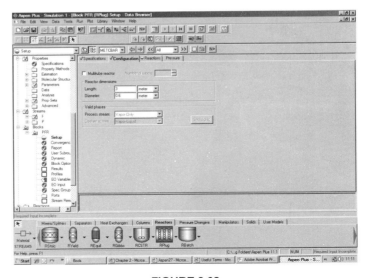

FIGURE 2.68

In the next, we define a reaction set for the present simulation. The default name R-1 is *OK*. Then select LHHW kinetics and obtain the screen, exhibited in Figure 2.69.

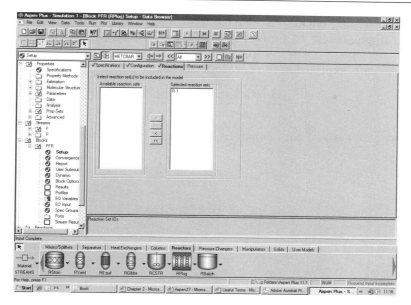

FIGURE 2.69

Specifying reaction information

Press *Next* knob and then click on *New*. Under *Reactants*, select 'ACETONE' from the *Component* dropdown menu and set the coefficient to −1. Similarly under *Products*, select 'KETENE' and 'METHANE', and set both coefficients to 1 (see Figure 2.70).

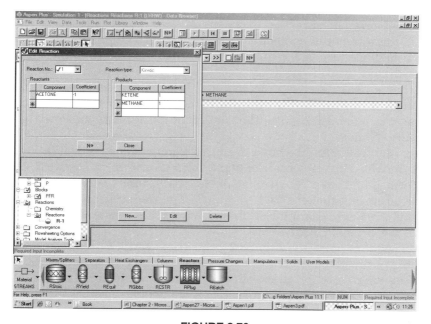

FIGURE 2.70

Hitting on *Next* and clicking *Kinetic* button, we get the kinetics input form. A little description is given below to understand the use of LHHW kinetic model in Aspen simulator.

The LHHW rate expression is represented by:

$$r = \frac{\text{(kinetic factor) (driving force)}}{\text{(adsorption expression)}} \tag{2.7}$$

The kinetic factor (reaction rate constant) has the following form:

$$K = k \left(\frac{T}{T_0}\right)^n \exp\left[-\frac{E}{R}\left(\frac{1}{T} - \frac{1}{T_0}\right)\right] \tag{2.8}$$

If T_0 is ignored, Eq. (2.3) replaces the above expression. Note that all the notations used in Eq. (2.8) have been defined earlier.

The driving force is expressed by:

$$K_1 \left(\prod_{i=1}^{N} C_i^{\alpha i}\right) - K_2 \left(\prod_{j=1}^{N} C_j^{\alpha j}\right)$$

and the adsorption is modeled as:

$$\left[\sum_{i=1}^{M} K_i \left(\prod_{j=1}^{N} C_j^{\alpha j}\right)\right]^m$$

where

$$\ln(K_i) = A_i + B_i/T + C_i \ln(T) + D_i T \tag{2.9}$$

Here, m is the adsorption expression exponent, M the number of terms in the adsorption expression, N the number of components, α the concentration exponent, K_1, K_2, K_i the equilibrium constants [Eq. (2.9)], A_i, B_i, C_i, D_i the coefficients and Σ the summation operator. Notice that the concentration term C used in the above discussion is dependent on the $[C_i]$ basis. Say for example, when $[C_i]$ basis is selected as molarity, the concentration term represents the component molar concentration ($kmol/m^3$); similarly when $[C_i]$ basis is partial pressure, the concentration term represents the component partial pressure (N/m^2).

Providing required data, we have the filled kinetic sheet, shown in Figure 2.71.

Click on *Driving Force* to obtain a blank form as shown in Figure 2.72.

Select 'Term 1' and then 'Molarity' as $[C_i]$ basis. Under *Concentration exponents for reactants*, set acetone exponent to 1. Similarly for products, set ketene and methane exponents to 0. Also enter zero for all four driving force constants (see Figure 2.73) as mentioned in the problem statement.

In the subsequent step (see Figure 2.74), select 'Term 2' from the pulldown *Enter term* menu. Since the given reaction is first-order with respect to acetone and there is no second term, enter zero for all exponents and B, C, and D coefficients. Owing to the method Aspen Plus uses to specify a reaction, we should insert a very large negative value for coefficient A (say, -10^6) to make Term 2 essentially zero [see Eq. (2.9)]. Finally, click on *Next* icon.

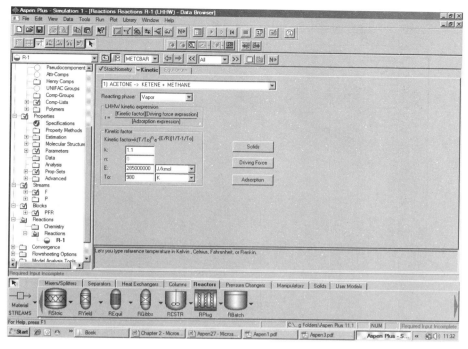

FIGURE 2.71

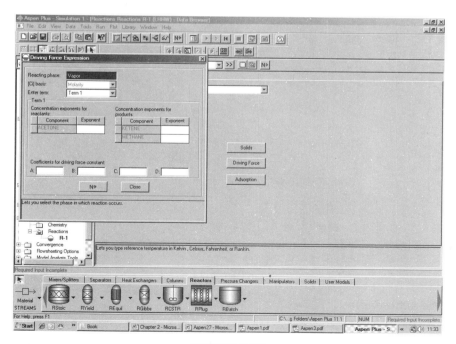

FIGURE 2.72

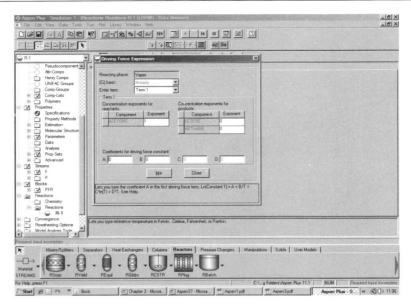

FIGURE 2.73

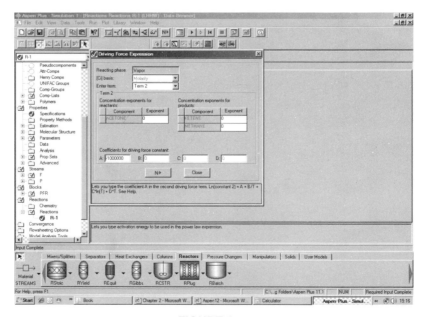

FIGURE 2.74

Running the simulation

The *Status bar* displays a message of *Required Input Complete* in the bottom right corner of the window shown in Figure 2.74. Subsequently, run the simulation and obtain the status report as displayed in Figure 2.75.

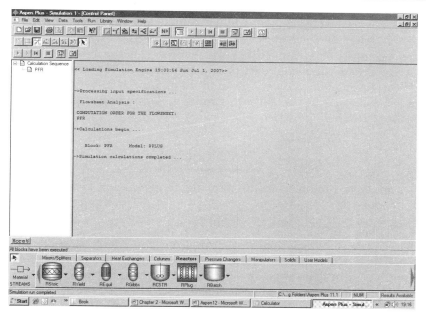

FIGURE 2.75

Viewing results

Pressing *Solver Settings* knob and selecting *Results Summary / Streams*, we obtain the final results as reported in Figure 2.76.

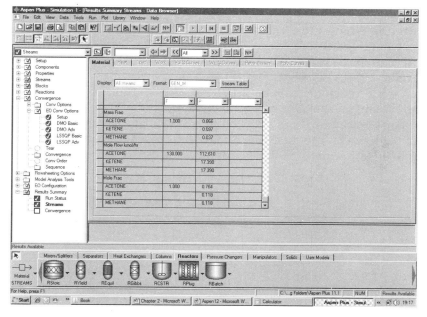

FIGURE 2.76

SUMMARY AND CONCLUSIONS

This chapter presents the simulation of several reactor models. Here, we have considered a variety of chemical reactions in the Aspen Plus simulator. Probably the most useful kinetic models, Power law and Langmuir-Hinshelwood-Hougen-Watson (LHHW), have been used in the solved examples. A number of problems are given in the exercise for extensive practice.

PROBLEMS

2.1 Ethyl acetate is produced in an esterification reaction between acetic acid and ethyl alcohol.

$$\text{acetic acid} + \text{ethyl alcohol} \leftrightarrow \text{ethyl acetate} + \text{water}$$

The feed mixture, consisting of 52.5 mole% acetic acid, 45 mole% ethyl alcohol and 2.5 mole% water, enters the RCSTR model with a flow rate of 400 kmol/h at 75°C and 1.1 atm. The reactor operates at 70°C and 1 atm. Both the reactions are first-order with respect to each of the reactants (i.e., overall second-order). For these liquid-phase reactions, the kinetic data for the Arrhenius law are as follows:

$$\text{Forward reaction: } k = 2.0 \times 10^8 \text{ m}^3/\text{kmol} \cdot \text{s}$$

$$E = 6.0 \times 10^7 \text{ J/kmol}$$

$$\text{Reverse reaction: } k = 5.0 \times 10^7 \text{ m}^3/\text{kmol} \cdot \text{s}$$

$$E = 6.0 \times 10^7 \text{ J/kmol}$$

$$[C_i] \text{ basis} = \text{Molarity}$$

Perform the Aspen Plus simulation using the NRTL thermodynamic model and reactor volume of 0.15 m^3.

2.2 Repeat the above problem replacing RCSTR model by RStoic model with 80% conversion of ethyl alcohol.

2.3 Simulate the reactor (Problem 2.1) for the case of an RGibbs model.

2.4 An input stream, consisting of 90 mole% di-*tert*-butyl peroxide, 5 mole% ethane and 5 mole% acetone, is introduced in a CSTR at 10 atm and 125°C and a flow rate of 0.2 kmol/h. The following elementary irreversible vapor-phase reaction is performed isothermally with no pressure drop.

$$(CH_3)_3COOC(CH_3)_3 \rightarrow C_2H_6 + 2CH_3COCH_3$$

Fake kinetic data for the Arrhenius formula are given as:

$$k = 1.67 \times 10^4 \text{ kmol/m}^3 \cdot \text{s} \cdot (\text{N/m}^2)$$

$$E = 85 \times 10^3 \text{ kJ/kmol}$$

$$[C_i] \text{ basis} = \text{Partial pressure}$$

The reactor operates at 50°C and its volume is 6 m^3. Using the SYSOP0 thermodynamic method, simulate the CSTR model and compute the component mole fractions in the product stream.

2.5 A feed stream, consisting of di-*tert*-butyl peroxide, ethane and acetone, enters a RYield model at 10 atm and 125°C. The reactor operates at 10 atm and 50°C. Use the SYSOP0 property method and assume the following component-wise flow rates in the feed and product streams (see Table 2.3).

TABLE 2.3

Component	Feed flow rate (kg/h)	Product flow rate (kg/h)
di-*tert*-butyl peroxide	26.321	1.949
ethane	0.301	5.314
acetone	0.581	19.94

Simulate the RYield reactor and compare the results (mole fractions in the product) with those obtained for Problem 2.4.

2.6 As stated in Problem 2.1, the reaction between acetic acid and ethanol gives ethyl acetate and water.

$$CH_3COOH + C_2H_5OH \leftrightarrow CH_3COOC_2H_5 + H_2O$$

The inlet stream, consisting of 50 mole% acetic acid, 45 mole% ethanol and 5 mole% water, is fed to a REquil model with a flow rate of 400 kmol/h at 75°C and 1.1 atm. The reactor operates at 80°C and 1 atm. Using the NRTL property method, simulate the reactor model and report the compositions of the product streams.

2.7 Ethylene is produced by cracking of ethane in a plug flow reactor. The irreversible elementary vapor-phase reaction is given as:

$$C_2H_6 \rightarrow C_2H_4 + H_2$$

ethane   ethylene   hydrogen

Pure ethane feed is introduced with a flow rate of 750 kmol/h at 800°C and 5.5 atm. The reactor is operated isothermally at inlet temperature. The kinetic data for the LHHW model are given below (Fogler, 2005).

$$k = 0.072 \text{ s}^{-1}$$

$$E = 82 \times 10^3 \text{ cal/mol}$$

$$T_0 = 1000 \text{ K}$$

$$[C_i] \text{ basis} = \text{Molarity}$$

The reactor length is 3 m and diameter is 0.8 m. Using the SYSOP0 thermodynamic model, simulate the reactor.

2.8 Repeat the above problem replacing the PFR by a stoichiometric reactor with 80% conversion of ethane. If require, make the necessary assumptions.

2.9 In acetic anhydride manufacturing, the cracking of acetone occurs and produces ketene and methane according to the following irreversible vapor-phase reaction:

$$CH_3COCH_3 \xrightarrow{K} CH_2CO + CH_4$$

In the CSTR model, ketene is decomposed producing carbon monoxide and ethylene gas.

$$CH_2CO \xrightarrow{K'} CO + 0.5\ C_2H_4$$

where

$$-r_A = K P_A^{1.5}$$

$$-r_k = K'$$

$$K = \exp\left(22.8 - \frac{26586}{T}\right) \text{mol/lit} \cdot \text{s} \cdot \text{atm}^{1.5}$$

$$K' = \exp\left(19.62 - \frac{25589}{T}\right) \text{mol/lit} \cdot \text{s}$$

$[C_i]$ basis = Partial pressure

Here, $-r_A$ is the rate of disappearance of acetone (A), $-r_k$ the rate of disappearance of ketene (k), P_A the partial pressure of A, and K and K' the reaction rate constants. Pure acetone feed with a flow rate of 130 kmol/h enters the reactor at 725°C and 1.5 atm. The reactor with a volume of 1.4 m^3 operates at 700°C and 1.5 atm. Applying the SYSOP0 base method, compute the component mole fractions in the product stream.

REFERENCE

Fogler, H. Scott (2005), *Elements of Chemical Reaction Engineering*, Prentice-Hall of India, 3rd ed., New Delhi.

CHAPTER 3

Aspen Plus™ Simulation of Distillation Models

3.1 BUILT-IN DISTILLATION MODELS

The Aspen simulation package has nine built-in unit operation models for the separating column. In the Aspen terminology, these models are named as DSTWU, Distl, RadFrac, Extract, MultiFrac, SCFrac, PetroFrac, RateFrac and BatchFrac. Under these categories, again several model configurations are there. Note that Extract model is used for liquid-liquid extraction. Among the built-in column models, DSTWU, Distl and SCFrac represent the shortcut distillation and the rest of the distillation models perform rigorous calculations.

The DSTWU model uses Winn-Underwood-Gilliland method for a single-feed two-product fractionating column having either a partial or total condenser. It estimates minimum number of stages using Winn method and minimum reflux ratio using Underwood method. Moreover, it determines the actual reflux ratio for the specified number of stages or the actual number of stages for the specified reflux ratio, depending on which is entered using Gilliland correlation. It also calculates the optimal feed tray and reboiler as well as condenser duty. Remember that this model assumes constant molar overflow and relative volatilities.

The *Distl* model also includes a single feed and two products, and assumes constant molar overflow and relative volatilities. It uses Edmister approach to calculate product composition. We need to specify the number of stages, feed location, reflux ratio, pressure profile and distillate to feed (D/F) ratio. Actually, when all the data are provided, we can use this column model to verify the product results.

The *RadFrac* is a rigorous fractionating column model that can handle any number of feeds as well as side draws. It has a wide variety of applications, such as absorption, stripping, ordinary distillation, extractive and azeotropic distillation, reactive distillation, etc.

The *MultiFrac* is usually employed for any number of fractionating columns and any number of connections between the columns or within the columns. It has the ability to simulate the distillation columns integrated with flash towers, feed furnaces, side strippers, pumparounds, etc. This rigorous column model can be used as an alternative of PetroFrac, especially when the configuration is beyond the capabilities of PetroFrac.

As mentioned earlier, *SCFrac* is a shortcut column model. It simulates a distillation unit connected with a single feed, multiple products and one optional stripping steam. It is used to model refinery columns, such as atmospheric distillation unit (ADU) and vacuum distillation unit (VDU).

The *PetroFrac* is commonly employed to fractionate a petroleum feed. This rigorous model simulates the refinery columns, such as ADU, VDU, fluidized-bed catalytic cracking (FCC) fractionator, etc., equipped with a feed furnace, side strippers, pumparounds and so on.

The *RateFrac* is a rate-based nonequilibrium column model employed to simulate all types of vapour-liquid separation operations, such as absorption, desorption and distillation. It simulates single and interlinked columns with tray type as well as packed type arrangement.

The *BatchFrac* is a rigorous model used for simulating the batch distillation columns. It also includes the reactions occurred in any stage of the separator. BatchFrac model does not consider column hydraulics, and there is negligible vapor holdup and constant liquid holdup.

It is worthy to mention that for detailed information regarding any built-in Aspen Plus™ model, select that model icon in the *Model Library* toolbar and press F1.

In this chapter, we will simulate different distillation models, including a petroleum refining column and a reactive distillation, using the Aspen Plus software. Moreover, an absorption column will be analyzed. In addition to the steady state simulation, the process optimization will also be covered in the present study.

3.2 ASPEN PLUS SIMULATION OF THE BINARY DISTILLATION COLUMNS

3.2.1 Simulation of a DSTWU Model

Problem statement

A feed stream, consisting of 60 mole% ethane and 40 mole% ethylene, enters a DSTWU column having a flow rate of 200 lbmol/hr at 75°F and 15 psia. This feed is required to fractionate in a distillation column capable of recovering at least 99.6% of the light key component in the distillate and 99.9% of the heavy key component in the bottoms. The sample process operates at 15 psia with zero tray-to-tray pressure drop. The pressure in the reboiler as well as condenser is also 15 psia. In the simulation, use total 30 theoretical stages (including a total condenser and reboiler). Applying the RK-Soave property method, simulate the column and calculate the minimum reflux ratio, actual reflux ratio, minimum number of stages, actual number of stages, and feed location.

Simulation approach

From the desktop, select *Start* button, and then click on *Programs, AspenTech, Aspen Engineering Suite, Aspen Plus Version* and *Aspen Plus User Interface.* Then choose *Template* option in the *Aspen Plus Startup* dialog and hit *OK* (see Figure 3.1).

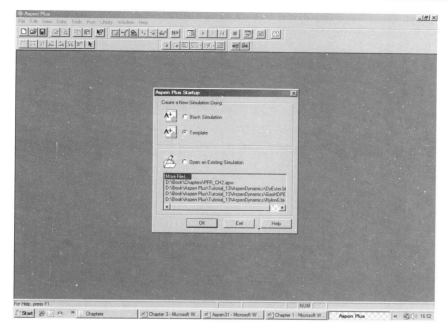

FIGURE 3.1

Select *General with English Units* as the next window appears (see Figure 3.2).

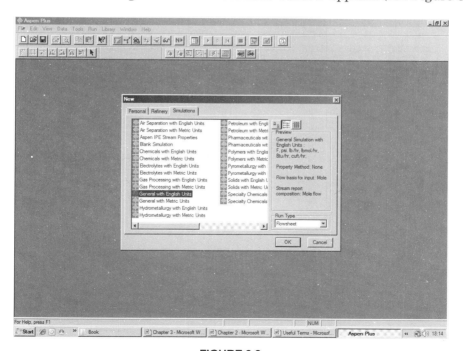

FIGURE 3.2

Again press *OK* to see the *Connect to Engine* dialog (see Figure 3.3). Here we choose 'Local PC' by scrolling down. Hit *OK* knob and move on to develop the process flow diagram.

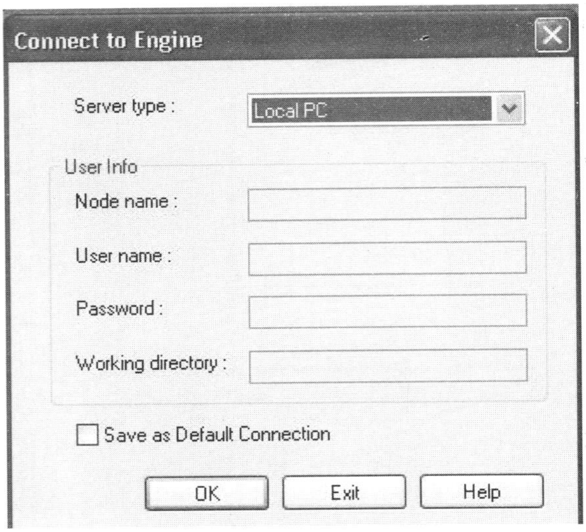

FIGURE 3.3

Creating flowsheet

As we select *Columns* tab in the bottom *Model Library* toolbar, Aspen Plus shows all built-in column models (Figure 3.4).

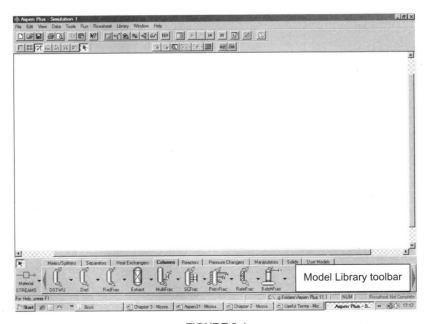

FIGURE 3.4

In the next, select DSTWU icon that represents a short-cut distillation process. Once we have selected the icon, place the icon on the flowsheet by clicking with the cross-hair somewhere on the flowsheet background. When finished, click on $\boxed{\nwarrow}$ symbol or right-click on the flowsheet background. By default, the column is named as B1 (see Figure 3.5).

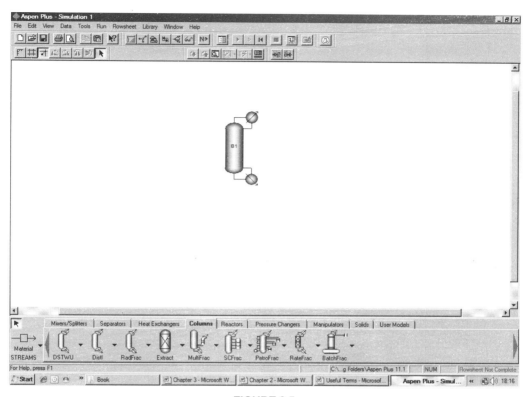

FIGURE 3.5

In the screen, shown in Figure 3.5, only the block is displayed; there are no incoming and outgoing streams connected with the block. Therefore, the *Status message* in the bottom right of the window includes *Flowsheet Not Complete*. Interestingly, after connecting all required streams with the unit, this message sometime may also be retained. This happens because of improper flowsheet connectivity.

To add a single-feed stream and two-product outlets (distillate and bottoms), click on *Material STREAMS* tab in the lower left-hand corner. As we move the cursor (a crosshair) onto the process flowsheet, suddenly three red arrows and one blue arrow appear around the block. These arrows typically indicate places to attach streams to the block. As we know, red arrows are required ports and blue arrows are optional ports. Click once on the connection point between the feed stream and the DSTWU block, enlarge the feed line and finally click again. By default, this stream is labeled as 1. In the similar fashion, we can add the two product streams, namely 2 and 3, to the distillation unit (see Figure 3.6). Recall that to make the stream flowline that is adhered to the block straight, right click highlighting the line and then click "Align Blocks" (or Ctrl+B).

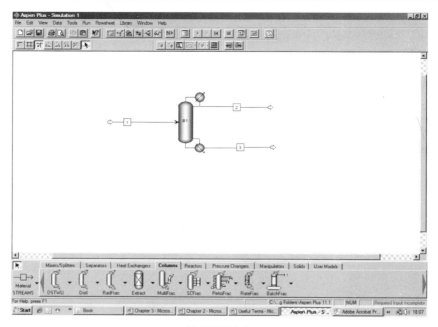

FIGURE 3.6

After renaming Stream 1 to F, Stream 2 to D, Stream 3 to B and Block B1 to DSTWU, the flowsheet finally looks like Figure 3.7.

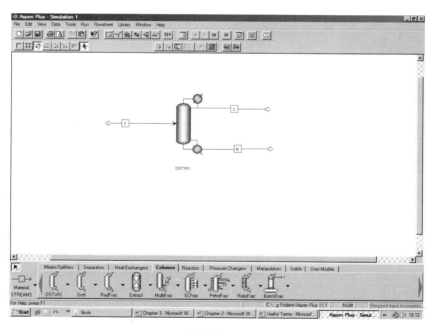

FIGURE 3.7

Now the *Status bar* in the window, shown in Figure 3.7, says *Required Input Incomplete* indicating that the flowsheet is complete and the input specifications are required to provide using available input forms for running the Aspen simulator.

Configuring settings

Recall that within the Aspen simulation software, the simplest way to find the next step is to use one of the following equivalent commands:

(a) press the *Next* button **N»**

(b) find 'Next' in the *Tools* menu

(c) use shortcut key *F4*

and obtain Figure 3.8.

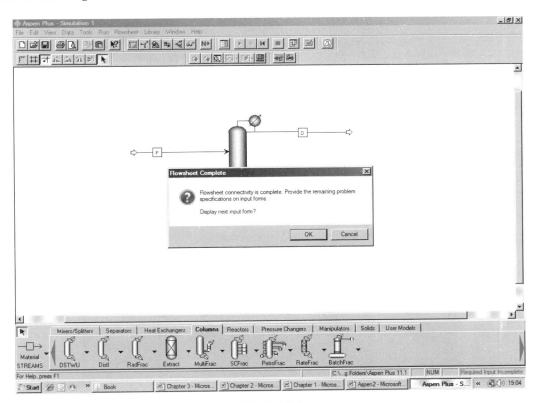

FIGURE 3.8

Hitting *OK* on the above message, we obtain the setup input form. Alternatively, select *Solver Settings* knob and choose *Setup/Specifications* in the list on the left (see Figure 3.9).

Although optional, it is a good practice to fill out the above form with a title and to provide the accounting information subsequently. The present project is titled as 'Simulation of a Shortcut Distillation Column' (see Figure 3.10).

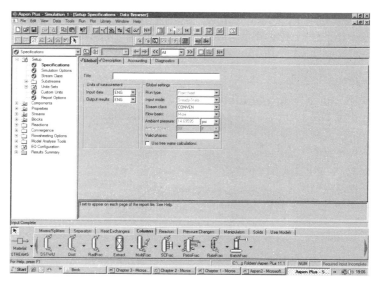

FIGURE 3.9

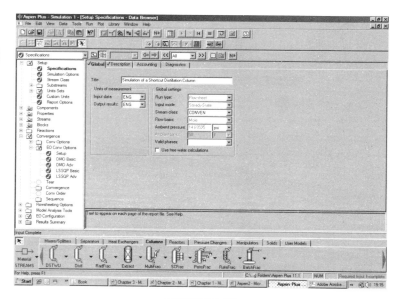

FIGURE 3.10

In the next (see Figure 3.11), the Aspen Plus *accounting* information (required at some installations) is given in the following way.

User name: AKJANA

Account number: 9

Project ID: ANY ID

Project name: YOU CHOOSE

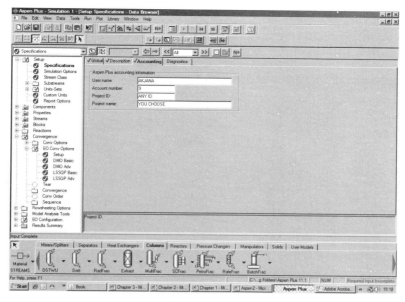

FIGURE 3.11

We may wish to have stream results summarized with mole fractions and/or some other basis that is not set by default. For this, we can use *Report Options* under *Setup* folder. In the subsequent step, open *Stream* sheet and then choose 'Mole' fraction basis. In this regard, a sample copy is shown in Figure 3.12, although this is not essential for the present problem.

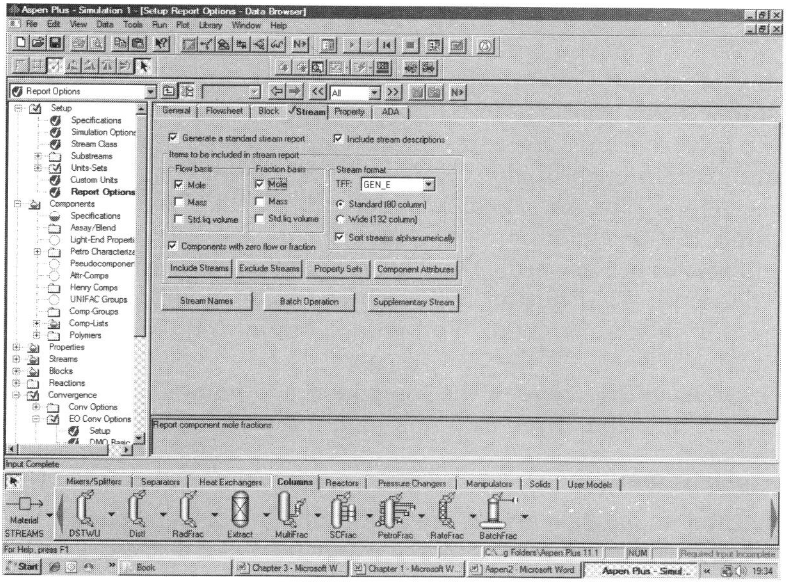

FIGURE 3.12

Specifying components

Use the Data Browser menu tree to navigate to the *Components / Specifications / Selection* sheet (see Figure 3.13).

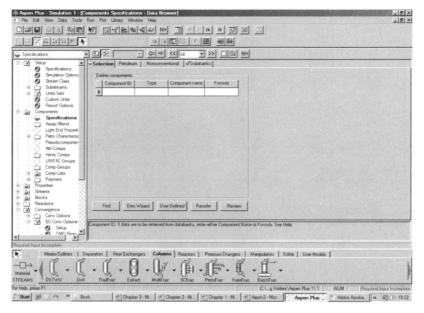

FIGURE 3.13

In Figure 3.13, the table has four columns; they are under the headings of *Component ID*, *Type*, *Component name* and *Formula*. Among them, the *Type* is a specification of how the Aspen software calculates the thermodynamic properties. For fluid processing of organic chemicals, it is generally suitable to use 'Conventional' option. Remember that component ID column should be filled out by the user. A *Component ID* is essentially an alias for a component. It is sufficient to use the chemical formulas or names of the components as their IDs. On the basis of these component IDs, Aspen Plus may spontaneously fill up the *Type*, *Component name* and *Formula* columns. If it does not happen, it means that Aspen Plus fails to find an exact match in its library. In other words, the Aspen Plus does not recognize the components by their IDs. Then use *Find* button to search the components. Select the components from the lists and then *Add* them. For more details, follow the approach given in Subsection 1.3.2.

Here, we have used the name of the components, ethane and ethylene, as their IDs (see Figure 3.14). The other three columns have been automatically filled out.

Specifying property method

A property method includes various methods and models to compute the physical properties. In order to obtain the property input form, either hit *Next* icon or choose *Properties / Specifications* in the left pane of the Data Browser window. Set RK-Soave property method by scrolling down (see Figure 3.15).

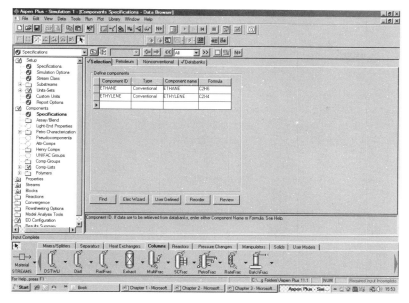

FIGURE 3.14

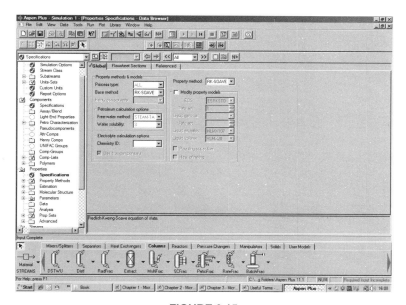

FIGURE 3.15

Specifying stream information

The *Streams/F/Input/Specifications* sheet appears with the Data Browser menu tree in the left pane (see Figure 3.16). Here, we have to provide the values for all state variables (temperature, pressure and total flow) and composition (component mole fractions).

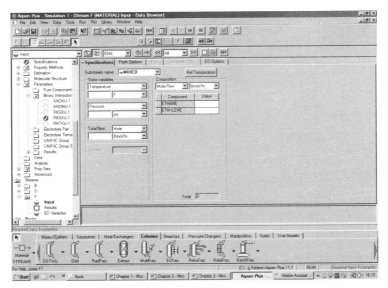

FIGURE 3.16

Filling out the form, shown in Figure 3.16, with the data given in the problem statement, one obtains Figure 3.17.

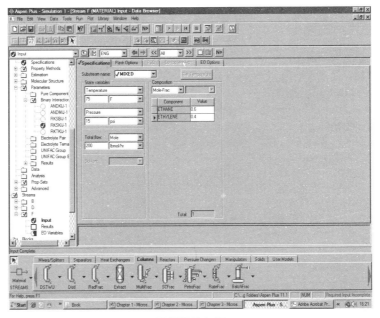

FIGURE 3.17

Specifying block information

In the column at the left side, choose *DSTWU/Input* under *Blocks* folder. As a result, a blank block input form is displayed (see Figure 3.18).

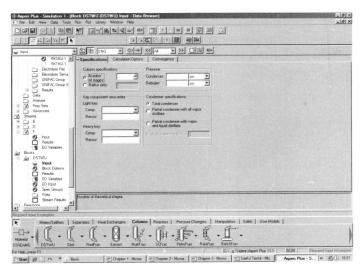

FIGURE 3.18

Under *Column specifications* option, here we enter the number of stages that is 30. It is clear that we can alternatively specify the reflux ratio when the number of stages is asked to compute. Note that ethylene is the light key and naturally ethane is the heavy key. As mentioned in the problem statement, recovery of the light key component in the distillate (= moles of light key in the distillate/moles of light key in the feed) is 0.996 and recovery of the heavy key component in the distillate (= moles of heavy key in the distillate/moles of heavy key in the feed) is 0.001. In addition, the pressure of the total condenser and reboiler is given as 15 psia. Entering all these information, one obtains Figure 3.19.

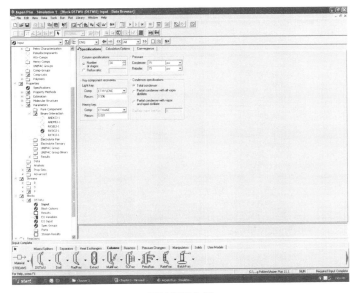

FIGURE 3.19

Running the simulation

The *Status message* includes *Required Input Complete* indicating that we are in a position to run the simulation. Simply press *Next* button and receive a message regarding the present status (see Figure 3.20).

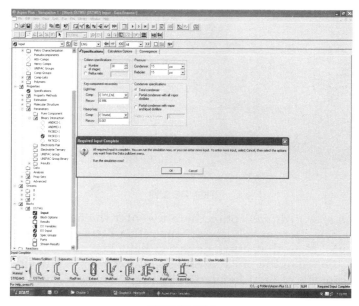

FIGURE 3.20

Click *OK* on the above message and obtain the *Control Panel* window (see Figure 3.21) that shows the progress of the simulation.

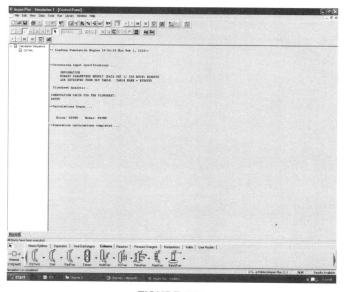

FIGURE 3.21

Hitting *Next* followed by *OK*, we have the *Run Status* screen (see Figure 3.22).

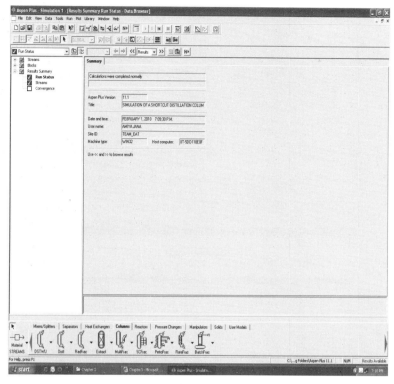

FIGURE 3.22

Viewing results

In the next, select *Blocks/DSTWU/Results* from the Data Browser. From Figure 3.23, we get the answers as:

      Minimum reflux ratio = 4.149

      Actual reflux ratio = 4.856

      Minimum number of stages = 15.273

      Actual number of stages = 30

      Feed location = 18.719

Save the work by choosing *File/Save As/...*in the menu list on the top. We can name the file whatever we like. Remember that a backup file (*.bkp) takes much less space than a normal Aspen Plus documents file (*.apw).

Viewing input summary

If we wish to have the input information, press Ctrl + Alt + I on the keyboard or select *Input Summary* from the *View* pulldown menu (see Figure 3.24).

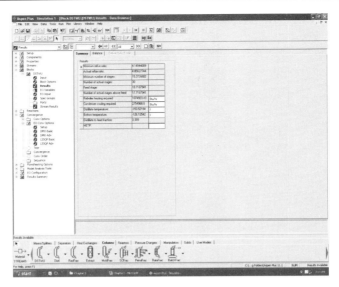

FIGURE 3.23

FIGURE 3.24

3.2.2 Simulation of a RadFrac Model

Problem statement

We will continue the above problem (simulated in Subsection 3.2.1) with few modifications. A hydrocarbon stream, consisting of 60 mole% ethane and 40 mole% ethylene, enters a RadFrac column having a flow rate of 200 lbmol/h at 75°F and 15 psia. The distillation process that has total 30 theoretical stages (including a total condenser and reboiler) operates at 15 psia with zero pressure drop throughout. The distillate rate, reflux ratio and feed tray location are obtained in Subsection 3.2.1 as 79.8 lbmol/h, 4.856 (mole basis) and 19 (above-stage), respectively. Consider the RK-Soave property method.

(a) Simulate the column and compute the compositions of top as well as bottom products.
(b) Is there any discrepancy in product compositions obtained from RadFrac and DSTWU columns? If yes, what is the main reason?

Simulation approach

(a) Start with the *General with English Units* Template, as shown in Figures 3.25(a) and 3.25(b).

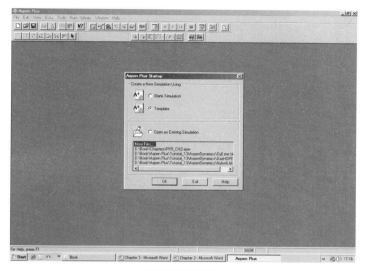

FIGURE 3.25(a)

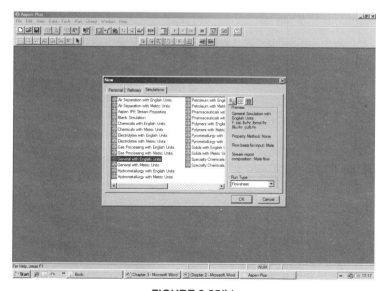

FIGURE 3.25(b)

Click *OK* in the screen, shown in Figure 3.25(b). When the *Connect to Engine* dialog pops up, again press *OK* button to obtain a blank *Process Flowsheet Window*.

Creating flowsheet

Among the built-in columns in the *Model Library* of Aspen simulator, select RadFrac and place it in the flowsheet window. Connecting feed, distillate and bottom product streams with the distillation column, and changing the default names of the block and all streams, finally we get Figure 3.26.

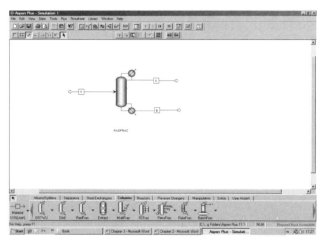

FIGURE 3.26

Configuring settings

In the subsequent step, simply hit *Next* button followed by *OK* to open a setup input form. These two windows, shown in Figures 3.27(a) and (b), include the *Global* and *Accounting* information for the present project.

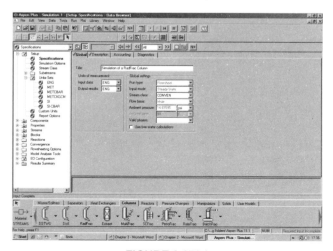

FIGURE 3.27(a)

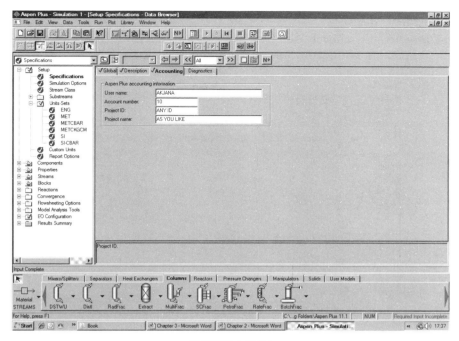

FIGURE 3.27(b)

In the *Setup/Report Options/Stream* sheet, select 'Mole' as well as 'Mass' fraction basis as shown in Figure 3.28.

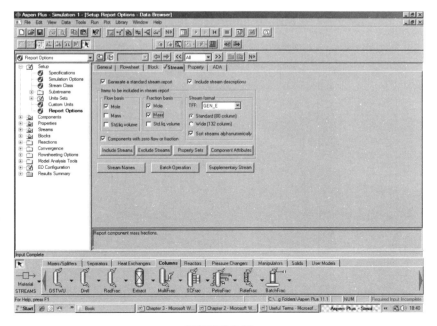

FIGURE 3.28

Specifying components

In the list on the left, choose *Components / Specifications* to define the components. Using the component names, ethane and ethylene, as their IDs, we obtain the filled table as shown in Figure 3.29.

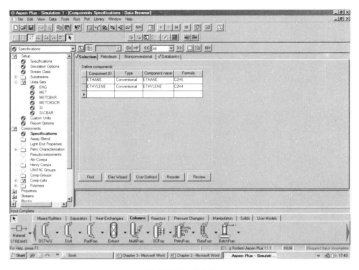

FIGURE 3.29

Specifying property method

From the Data Browser, select *Specifications* under *Properties* folder and then set RK-Soave base method to compute the physical properties (see Figure 3.30).

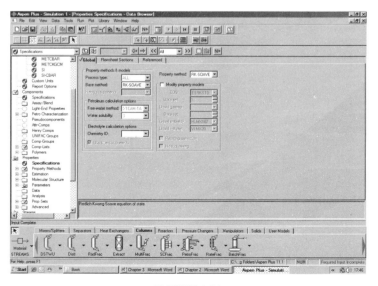

FIGURE 3.30

Specifying stream information

Use the Data Browser menu tree to navigate to the *Streams / F / Input / Specifications* sheet. Inserting the given values for the feed stream, Figure 3.31 is obtained.

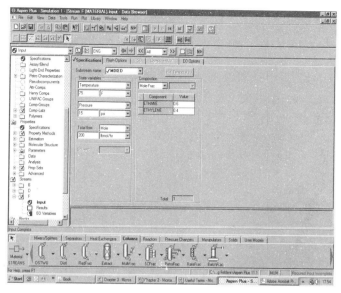

FIGURE 3.31

Specifying block information

In the left pane of the Data Browser window, select *Blocks / RADFRAC / Setup*. Fill up the *Configuration* sheet as shown in Figure 3.32.

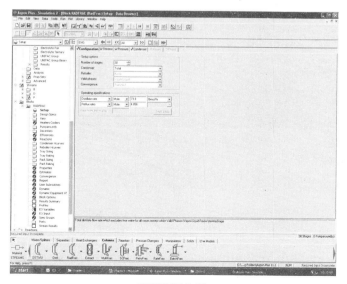

FIGURE 3.32

Here we consider *Total* condenser as asked in the problem statement. It does not matter whether we select the *Kettle* type or *Thermosyphon* reboiler because both are partial reboilers (the outgoing liquid and vapor streams from the reboiler are in equilibrium). Regarding convergence, the *Standard* method works well for the representative hydrocarbon system. Alternative methods should be used for highly nonideal systems.

Under *Setup* subfolder, the filled *Streams* sheet looks like Figure 3.33.

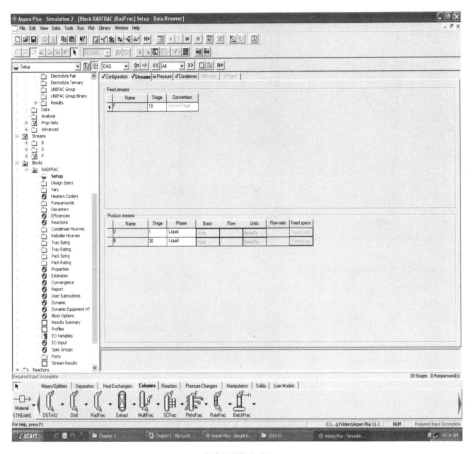

FIGURE 3.33

In the next, simply input 15 psi under *Stage 1/Condenser pressure* (see Figure 3.34). Aspen simulator assumes that the column operates isobarically if no additional pressure information is provided.

Running the simulation

To run the simulation, hit *Next* and then *OK* to observe the progress of the simulation in the *Control Panel* window, shown in Figure 3.35.

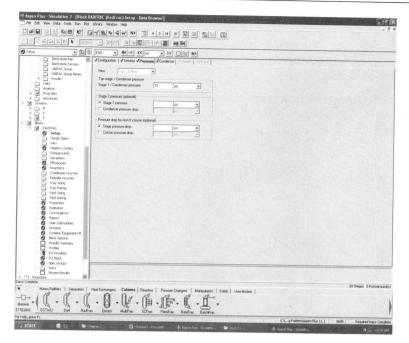

FIGURE 3.34

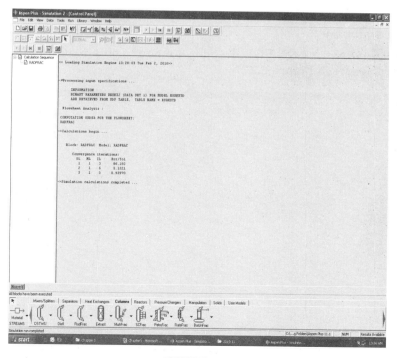

FIGURE 3.35

Viewing results

Click on *Solver Settings* followed by *Results Summary* and *Streams*, we have the table, shown in Figure 3.36, accompanying the results of all individual streams. Save the work in a folder as a file.

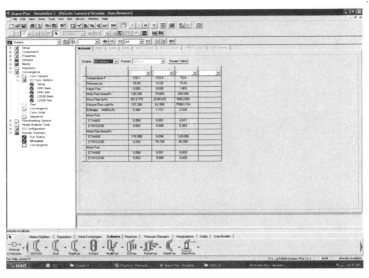

FIGURE 3.36

Viewing input summary

Select *Input Summary* from the *View* dropdown menu to obtain all input information of the present problem (see Figure 3.37).

FIGURE 3.37

(b) Results of the RadFrac column

TABLE 3.1

	Composition (mole fraction)	
Component	B	D
ethane	0.998	0.001
ethylene	0.002	0.999

Results of the DSTWU column

TABLE 3.2

	Composition (mole fraction)	
Component	B	D
ethane	0.997	0.002
ethylene	0.003	0.998

From Tables 3.1 and 3.2, it is obvious that there is a little difference between the product compositions. However, the main reason behind this fact is that the RadFrac performs rigorous calculations, whereas the DSTWU is a shortcut model. Another possibility is the round-off error associated in the reflux ratio and feed tray position.

3.3 ASPEN PLUS SIMULATION OF THE MULTICOMPONENT DISTILLATION COLUMNS

3.3.1 Simulation of a RadFrac Model

Problem statement

A multicomponent distillation column, specified in Figure 3.38, has total 20 stages (including condenser and reboiler) with 60% Murphree efficiency. A hydrocarbon feed

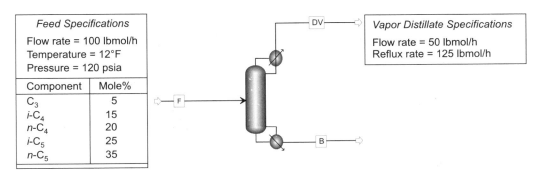

FIGURE 3.38 A flowsheet of a distillation column.

mixture enters above tray 10 of the RadFrac column. Apply the Peng-Robinson correlation and consider 120 psia pressure throughout the column.

(a) Simulate the model and calculate the product compositions,
(b) Produce a 'Temperature' (°F) vs. 'Stage' plot, and
(c) Discuss the column sizing of the example multicomponent process

Simulation approach

(a) As we start Aspen Plus from the *Start* menu or by double-clicking the Aspen Plus icon on our desktop, the *Aspen Plus Startup* dialog appears (see Figure 3.39). Select *Template* option.

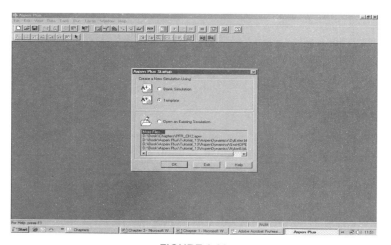

FIGURE 3.39

As Aspen Plus presents the window after clicking *OK* in Figure 3.39, choose *General with English Units*. Then hit *OK* in Figure 3.40.

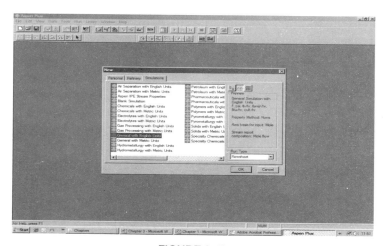

FIGURE 3.40

Click *OK* when the Aspen Plus engine window is displayed. Remember that this step is specific to the installation.

Creating flowsheet

At present, we have a blank *Process Flowsheet Window*. So, we start to develop the process flow diagram by adding a RadFrac column from the *Model Library* toolbar and drawing the inlet and product streams by the help of *Material STREAMS*.

Now the process flowsheet is complete. The *Status bar* in the bottom right of the screen, shown in Figure 3.41, displays a message of *Required Input Incomplete* indicating that input data are required to enter to continue the simulation.

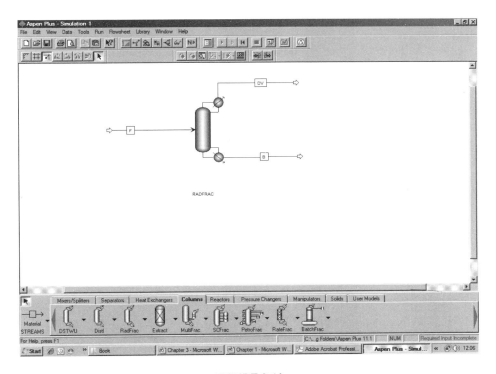

FIGURE 3.41

Configuring settings

Hitting *Next* knob and then clicking *OK*, we get the setup input form. In Figures 3.42(a) and (b), the *Title* of the problem ('Simulation of a Multicomponent Column') followed by the Aspen Plus *accounting* information (AKJANA/11/ANY ID/FINE) are provided.

Include two more bases in *Report Options/Stream* sheet under *Setup* folder (see Figure 3.43).

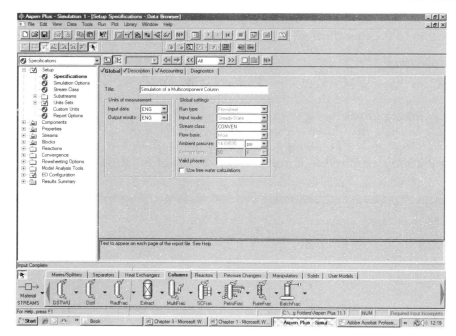

FIGURE 3.42(a)

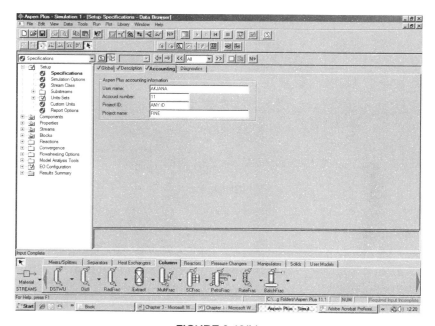

FIGURE 3.42(b)

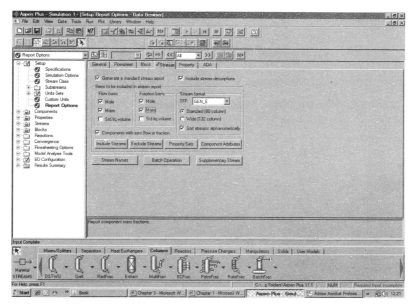

FIGURE 3.43

Specifying components

In the left pane of the Data Browser window, select *Components / Specifications*. Filling out the *Component ID* column, we finally obtain the table as shown in Figure 3.44.

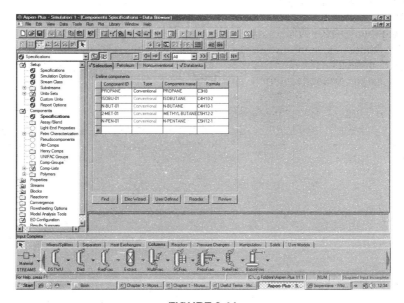

FIGURE 3.44

Specifying property method

In order to define the base property method, press *Next* icon or select *Properties/ Specifications* in the column at the left side (Figure 3.45). From the *Property method* pulldown menu, select PENG-ROB. This equation of state model is chosen for thermodynamic property predictions.

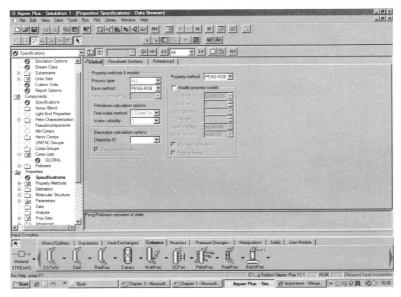

FIGURE 3.45

Specifying stream information

In the next, use the Data Browser menu tree to navigate to the *Streams/F/Input/ Specifications* sheet. Entering the values of all state variables and component mole fractions, we get Figure 3.46.

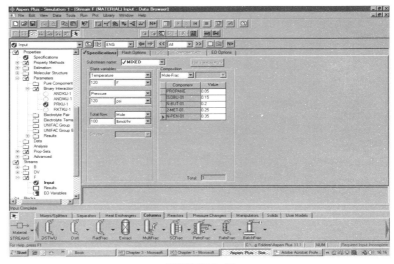

FIGURE 3.46

Specifying block information

Open the *Configuration* sheet choosing *Blocks/RADFRAC* in the list on the left. Then the information on number of stages, condenser type, vapor distillate flow rate and reflux rate are entered in Figure 3.47.

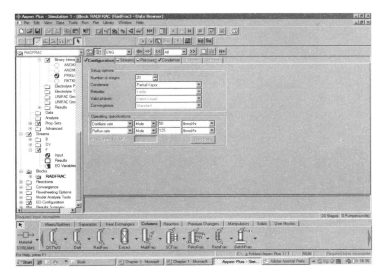

FIGURE 3.47

In the subsequent step, specify the feed tray location in the *Streams* sheet as shown in Figure 3.48.

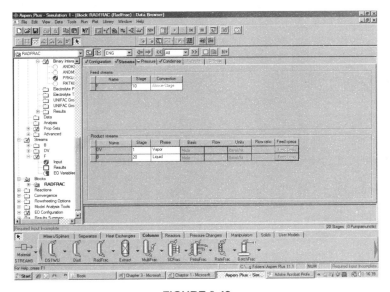

FIGURE 3.48

Enter the column pressure of 120 psi and get Figure 3.49.

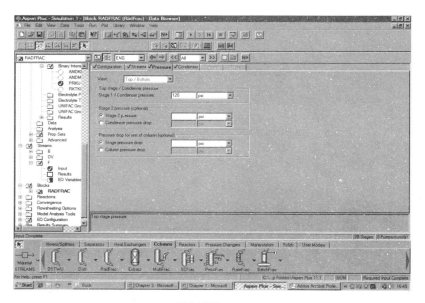

FIGURE 3.49

The *Blocks/RADFRAC/Efficiencies/Options* sheet appears with the Data Browser menu tree in the left pane. To input the Murphree efficiency value for all trays (excluding the condenser and reboiler), we have the screen, shown in Figure 3.50 first.

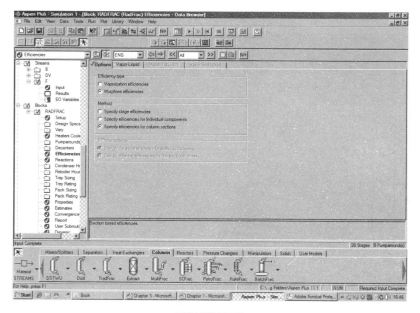

FIGURE 3.50

Press the knob to open the *Vapor-Liquid* sheet (see Figure 3.51).

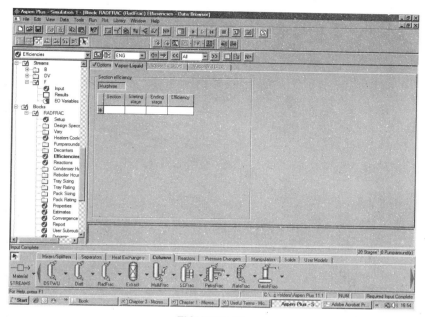

FIGURE 3.51

Assume the rectifying along with the stripping zone as Section 1 and fill up the table, shown in Figure 3.52.

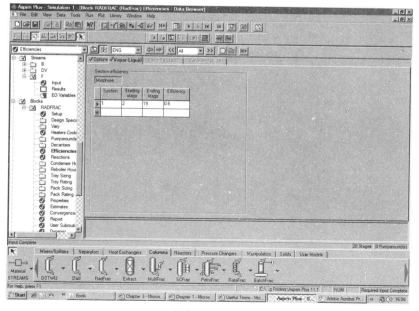

FIGURE 3.52

Running the simulation

Hit *Next* button followed by *OK* and observe the progress of the simulation in the *Control Panel* window as shown in Figure 3.53.

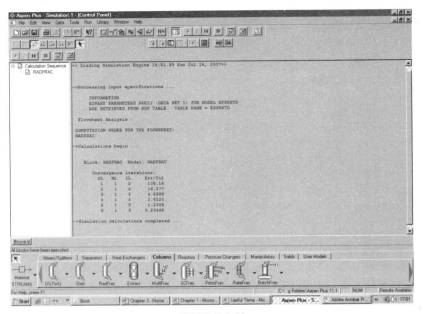

FIGURE 3.53

Viewing results

Click on *Solver Settings* knob and then choose *Results Summary/Streams* to obtain the product compositions (see Figure 3.54).

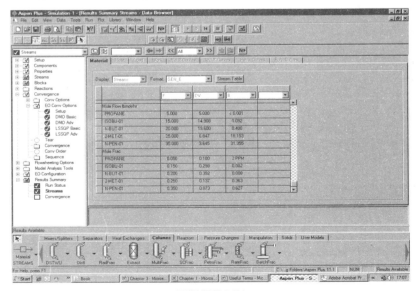

FIGURE 3.54

It is a good habit to save the work done at least at this moment. If we wish to see the tabulated results with the process flow diagram in a single sheet (see Figure 3.55), simply hit *Stream Table* button just above the results table in Figure 3.54.

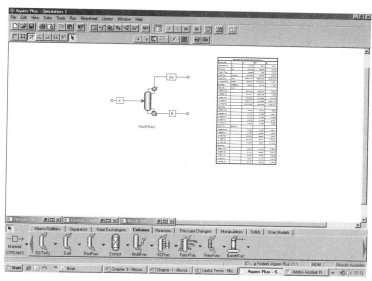

FIGURE 3.55

Viewing input summary

As stated previously, to obtain the input information, press Ctrl+Alt+I or select *Input Summary* from the *View* pulldown menu (see Figure 3.56).

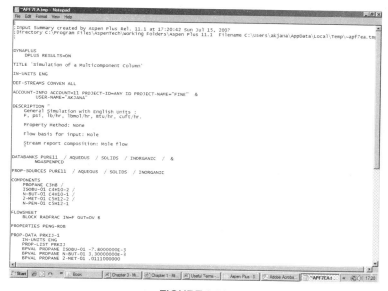

FIGURE 3.56

(b) First, choose *Blocks*/*RADFRAC*/*Profiles* in the column at the left side. Accordingly, we have the stage-wise data as shown in Figure 3.57.

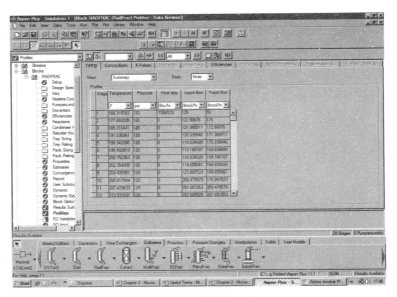

FIGURE 3.57

In the next, select *Plot Wizard* from the *Plot* dropdown menu or press Ctrl + Alt + W on the keyboard to get Figure 3.58.

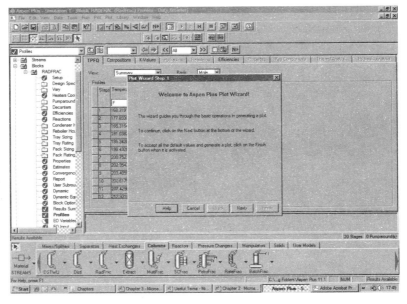

FIGURE 3.58

Click on *Next* button in the *Plot Wizard Step* 1 dialog and get a variety of plot types, shown in Figure 3.59.

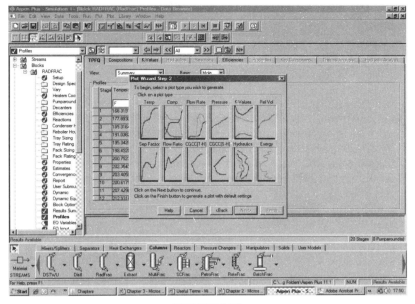

FIGURE 3.59

Select the plot type under the heading of *Temp* and press *Finish* button to obtain a plot of 'Temperature' (°F) vs. 'Stage' (see Figure 3.60).

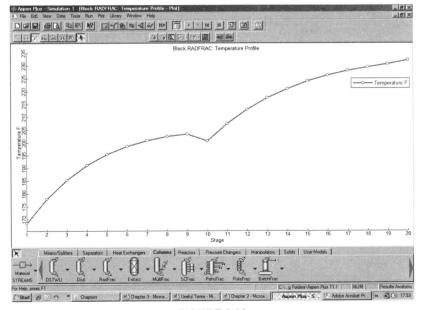

FIGURE 3.60

Recall that the above plot window can be edited by right clicking on that window and selecting *Properties*. Then the user can easily modify the title, axis scale, font and colour of the plot.

(c) Column sizing

(i) Column length (L_C): Typically the tray spacing has been considered as 2 ft. With known total number of stages (NT), the column height with a tray spacing of 2 ft is:

$$L_C = 2(NT - 2) \tag{3.1}$$

Note that in the Aspen simulator, the reflux condenser and reboiler are counted as one stage each. It is a common practice to include some additional space at the top of the distillation tower as a vapor-liquid disengaging space and at the feed stage for feed distribution. Similarly, at the bottom, additional space must be provided for a liquid sump. Enough space is also needed aiming to maintain the liquid level in the column base above the elevation of the bottoms pump to provide the necessary net positive suction head (NPSH) requirements for this pump. Hence, a design heuristic is to include a 20% allowance for this excess space. Accordingly, it is suggested to use the following correlation for calculating the length of the column as:

$$\begin{aligned} L_C &= 1.2 \times 2(NT - 2) = 2.4\ (NT - 2) \\ &= 43.2\ \text{ft} \end{aligned} \tag{3.2}$$

(ii) Column diameter (D_C): Using the Aspen package, it is easy to determine the distillation column diameter. In the list on the left, select *Blocks / RADFRAC / Tray Sizing*. Press the *New* button, accept the default tray sizing section number (i.e., 1) and click *OK* to get the following window (Figure 3.61):

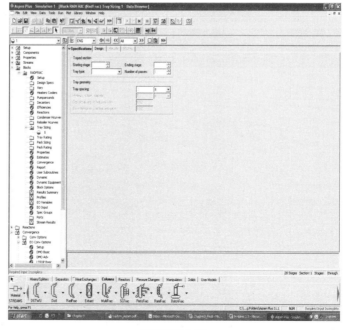

FIGURE 3.61

Obviously, the starting stage is 2 and the ending stage is 19. Specifying tray type (i.e., Sieve tray), we get Figure 3.62.

FIGURE 3.62

After running the simulator, the *Results* sheet under *Blocks/RADFRAC/Tray Sizing/1* shows the following window (Figure 3.63) with a column diameter of 2 ft. Remember that for large diameter distillation columns, it is recommended to use more than one 'Number of passes' mainly to avoid large liquid gradients across the tray.

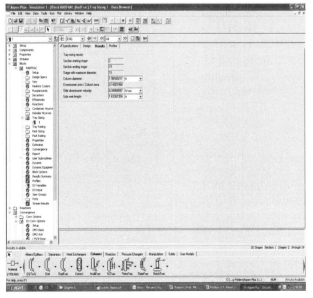

FIGURE 3.63

Verification: It is a good practice to verify the column diameter calculated using the Aspen simulator. For this purpose, it can be assumed an *F*-factor of 1 (in English engineering units). It implies,

$$F = V_{\max} \rho_V^{0.5} = 1.0 \tag{3.3}$$

In the above expression, $V_{\max}$ designates the maximum vapor velocity (ft/s) and ρ_V the vapor density (lb/ft^3). Since the vapor flow rates vary from tray to tray in a nonequimolal overflow system, the tray with the maximum vapor velocity will set the minimum distillation column diameter.

It is obvious that if ρ_V is known, we can determine $V_{\max}$ from Eq. (3.3). Open the window by clicking *Blocks / RADFRAC / Report* and then under *Property Options*, select *Include hydraulic parameters* (Figure 3.64). In the subsequent step, run the program, open *Hydraulics* window under *Blocks / RADFRAC / Profiles* and obtain Figure 3.65 that provides many important information on liquid and vapor streams.

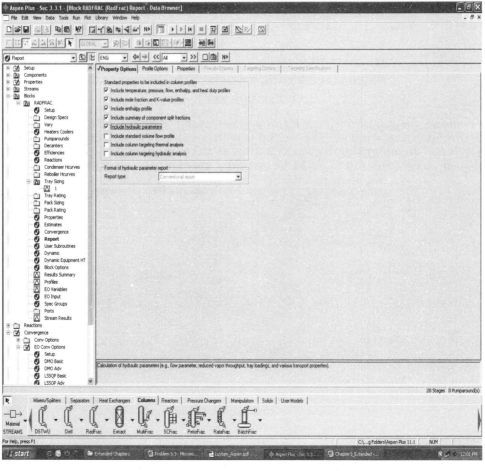

FIGURE 3.64

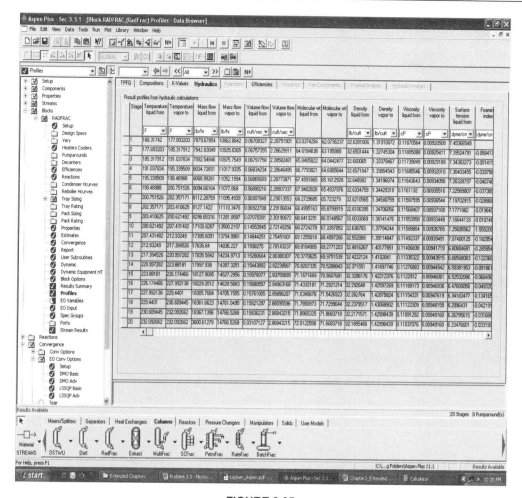

FIGURE 3.65

From Figure 3.65, we get the maximum vapor volumetric flow rate of 2.8684 ft³/s and it occurs on Stage 20. In the same figure, the vapor density (ρ_V) on 20th stage is reported as 1.42998 lb/ft³. According to the heuristic with an F-factor of 1, the maximum velocity ($V_{\max}$) is calculated as 0.836 ft/s. Hence, the cross-sectional area is:

$$\frac{\pi}{4} D_C^2 = \frac{2.8684}{0.836} = 3.431 \text{ ft}^2 \tag{3.4}$$

Solving, we get the column diameter of 2.09 ft which is very close to the value (= 2 ft) obtained from the Aspen simulator.

Douglas (1988) suggested a design guideline that the distillation column height-to-diameter ratio should be less than 20 to 30. For the representative distillation process, this ratio is 43.2/2 = 21.6.

(iii) Reflux condenser: A partial condenser is used in the example multicomponent distillation column. Here, the condenser duty (Q_c) refers to the amount of heat required for partial condensation of the vapor passing overhead. The energy balance for a partial reflux condenser gives:

$$Q_c = U_c A_c \Delta T_{LMTDc} = m_{cw} C_{pcw} \Delta T_{cw} \qquad (3.5)$$

with

$$\Delta T_{LMTDc} = \frac{120 - 90}{\ln[(T_T - 90)/(T_T - 120)]}$$

$$\Delta T_{cw} = (120 - 90)$$

Here, U_c denotes the overall heat transfer coefficient [Btu/(h.ft².°F)], A_c the heat transfer area (ft²), ΔT_{LMTDc} the log-mean temperature difference (°F), m_{cw} the flow rate of cooling water (cw) (lb/h), C_{pcw} the specific heat capacity of *cw* [Btu/(lb.°F)] and T_T the temperature of overhead vapor (°F). It is supposed that the cooling water is available at 90°F and it leaves the condenser at 120°F. So, ΔT_{cw} is 30°F.

For the condenser, U_c of 150 Btu/(h.ft².°F) gives reasonable results. Apart from the assumption of U_c value, from the Aspen simulation, we can have the following information:

$$Q_c = 1050116.6 \text{ Btu/h}$$

$$T_T = 177.89°F$$

Accordingly, Eq. (3.5) yields,

$$Q_c = U_c A_c \frac{120 - 90}{\ln[(177.89 - 90)/(177.89 - 120)]} \qquad (3.6)$$

$$\Rightarrow 1050116.6 = 150 \times A_c \times 71.85$$

$$\Rightarrow A_c = 97.44 \text{ ft}^2$$

Calculation of coolant flow rate (m_{cw}). Next we wish to estimate the required coolant rate using Eq. (3.5) as:

$$Q_c = m_{cw} C_{pcw} \Delta T_{cw} \qquad (3.7)$$

Since $C_{pcw} = 1$ Btu/(lb.°F), we have

$$m_{cw} = \frac{Q_c}{C_{pcw} DT_{cw}} = \frac{1050116.6}{1 \times 30} = 35003.9 \text{ lb/h}$$

(iv) Bottom reboiler: In the reboiler, let us assume that the steam provides heat (Q_r) for producing vapor at a boil-up rate of V_B (lb/h). It is well-known fact that this vapor leaves the reboiler and enters the bottom tray. Like the reflux condenser, a simple energy balance around the bottom reboiler gives,

$$Q_r = \Delta H_B V_B = U_r A_r \Delta T_r = m_s \Delta H_s \qquad (3.8)$$

Here, ΔH_B denotes the heat of vaporization (Btu/lb). It is reasonable to assume the overall heat transfer coefficient (U_r) of 250 Btu/(h.ft².°F) and ΔT_r of 35°F. Also, we have the Aspen simulation results:

$$Q_r = 1772257 \text{ Btu/h}$$

From Eq. (3.8), we can easily calculate the heat transfer area of reboiler (A_r) as:

$$1772257 = 250 \times A_r \times 35$$

$$\Rightarrow A_r = 202.54 \text{ ft}^2$$

Calculation of steam flow rate (m_s): From the Aspen simulation, the temperature of the bottom stream (T_B) is directly obtained as 230.6°F. Hence, the used steam temperature would be 265.6°F (= $T_B + \Delta T_r$). From the steam table, we get the corresponding ΔH_s of 935 Btu/lb. Using Eq. (3.8),

$$m_s = \frac{Q_r}{\Delta H_s} = \frac{1772257}{935} = 1895.5 \text{ lb/h}$$

(v) Size of reflux drum and column base: The volumetric flow rates of liquid into the reflux accumulator (Stage 1, the *condenser* in Aspen terminology) as well as the column base (the last stage, the *sump* in Aspen terminology) can be used to size the two vessels by using the heuristic of a 5–10 minute holdup time.

For the reflux drum of a column equipped with a partial condenser, let L_R be the volumetric flow rate of liquid (ft³/min) leaving the drum (reflux[1]). Considering a 10 minute holdup time, the reflux drum volume (V_R, ft³) is:

$$V_R = 10 \times L_R \tag{3.9}$$

If we consider an aspect ratio (length to diameter, L/D) of 2, the diameter of the reflux drum is:

$$D = \left(\frac{2V_R}{\pi}\right)^{1/3} \tag{3.10}$$

From Figure 3.65, we take L_R = 4.025 ft³/min. It gives $V_R = 10 \times L_R$ = 40.25 ft³. Using Eq. (3.10), one obtains D = 2.95 ft. So, L = 5.9 ft.

By the similar way, we can easily determine the dimensions of a bottom vessel. It is left for the students.

3.3.2 Simulation of a PetroFrac Model

Problem statement

An artificial petroleum refining column (PRC), shown in Figure 3.66, consists of a feed furnace and a distillation tower. The tower has two pumparound circuits, a partial condenser and three side strippers. The furnace (single stage flash type) operates at 25 psia and provides a fractional overflash of 40% (StdVol basis) in the tower. The outlet stream of the furnace goes to the tower on Stage 22. The tower has 26 stages with a Murphree stage efficiency equal to 90%. A steam stream, STEAM, is introduced at the bottom of the fractionator (26th stage with on-stage convention). There are another three steam streams, STM1, STM2 and STM3, used in the side strippers. The condenser runs at 15.7 psia with a pressure drop of 5 psi. The tower pressure drop is equal to 4 psi. The distillate rate is 10000 bbl/day and the distillate vapor fraction in the condenser is 0.2 (StdVol basis).

---

[1] For the case with a total condenser, L_R includes both the distillate and reflux flows.

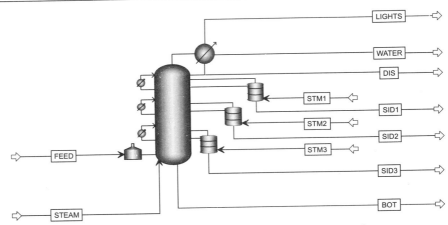

FIGURE 3.66 A flowsheet of a petroleum refining column.

A hydrocarbon mixture with the following component-wise flow rates enters the furnace at 117°F and 44.7 psia (see Table 3.3).

TABLE 3.3

Component	Flow rate (bbl/day)
C_1	3
C_2	65
C_3	575
$i\text{-}C_4$	1820
$n\text{-}C_4$	7500
$i\text{-}C_5$	30000
$n\text{-}C_5$	42000
H_2O	250

In Table 3.4, two pumparound circuits and three side strippers are specified.

TABLE 3.4

Pumparound (drawoff type)	Location Draw stage	Location Return stage	Specifications Flow rate (bbl/day)	Specifications Heat duty (MMBtu/h)
1 (partial)	8	6	49000	−40 (for cooling)
2 (partial)	14	12	1000	−17 (for cooling)

Stripper	No. of stages	Stripper product	Location Draw stage	Location Return stage	Stripping steam	Bottom product flow rate (bbl/day)
1	5	SID1	6	5	STM1	11000
2	4	SID2	12	11	STM2	15000
3	3	SID3	19	18	STM3	8000

Four steam streams used in the column model are described in Table 3.5.

TABLE 3.5

		Specifications		
Steam stream	Location	Temperature (°F)	Pressure (psia)	Flow rate (lb/h)
STEAM	Main tower	350	50	11500
STM1	SID1 stripper	350	50	4000
STM2	SID2 stripper	350	50	1500
STM3	SID3 stripper	350	50	1000

Considering the 'BK10' base method under 'REFINERY' process type, simulate the PetroFrac column and report the flow rates (bbl/day) of all product streams.

Simulation approach

Select *Aspen Plus User Interface*. When the *Aspen Plus Startup* dialog appears, choose *Template* and click on *OK* (see Figure 3.67).

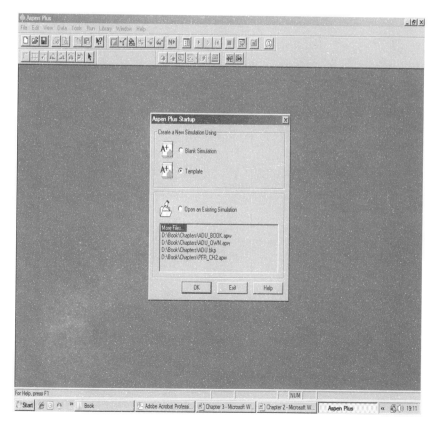

FIGURE 3.67

As the next window pops up (see Figure 3.68), select *Petroleum with English Units* and press *OK* knob.

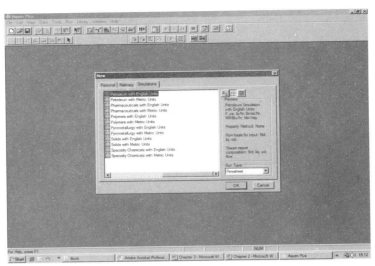

FIGURE 3.68

Click *OK* when the *Connect to Engine* dialog appears. The next screen presents a blank process flowsheet.

Creating flowsheet

Select the *Columns* tab from the *Model Library* toolbar. As we expand the PetroFrac block icon, a variety of models is displayed as shown in Figure 3.69. Select a model icon and press F1 to know more about that.

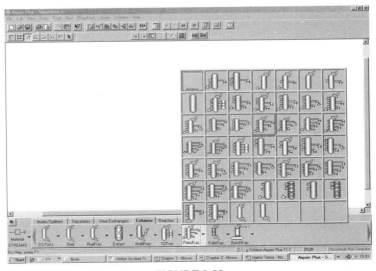

FIGURE 3.69

As the distillation tower described in the problem statement, it is appropriate to choose CDU10F PetroFrac model. Then place it in the flowsheet window. Adding all incoming and outgoing streams and renaming the streams as well as block, the process flow diagram takes the shape as shown in Figure 3.70.

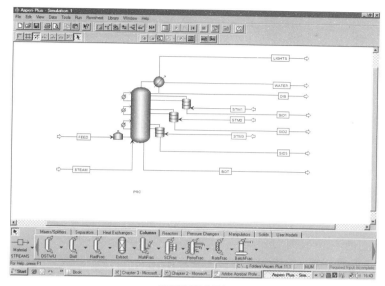

FIGURE 3.70

Configuring settings

Click *Next* to continue the simulation In the *Title* field, we enter 'Simulation of a Petroleum Refining Column'. and keep untouched the other global defaults set by Aspen Plus (see Figure 3.71).

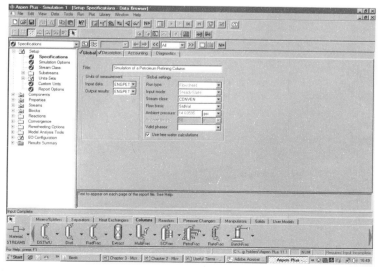

FIGURE 3.71

In the form, shown in Figure 3.72, the Aspen Plus *accounting* information is inserted.

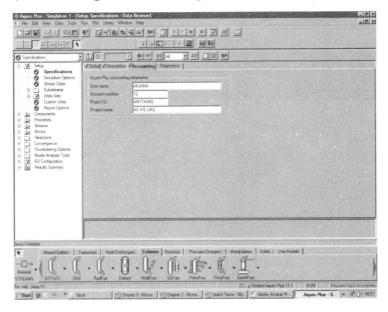

FIGURE 3.72

Specifying components

In the subsequent step, use the Data Browser menu tree to navigate to the *Components/ Specifications* sheet. Filling out the component input form, we have Figure 3.73.

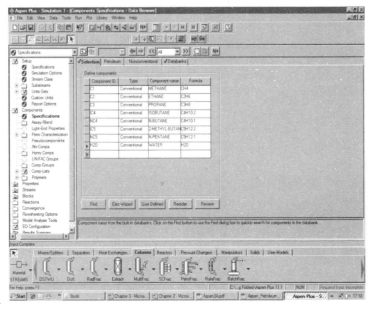

FIGURE 3.73

Specifying property method

We know that the thermodynamic models calculate the properties, such as vapor-liquid equilibrium coefficient, enthalpy and density. In the list on the left, choose *Properties/ Specifications* to open the property input form. As depicted in Figure 3.74, in the *Process type* field, select 'REFINERY' and in the *Base method* field, select 'BK10' (Braun K-10 method).

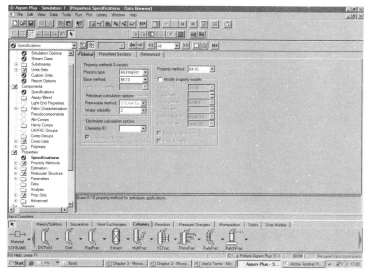

FIGURE 3.74

Specifying stream information

Next the *Streams/FEED/Input/Specifications* sheet appears with the Data Browser menu tree in the left pane. Entering the feed data, Figure 3.75 is obtained.

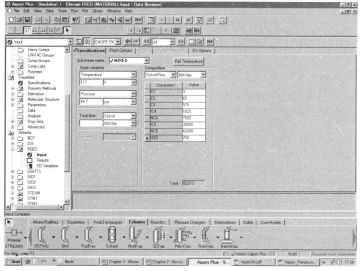

FIGURE 3.75

As we hit *Next* icon, an input form for Stream STEAM opens up. After filling out, it looks like Figure 3.76.

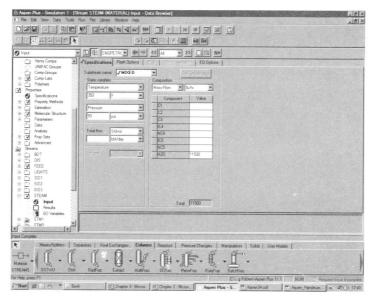

FIGURE 3.76

In Figures 3.77(a), (b) and (c), three filled input forms are shown for STM1, STM2 and STM3 streams.

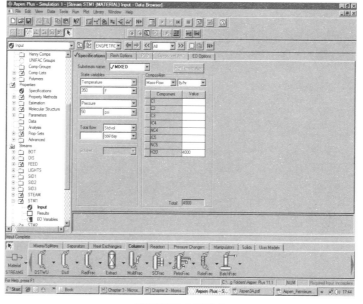

FIGURE 3.77(a)

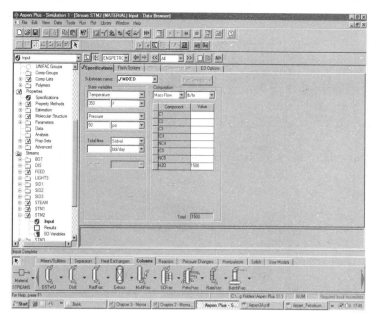

FIGURE 3.77(b)

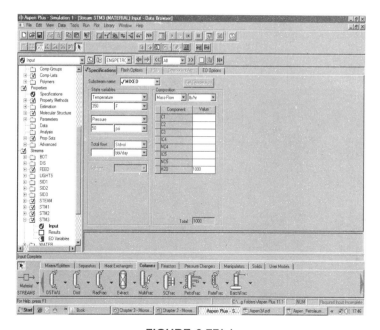

FIGURE 3.77(c)

Specifying block information

From the Data Browser, open *Blocks/PRC/Setup/Configuration* sheet and fill it up (see Figure 3.78).

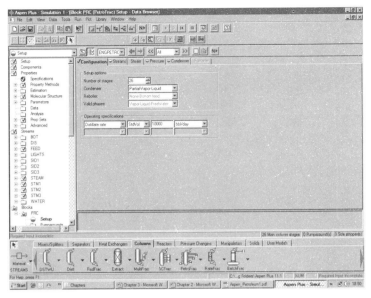

FIGURE 3.78

As we press *Next* icon, the *Blocks/PRC/Setup/Streams* sheet appears as shown in Figure 3.79.

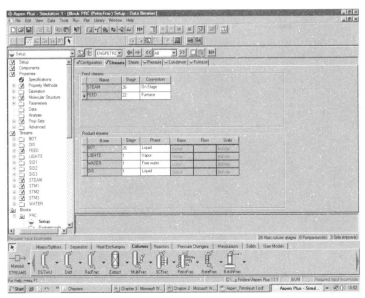

FIGURE 3.79

In Figure 3.80, the *pressure* sheet shows the condenser pressure along with the top as well as bottom stage pressure of the distillation tower.

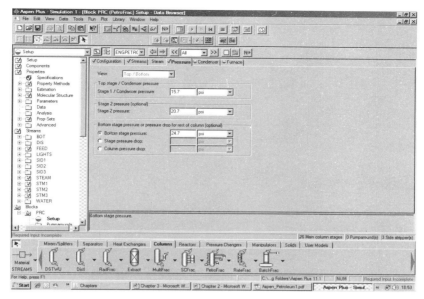

FIGURE 3.80

As given in the problem statement, enter 0.2 in the *Distillate vapour fraction* field under the heading of *Condenser specification* (see Figure 3.81).

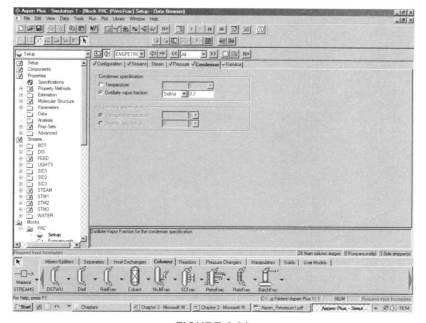

FIGURE 3.81

In the next step, the feed furnace is specified by selecting the type of furnace and giving the values of pressure and fractional overflash (see Figure 3.82).

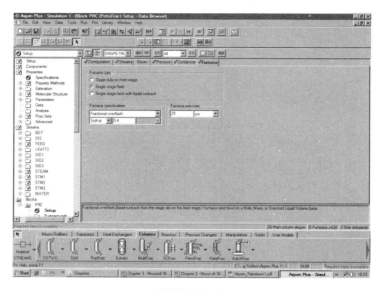

FIGURE 3.82

In the left pane of the Data Browser window, select *Blocks/PRC/Efficiencies* and provide 90% Murphree tray efficiency (see Figure 3.83).

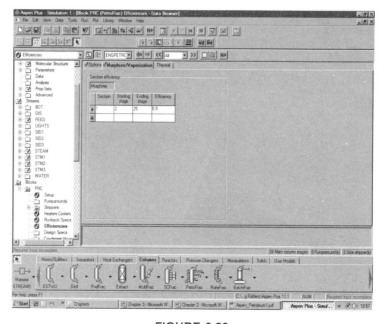

FIGURE 3.83

The three windows, shown in Figures 3.84(a), (b) and (c), specify the side strippers based on the given input data.

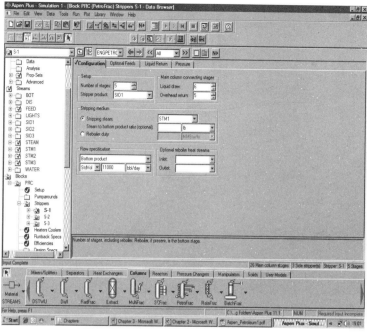

FIGURE 3.84(a)

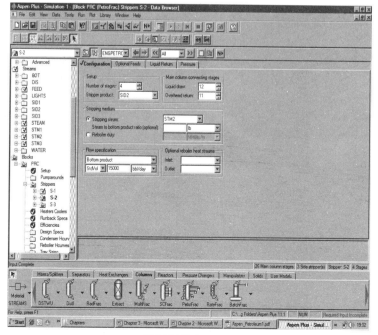

FIGURE 3.84(b)

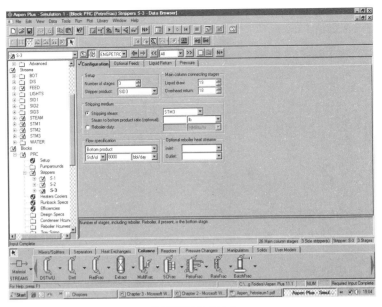

FIGURE 3.84(c)

Although the *Status bar* says *Required Input Complete*, we have to specify the two pumparound circuits connected with the main fractionator. Select *Blocks/PRC/Pumparounds* in the list on the left. Click on *New* as the object manager appears. We may accept the default *Pumparound ID* 'P-1'. Then specify the first pumparound circuit (see Figure 3.85).

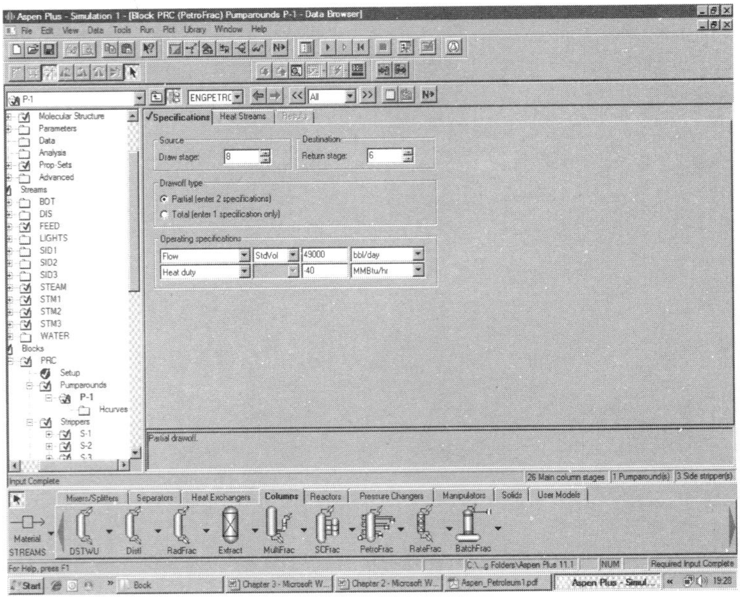

FIGURE 3.85

Select again *Blocks/PRC/Pumparounds* to reopen the pumparounds object manager. By the same way, fill out the form for second pumparound circuit, shown in Figure 3.86.

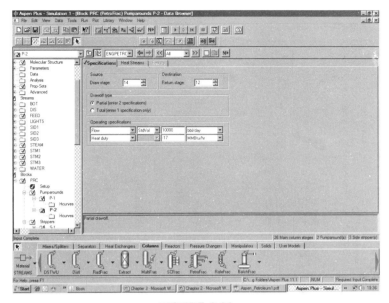

FIGURE 3.86

Running the simulation

Hit *Next* icon and click *OK* to run the simulation. The *Control Panel* window is presented in Figure 3.87.

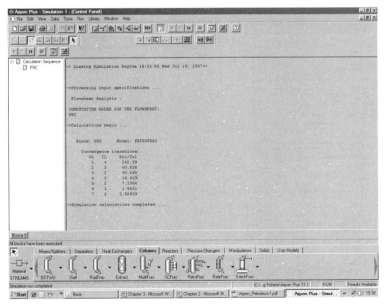

FIGURE 3.87

Viewing results

From the Data Browser, choose *Results Summary/Streams* and obtain the table, shown in Figure 3.88, that includes the flow rates (bbl/day) of all product streams. Save the work done.

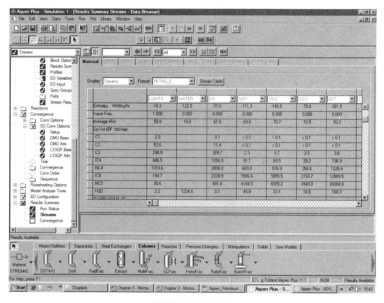

FIGURE 3.88

To obtain the input information (see Figure 3.89), select *Input Summary* from the *View* pulldown menu.

FIGURE 3.89

3.4 SIMULATION AND ANALYSIS OF AN ABSORPTION COLUMN

Problem statement

A hydrocarbon vapor enters an absorption column below the bottom stage and the absorbent enters above the top stage. The column operates at 75 psia with no pressure drop and it has four equilibrium stages. The absorber is specified in Figure 3.90.

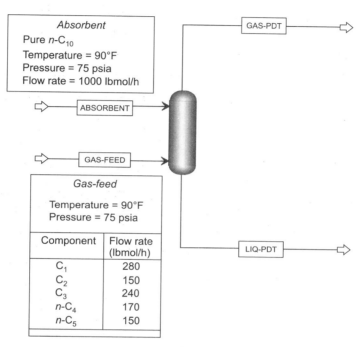

FIGURE 3.90 A flowsheet of an absorption column.

Apply the Peng-Robinson equation of state model in the simulation.

(a) Simulate the absorber model (ABSBR2 under RadFrac) and compute the product compositions.

(b) Perform the sensitivity analysis by examining the effect of absorbent flow rate on the exiting C_3 concentration in the top product.

(c) Compute the absorbent flow rate to keep 15 mole% of C_3 in the gas product (GAS-PDT).

Simulation approach

(a) Double-click *Aspen Plus User Interface* icon on the desktop. When Aspen Plus window pops up, select *General with English Units* Template as shown in Figures 3.91(a) and (b).

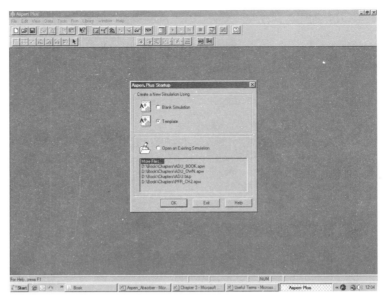

FIGURE 3.91(a)

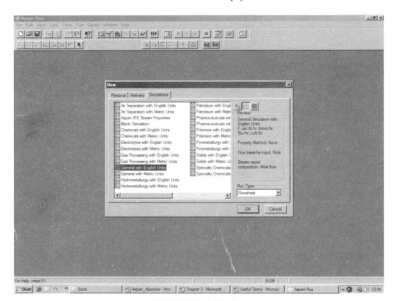

FIGURE 3.91(b)

Click *OK* when the *Connect to Engine* dialog is displayed and proceed to develop the process flow diagram.

Creating flowsheet

Select the *Columns* tab from the bottom toolbar. Among the available RadFrac models, select *ABSBR2* and then place it on the flowsheet by clicking with the cross hair somewhere on the flowsheet background. Right-click to de-select the block. Connecting the inlet and outlet streams and changing the all default labels, we have Figure 3.92.

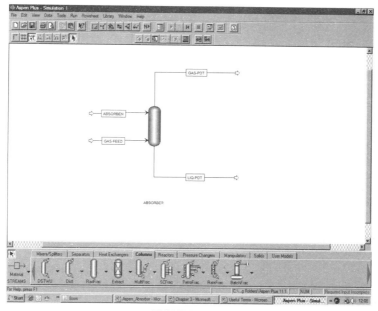

FIGURE 3.92

Configuring settings

In the subsequent step, hit *Next* symbol and fill up the three setup input forms as shown in Figures 3.93(a), (b) and (c).

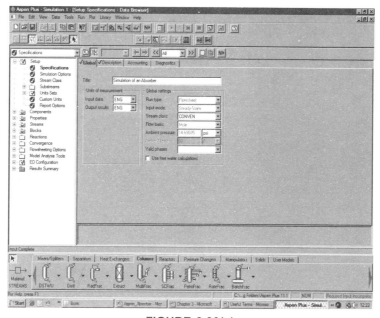

FIGURE 3.93(a)

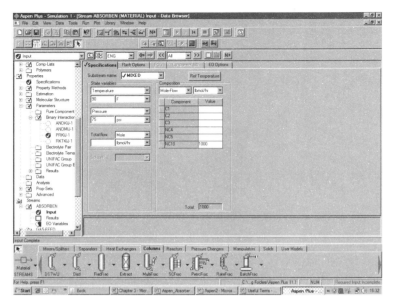

FIGURE 3.96(a)

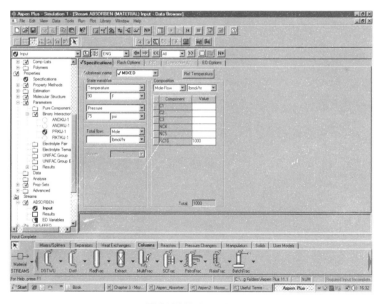

FIGURE 3.96(b)

Specifying block information

Use the Data Browser menu tree to navigate to the *Blocks/ABSORBER/Setup/Configuration* sheet (see Figure 3.97).

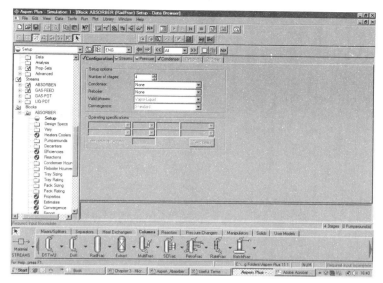

FIGURE 3.97

Select the *Streams* tab to specify stream location. Under *Convention*, there are two feeding options: On-Stage and Above-Stage. In the present problem, the top stage is the first stage and the bottom stage is the fourth one. Therefore the absorbent is fed above Stage 1 and the gas feed is introduced above Stage 5 (see Figure 3.98).

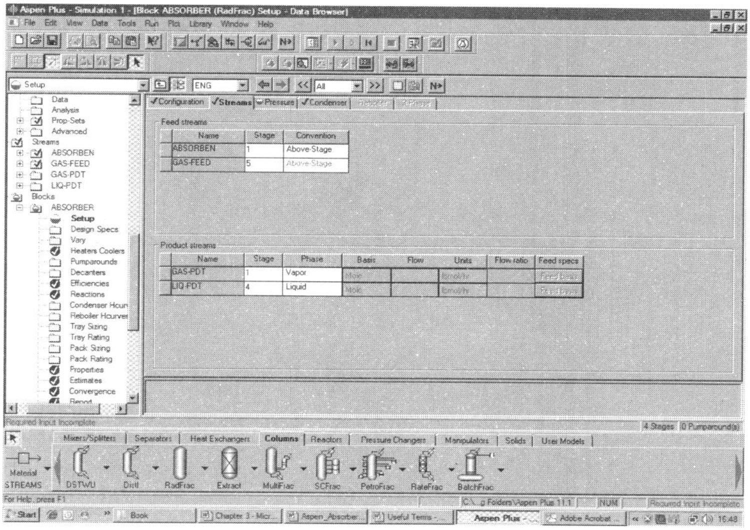

FIGURE 3.98

In the next step (see Figure 3.99), select *Pressure* tab to specify the pressure profile across the absorption column. In this case, the column is operated isobarically at 75 psia. Under *Top stage / Condenser pressure*, enter 75 psi. Aspen software assumes that the column operates isobarically if no additional information is provided.

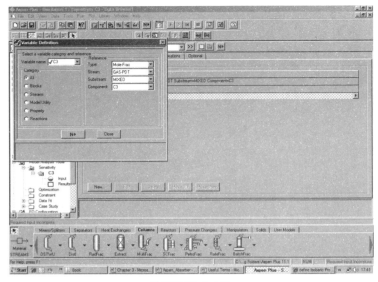

FIGURE 3.103

Type: Mole-Frac

Stream: GAS-PDT

Substream: MIXED

Component: C3

Hit *Next* and select the *Vary* tab. The manipulated variable is specified with the following data (see Figure 3.104):

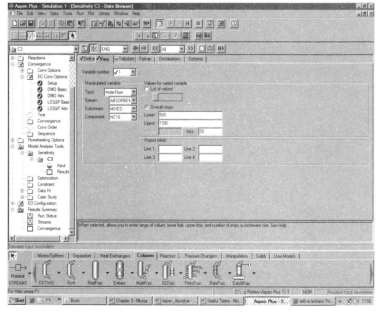

FIGURE 3.104

Type: Mole-Flow

Stream: ABSORBEN

Substream: MIXED

Component: NC10

Overall range

Lower: 500

Upper: 1500

Increment: 50

In the subsequent step (see Figure 3.105), select the *Tabulate* tab. This screen is used by Aspen to set up tables. Insert '1' under *Column No.* Then right click on the adjacent cell under *Tabulated variable or expression*. Select *Variable List* and drag and drop the variable name (C3) into the cell. We may also directly type 'C3' in the cell.

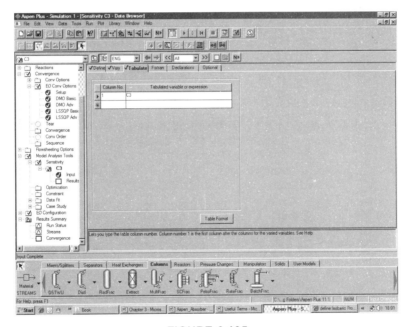

FIGURE 3.105

Then run the simulation and get the screen, shown in Figure 3.106.

Select *New* under *Define* tab. Then enter 'C3' as a variable name and press *OK*. In the next step (see Figure 3.110), the following information are required to input:

Type: Mole-Frac
Stream: GAS-PDT
Substream: MIXED
Component: C3

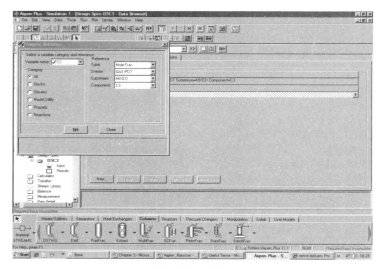

FIGURE 3.110

In the subsequent step (see Figure 3.111), select the *Spec* tab. Design specification data are noted below:

Spec: C3
Target: 0.15
Tolerance: 0.001

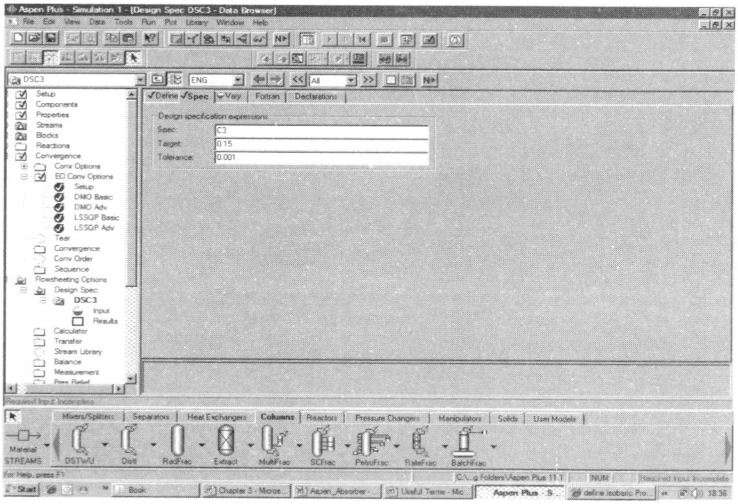

FIGURE 3.111

Finally, select the *Vary* tab and enter the following information in Figure 3.112:

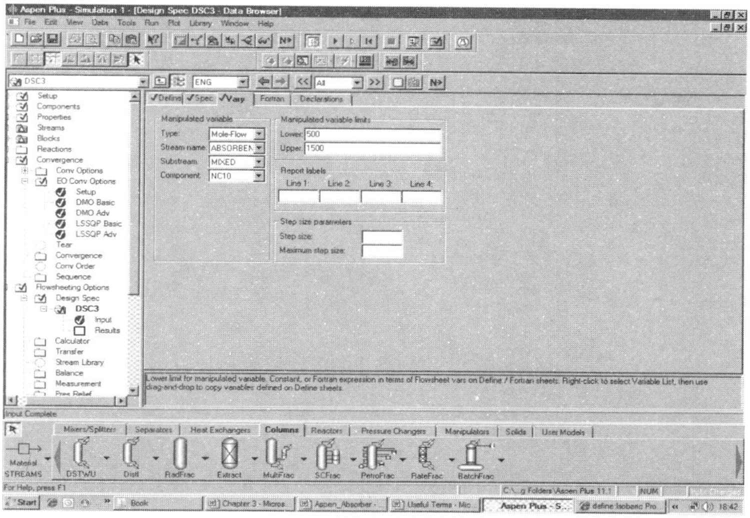

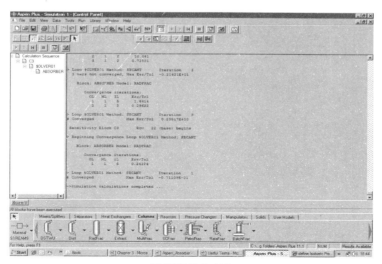

FIGURE 3.112

Type: Mole-Flow
Stream: ABSORBEN
Substream: MIXED
Component: NC10
Manipulated variable limits

Lower: 500
Upper: 1500

As we run the simulation, we get the screen, shown in Figure 3.113.

FIGURE 3.113

Right-click in the empty cell with selecting *Maximize* option. Then select *Variable List* and drag and drop the variable name (C3) into the cell (see **Figure 3.117**). We can also simply type C3 in the field.

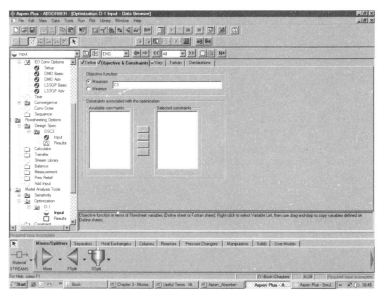

FIGURE 3.117

In the subsequent step, select the *Vary* tab. Under *Variable number*, as we choose 'New', automatically the number '1' appears. Filling out the form, we have the window as shown in Figure 3.118.

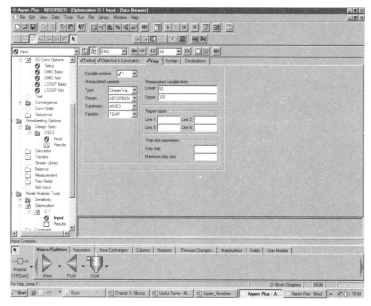

FIGURE 3.118

Pressing *Next* symbol and running the simulation, we get the answer (see Figure 3.119). The maximum C_3 mole fraction of 0.259 is obtained at absorbent inlet temperature of 179.8°F.

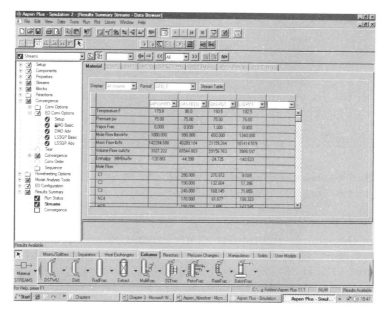

FIGURE 3.119

3.6 SIMULATION OF A REACTIVE DISTILLATION COLUMN

In a reactive distillation (RD) column, shown in Figure 3.120, the decomposition of HI to H_2 and I_2 is considered. The column has total 8 stages (including a partial condenser and reboiler) and it operates at 22 bara pressure.

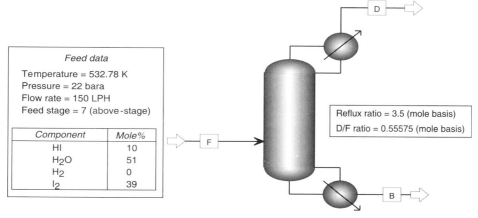

Feed data

Temperature = 532.78 K
Pressure = 22 bara
Flow rate = 150 LPH
Feed stage = 7 (above-stage)

Component	Mole%
HI	10
H_2O	51
H_2	0
I_2	39

Reflux ratio = 3.5 (mole basis)
D/F ratio = 0.55575 (mole basis)

FIGURE 3.120 A schematic of a reactive distillation column.

The HI decomposition reaction (Belaissaoui et al., 2008)

$$2HI \leftrightarrow H_2 + I_2$$

is an equilibrium limited reaction taking place in the vapor phase. The chemical equilibrium constant (K_{eq}) is expressed as the ratio of the partial pressure P_i of the species i:

$$K_{eq} = \frac{P_{H_2} P_{I_2}}{P_{HI}^2} \tag{3.11}$$

The expression of the HI decomposition chemical equilibrium constant as a function of the temperature (T) is given as follows:

$$\ln K_{eq}(T) = -0.03684 - \frac{2772.7729}{T} \tag{3.12}$$

with T in K. The reaction occurs from 2nd to 6th stage. Using the NRTL property method, simulate the reactive column (RadFrac).

Simulation approach

The flowsheet of the sample RD column is drawn and shown in Figure 3.121.

In the subsequent step, filling the *Global* and *Accounting* sheets, specify the components, namely HI, H_2, I_2 and H_2O. Use the NRTL thermodynamic property method to represent the vapor-liquid equilibrium.

In order to obtain the property data (e.g., surface tension, viscosity and conductivity), press *Property Sets* button in *Stream* sheet under *Setup/Report Options* (see Figure 3.122).

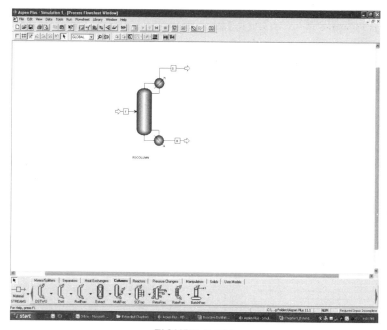

FIGURE 3.121

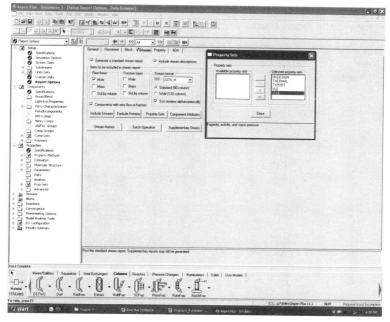

FIGURE 3.122

Use the Data Browser menu tree to navigate to the *Streams/F/Input/Specifications* sheet. Inserting the given values for the feed stream, Figure 3.123 is obtained.

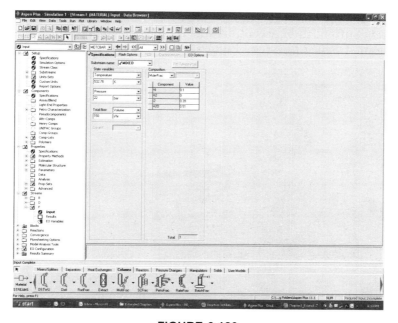

FIGURE 3.123

In the left pane of the Data Browser window, select *Blocks / RDCOLUMN / Setup*. Fill up the *Configuration* sheet and obtain Figure 3.124.

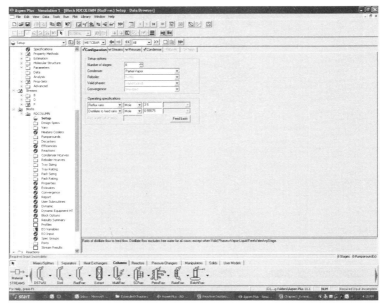

FIGURE 3.124

As defined in the problem statement, consider the feeding point at 7th stage (above-stage) and the column operating pressure of 22 bara in the Aspen simulation.

In the list on the left, select *Reactions* under *Reactions* folder. As the *Object manager* appears, hit *New* button, then accept default ID R-1. For equilibrium reactions, choose REAC-DIST against *Select type* (see Figure 3.125).

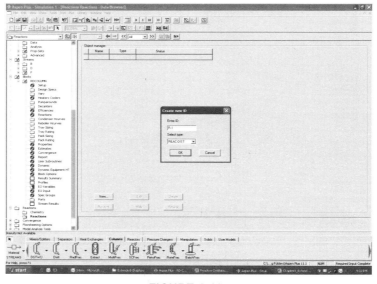

FIGURE 3.125

In the next, press *New* knob and choose *Kinetic/Equilibrium/Conversion*, and then accept *Reaction No.* 1 (see Figure 3.126).

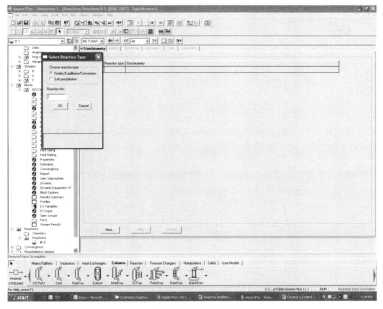

FIGURE 3.126

After inserting the information as asked in the *Edit Reaction* sheet, Figure 3.127 is obtained.

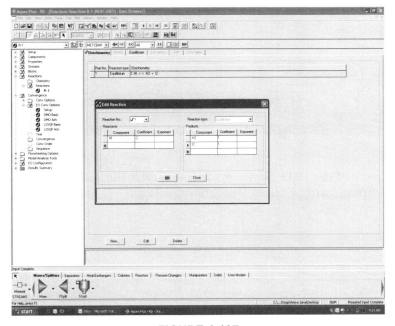

FIGURE 3.127

Open *Equilibrium* sheet under R-1 subfolder. Obtain Figure 3.128 providing the following data:

Reacting phase: Vapor
Keq basis: Partial pressure
Temperature approach[2] *to equilibrium*: 0
Then use Eq. (3.12) for the following parameter values:

A = −0.03684
B = −2772.7729
C = D = 0

Note that the Aspen simulation package can compute K_{eq} from Gibbs energies when the equilibrium constants are not defined.

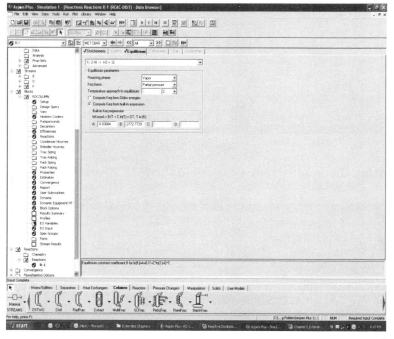

FIGURE 3.128

In the left pane of the Data Browser window, select *Blocks / RDCOLUMN / Reactions* and fill up the *Specifications* sheet (see Figure 3.129).

---

[2] Temperature approach is the number of degrees above the stage temperature at which chemical equilibrium constant is determined.

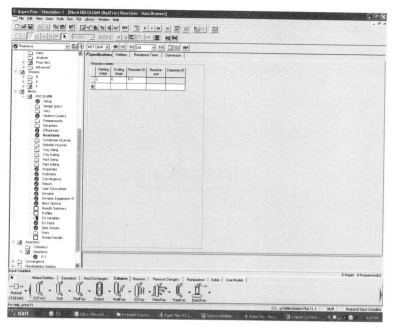

FIGURE 3.129

Running the simulator, we get the following picture (Figure 3.130) that includes the results, including the physical properties.

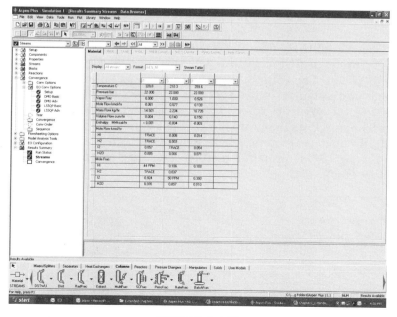

FIGURE 3.130

SUMMARY AND CONCLUSIONS

At the beginning of this chapter, a brief of all built-in column models of Aspen software has been presented. Several separating columns, including a petroleum refining column, an absorber and a reactive distillation, have been simulated using Aspen Plus. The process optimization has also been discussed with an example. The present study covers both the binary as well as multicomponent systems. Interested readers may try to simulate the models given in the exercise.

PROBLEMS

3.1 A feed mixture, consisting of 60 mole% ethanol and 40 mole% water, is to be separated by using a DSTWU model having a flow rate of 100 kmol/h at 40°C and 1 atm so as to recover at least 85% of the light key component in the liquid distillate and 80% of the heavy key component in the bottoms. The column operates at 1 atm with no pressure drop throughout. In the simulation, consider the reflux ratio of 1.5 and a total condenser. Applying the Wilson property method, simulate the column and find out the minimum number of stages, actual number of stages, and feed position.

3.2 A feed stream, consisting of 50 mole% ethane and 50 mole% ethylene, enters a Distl column having a flow rate of 200 lbmol/h at 75°F and 15 psia. This separator runs at 300 psia with no tray-to-tray pressure drop. The pressure in the reboiler as well as condenser is also 300 psia. The feed enters the model at 6th stage and the column has total 15 theoretical stages (including condenser and reboiler) and a total condenser. If the reflux ratio is 7 and the distillate to feed ratio is 0.8, compute the mole fraction of ethane in both the product streams with applying the RK-Soave equation of state model.

3.3 A feed mixture specified in Figure 3.131 is to be distilled by a rigorous RadFrac model (FRACT2). The column consists of total 24 equilibrium stages (including condenser and reboiler) with a stage pressure drop of 2 kPa. Consider the condenser (total) pressure of 125 kPa and the top stage (Stage no. 2) pressure of 130 kPa. The distillate flow rate is 120 kmol/h and the reflux ratio (mole basis) is 2. A side product (vapor) is withdrawn from 14th stage. Applying the Soave-Redlich-Kwong (SRK) property method, simulate the column model and report the product compositions.

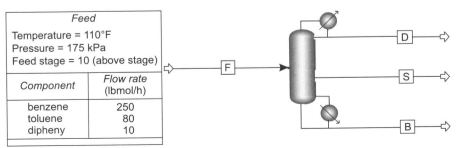

Feed	
Temperature = 110°F	
Pressure = 175 kPa	
Feed stage = 10 (above stage)	
Component	Flow rate (lbmol/h)
benzene	250
toluene	80
dipheny	10

FIGURE 3.131 A flowsheet of a distillation column.

3.4 A reboiled stripper is to be employed to remove mainly propane and lighter components from a feed stream, shown in Figure 3.132. It has total 6 stages (including condenser and reboiler) and no condenser. The bottoms rate is 100 lbmol/h and the column pressure is 150 psia throughout. Using the Peng-Robinson thermodynamic method, simulate the RadFrac model (STRIP2) and find out the product compositions.

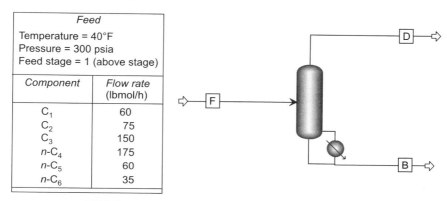

Feed	
Temperature = 40°F	
Pressure = 300 psia	
Feed stage = 1 (above stage)	

Component	Flow rate (lbmol/h)
C_1	60
C_2	75
C_3	150
$n\text{-}C_4$	175
$n\text{-}C_5$	60
$n\text{-}C_6$	35

FIGURE 3.132   A flowsheet of a stripping column.

3.5 A feed mixture of cyclopentane and cyclohexane is to be separated employing a liquid-liquid extraction unit at 25°C and 1 atm with the use of methanol as a solvent. The schematic diagram of the process with feed specifications is given in Figure 3.133. The process unit, having total five stages, is operated adiabatically. Applying the UNIQUAC property method, simulate the extraction model (ICON1) and note down the product compositions.

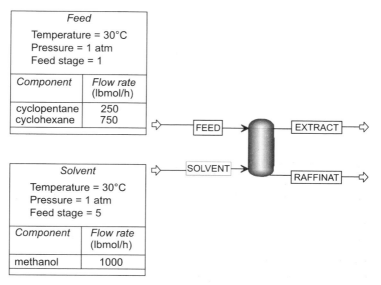

Feed	
Temperature = 30°C	
Pressure = 1 atm	
Feed stage = 1	

Component	Flow rate (lbmol/h)
cyclopentane	250
cyclohexane	750

Solvent	
Temperature = 30°C	
Pressure = 1 atm	
Feed stage = 5	

Component	Flow rate (lbmol/h)
methanol	1000

FIGURE 3.133   A flowsheet of an extraction column.

3.6 A gas consisting of 40 mole% ammonia, 60 mole% air at 20°C, 25 psia, flowing at the rate 120 kmol/h, is to be scrubbed counter-currently with water (pure) entering at 60°C and 30 psia at a rate 100 kmol/h. The column operates at 1 atm throughout and it has four stages. Using the UNIFAC thermodynamic model, (a) simulate the RadFrac absorber (ABSBR2) and determine the exiting ammonia concentration in the gas product, (b) Perform the sensitivity analysis by examining the effect of absorbent flow rate on the exiting ammonia concentration in the top product.

3.7 A distillation column (RadFrac) used for fractionating a ternary mixture (HI-I_2-H_2O) is shown in Figure 3.134.

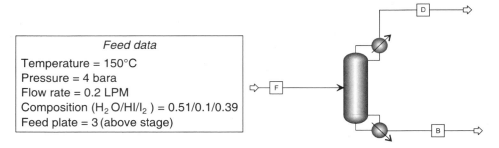

> *Feed data*
> Temperature = 150°C
> Pressure = 4 bara
> Flow rate = 0.2 LPM
> Composition (H_2O/HI/I_2) = 0.51/0.1/0.39
> Feed plate = 3 (above stage)

FIGURE 3.134 A flowsheet of a distillation column (RadFrac).

The column operates at 4 bara pressure and it has total five trays (including partial condenser and reboiler). In the simulation, the reflux ratio (mole) and distillate to feed ratio (mole) are taken as 0.94 and 0.6121, respectively.

(i) Simulate the model using the NRTL thermodynamic package.

(ii) Considering sieve tray type with tray spacing of 0.6096 m, report the Aspen generated column diameter.

(iii) Calculate the column diameter considering F-factor equals 1 (in English engineering units).

(iv) Compare the calculated column diameters with the value to be obtained using the following correlation:

$$D_c = \sqrt{\frac{V}{900\pi}\left(\frac{MWART}{P}\right)^{0.5}}$$

where, D_c denotes the column diameter (ft), T the average column temperature (°R), P the column pressure (psia), V the vapor flow rate (lbmol/h), and MWA the average molecular weight. Consider R equals 10.73 psia.ft³/lbmol.°R.

3.8 Repeat the above problem with the replacement of RadFrac model by the RateFrac column. The feed specifications are provided in Figure 3.135.

A hydrocarbon mixture with the given component-wise flow rates (Table 3.6) enters the furnace at 120°F and 45 psia.

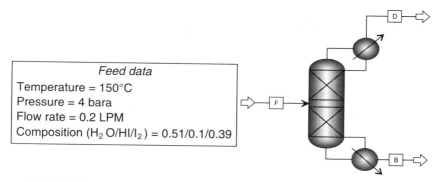

FIGURE 3.135 A flowsheet of a distillation column (RateFrac).

The column operates at 4 bara pressure and it has total four segments (including partial condenser and reboiler[3]). As mentioned in the previous problem statement, the reflux ratio (mole) and distillate to feed ratio (mole) are considered as 0.94 and 0.6121, respectively. The packing information is given as:

Arrangement: RANDOM

Type: RASCHIG

Material: GLASS

Dimension: 10-MM
Height per segment: 1.25 metre

Considering the column diameter of 60 mm and the default optional parameters (e.g., size of packing, packing factor, etc.),

(i) Simulate the column using the NRTL thermodynamic model.
(ii) Compare the simulation results with those obtained for the previous RadFrac column.
(iii) Perform a sensitivity test to investigate the effect of column pressure (SEG-PRES variable) on HI mole fraction in the vapor distillate.

3.9 An artificial petroleum refining column (PRC) shown in Figure 3.136 consists of a feed furnace and a fractionation tower. The tower includes one pumparound circuit, a partial condenser and one side stripper. The furnace (single stage flash type) operates at 20 psia and provides a fractional overflash of 50% (StdVol basis) in the tower. The outlet stream of the furnace enters the tower on stage 18. The column has total 20 stages. A steam stream, STEAM, is fed at the bottom of the fractionator (20th stage with on-stage convention). There is another steam stream, STEM1, used in the side stripper. The condenser runs at 15 psia with a pressure drop of 5 psi. The tower pressure drop is equal to 5 psi. The distillate rate is 12000 bbl/day and the distillate vapor fraction in the condenser is 0.25 (StdVol basis). The liquid product, SID1, is withdrawn from 5th stage with a flow rate of 2000 bbl/day.

---

[3] Condenser and reboiler are considered as individual equilibrium segment.

A hydrocarbon mixture with the given component-wise flow rates (Table 3.6) enters the furnace at 120°F and 45 psia.

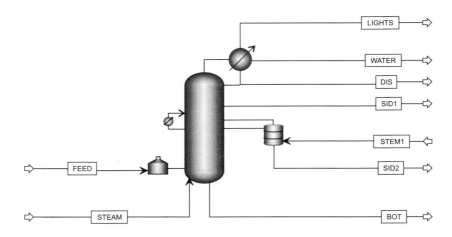

FIGURE 3.136 A flowsheet of a petroleum refining column.

TABLE 3.6

Component	Flow rate (bbl/day)
C_1	10
C_2	100
C_3	600
i-C_4	1800
n-C_4	7500
i-C_5	30000
n-C_5	42000
n-C_6	250
H_2O	250

The pumparound circuit (for cooling) and the side stripper are specified with the following information (see Table 3.7).

TABLE 3.7

Pumparound (drawoff type)	Location		Specifications	
	Draw stage	Return stage	Flow rate (bbl/day)	Temperature (°F)
1 (partial)	8	6	40000	20

Stripper	No. of stages	Stripper product	Location		Stripping steam	Bottom product flow rate (bbl/day)
			Draw stage	Return stage		
1	5	SID1	12	10	STEM1	15000

Two steam streams, used in the column model, are described in Table 3.8.

TABLE 3.8

		Specifications		
Steam stream	Location	Temperature (°F)	Pressure (psia)	Flow rate (lb/h)
STEAM	Main tower	350	50	12000
STEM1	Stripper	350	50	5000

Selecting the PENG-ROB base method under 'REFINERY' process type, simulate the model using a PetroFrac column and report the flow rates (bbl/day) of all product streams.

REFERENCES

Douglas, J. M. (1988). *Conceptual Design of Chemical Processes*, McGraw-Hill, New York.

Belaissaoui, B., Thery, R., Meyer, X. M., Meyer, M., Gerbaud, V., and Joulia, X. (2008). "Vapour Reactive Distillation Process for Hydrogen Production by HI Decomposition from $HI-I_2-H_2O$ Solutions," *Chem. Eng. Process.*, 47, 396–407.

Part II

Chemical Plant Simulation
using Aspen Plus™

CHAPTER 4

Aspen Plus™ Simulation of Chemical Plants

4.1 INTRODUCTION

In the last three chapters, we have studied in detail the simulation of individual processes, such as flash drum, dryer, chemical reactor, distillation column including petroleum refining process, absorber, stripper and liquid-liquid extraction unit, using the Aspen Plus™ software. Here, by a 'chemical plant' we mean a chemical process integrated with several single process units. The chemical process industries are typically constructed combining, for example, flash chamber, mixer, splitter, heat exchanger, pump, compressor, reactor, fractionator, filter and so on. It is easy to simulate even a large chemical plant by the use of Aspen software package.

In the present chapter, first the simulation of two chemical process flowsheets is discussed. They are a distillation train and a vinyl chloride monomer (VCM) manufacturing unit. Subsequently, we will also learn how to proceed for simulating a process unit with streams recycling by the use of the Aspen simulation package. After thoroughly reading this chapter and simulating the solved examples in hand, we will be able to use Aspen Plus flowsheet simulator for solving a wide variety of chemical plants. To improve the flowsheet simulation skills, it is recommended to solve the problems given in the exercise.

4.2 ASPEN PLUS SIMULATION OF A DISTILLATION TRAIN

Problem statement

A hydrocarbon stream H is supplied at 5°C and 2.5 atm. The pump P1 discharges the feed F at 10 atm. In Table 4.1 the component-wise flow rates are tabulated for stream H.

The schematic representation of the complete process integrated with a pump and five DSTWU column models (C1, C2, C3, C4 and C5) is shown in Figure 4.1.

TABLE 4.1

Component	Flow rate (kmol/h)
C_2	35
C_3	50
i-C_4	130
n-C_4	200
i-C_5	180
n-C_5	200
n-C_6	5

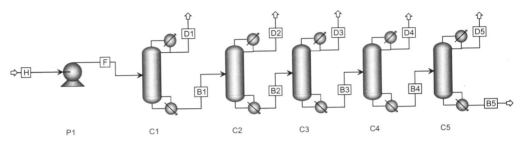

FIGURE 4.1 A flowsheet of a distillation train.

For Aspen Plus simulation of the distillation train, required information are given in Table 4.2.

TABLE 4.2

Column (abbreviation)	Condenser pressure (atm)	Reboiler pressure (atm)
Deethanizer (C1)	9	9
Depropanizer (C2)	5	5
Deisobutanizer (C3)	4	4
Debutanizer (C4)	3	3
Deisopentanizer (C5)	2	2

Each distillation model has total 20 theoretical stages (including a total condenser and a reboiler). For the light key (LK) and heavy key (HK), we expect 99.9% and 0.1% recovery, respectively, in the distillate of all columns. Using the Peng-Robinson property method, simulate the distillation train and report the compositions of all distillation products.

Simulation approach

From the desktop, select *Start* button followed by *Programs*, *AspenTech*, *Aspen Engineering Suite*, *Aspen Plus Version* and finally *Aspen Plus User Interface*. Then choose *Template* option in the *Aspen Plus Startup* dialog (see Figure 4.2).

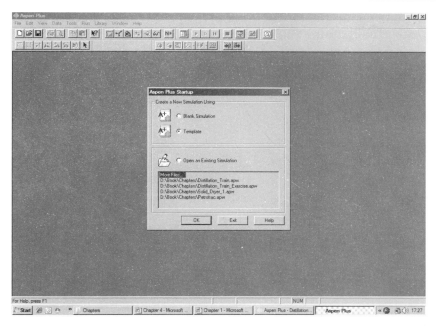

FIGURE 4.2

As we hit *OK* button, the following window appears (see Figure 4.3). Based on the units used in the problem statement, we select *General with Metric Units*.

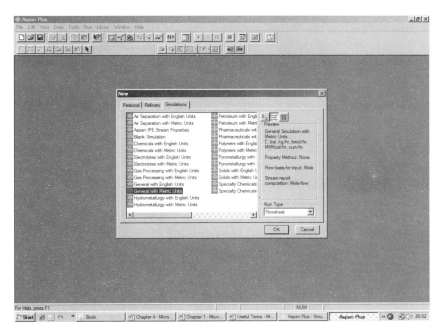

FIGURE 4.3

Press *OK* and obtain the *Connect to Engine* dialog. Select 'Local PC' as *Server type* and click *OK*. Actually, this step is specific to our installation (see Figure 4.4).

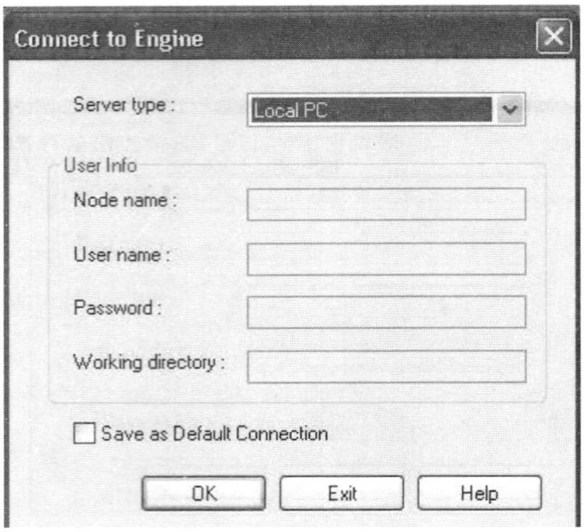

FIGURE 4.4

Creating flowsheet

The next screen represents a *Process Flowsheet Window*. Add a pump by selecting the *Pressure Changers* tab from the *Model Library* toolbar. Moreover, in the library, select the *Columns* tab and then choose DSTWU model to include five such columns consecutively on the flowsheet. Notice that to incorporate a block, click on the appropriate icon and then place the block on the process flowsheet by clicking with the cross hairs somewhere on the flowsheet background. Right click to de-select the block.

Now we need to interconnect the blocks and add the inlet as well as outlet streams. Select *Material STREAMS* on the left of the toolbar at the bottom. In the next, as we move the cursor to the process flowsheet window, several red and blue arrows appear around the blocks. The red arrows indicate required streams and the blue arrows are optional. In the previous chapters, we have learned how to connect the feed and product streams with a single block.

Let us observe Figure 4.5 to know how to interconnect the two blocks by a stream. Here, first we wish to interconnect the pump P1 with the column C1 using the feed stream F. Right-click with highlighting feed block, select *Reconnect Destination* and then move the cursor to click on an arrow that is fed to the column C1.

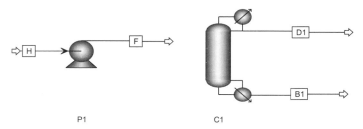

FIGURE 4.5

We can select *Reconnect Source* instead of *Reconnect Destination* if we modify Figure 4.5 to Figure 4.6.

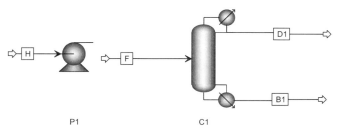

FIGURE 4.6

By the same way, interconnect remaining blocks. Renaming all blocks as well as incoming and outgoing streams, finally we have the screen shown in Figure 4.7. As mentioned earlier, to rename a particular stream (or block), first select it, then right-click, next select *Rename Stream* (or *Rename Block*) and finally enter the name as we like.

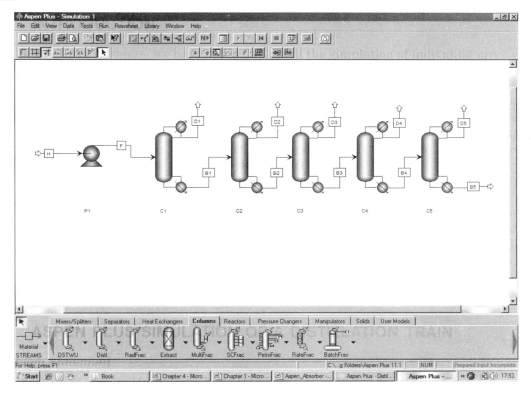

FIGURE 4.7

The status indicator in the bottom right of the window, shown in Figure 4.7, says *Required Input Incomplete* indicating that the process flowsheet is complete and input data are required to enter for running the simulation.

Configuring settings

As we hit *Next* icon and then click on *OK*, the following window pops up (see Figure 4.8). Remember that in the Data Browser, we need to enter information using data input forms at locations where there are red semicircles. As we finish a section, a blue checkmark appears.

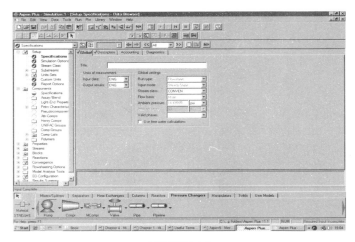

FIGURE 4.8

It is always a good practice to represent a simulation problem with entering a title. In the *Title* field, we may enter 'Simulation of a Distillation Train'. Note that we may change the input/output data units under *Units of measurement* (see Figure 4.9).

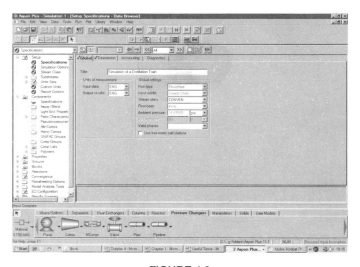

FIGURE 4.9

The next window (see Figure 4.10) includes the Aspen Plus *accounting* information, as given below, required at some installations.

User name: AKJANA
Account number: IIT-KGP
Project ID: CHEMICAL
Project name: DT

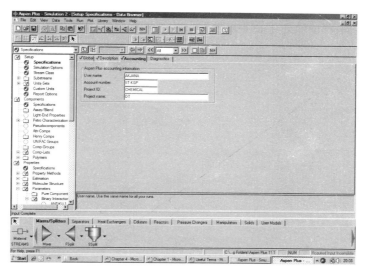

FIGURE 4.10

If we wish to have the streams results in terms of mole fraction, select *Report Options* under *Setup* folder to the left. Under the *Stream* tab, select 'Mole' as *Fraction basis* (see Figure 4.11).

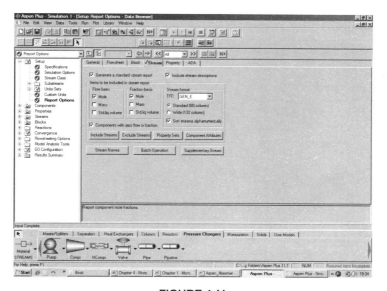

FIGURE 4.11

Specifying components

In the subsequent step, use the Data Browser menu tree to navigate to the *Components/Specifications* sheet. It is shown in Chapter 1 how to define components in the component input form. Here, we have this table as shown in Figure 4.12.

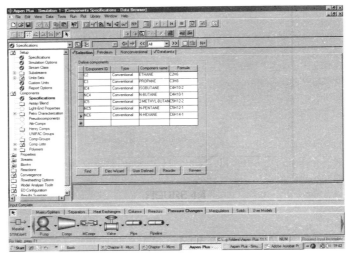

FIGURE 4.12

Specifying property method

In the list on the left, choose *Properties/Specifications* to obtain the property input form. As mentioned previously, *a property method* includes the models and methods to calculate the physical properties, such as vapor-liquid equilibrium coefficient, enthalpy and density. For the example plant, set PENG-ROB base property method by scrolling down (Figure 4.13).

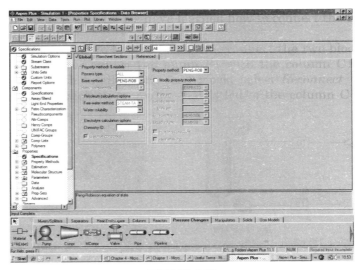

FIGURE 4.13

Note that there is no compulsion to use only a single thermodynamic property method for all processes in a chemical plant. Aspen software provides an option to choose different property methods for different processes. To do so, select *Block Options/ Properties* under a particular model of *Blocks* folder in the list on the left and then choose the suitable property method.

Specifying stream information

The *Streams/H/Input/Specifications* sheet appears with the Data Browser menu tree in the left pane. Entering the given data for stream H, we obtain the sheet as shown in Figure 4.14.

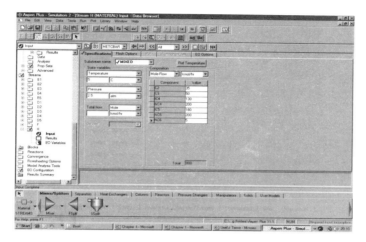

FIGURE 4.14

Specifying block information

As we hit *Next* button, the block input form appears. The deethanizer column is specified with the given data as shown in Figure 4.15.

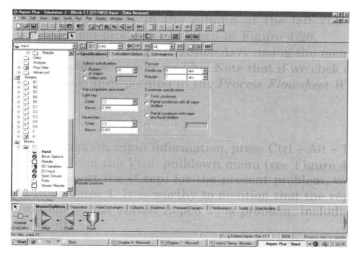

FIGURE 4.15

Subsequently, the filled input forms are shown in Figures 4.16(a), (b), (c) and (d) for other four DSTWU columns.

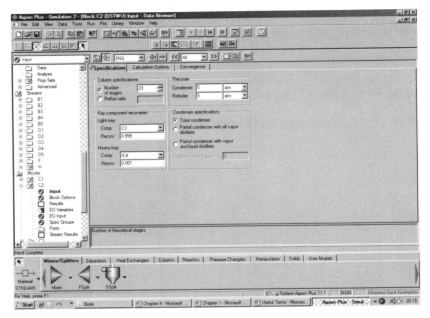

FIGURE 4.16(a)

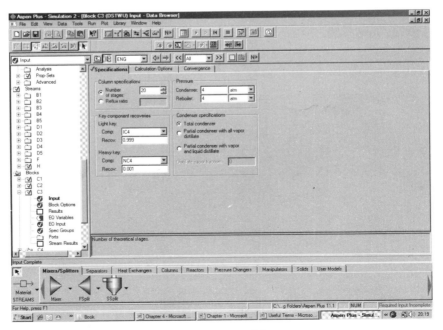

FIGURE 4.16(b)

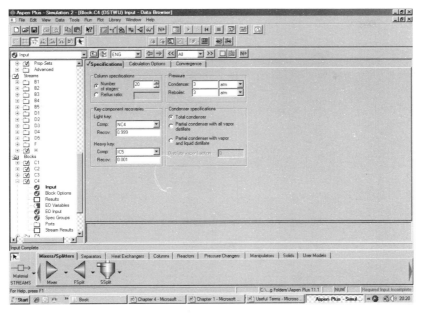

FIGURE 4.16(c)

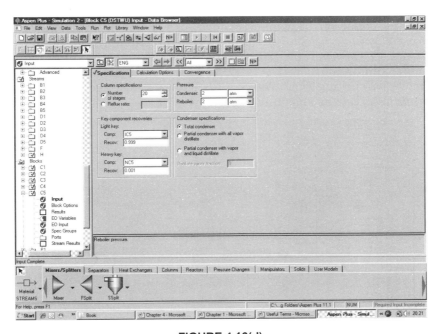

FIGURE 4.16(d)

Click on *Next* and specify the pump (P1) outlet by providing the discharge pressure of 10 atm (see Figure 4.17).

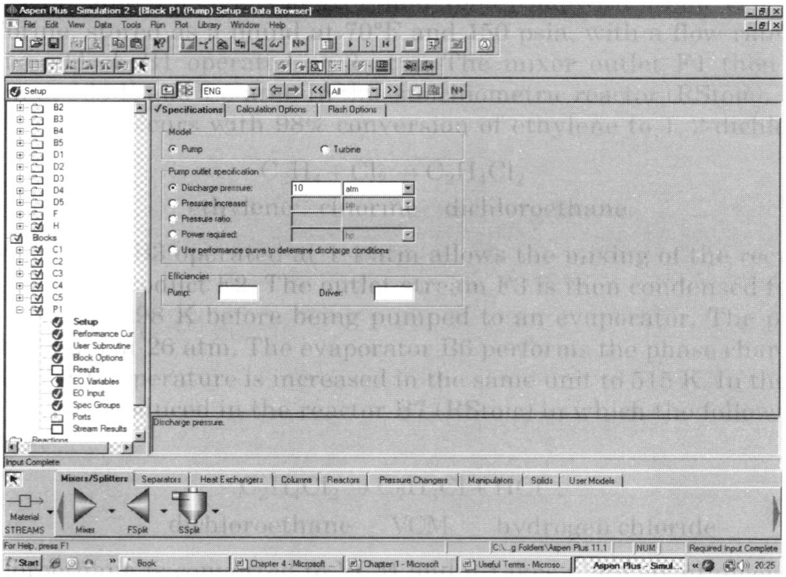

FIGURE 4.17

At this moment, we can observe that all the red semicircles in the left became blue, indicating that the data provided for running the Aspen simulator are sufficient. In addition, the status bar in the window, shown in Figure 4.17, includes a message of *Required Input Complete*; it confirms that to run the simulator, minimum input data have been supplied. However, provision is there for putting more input information for different kind of simulations.

As we move on clicking *Next*, Aspen Plus shows a message under the heading of *Required Input Complete* as shown in Figure 4.18.

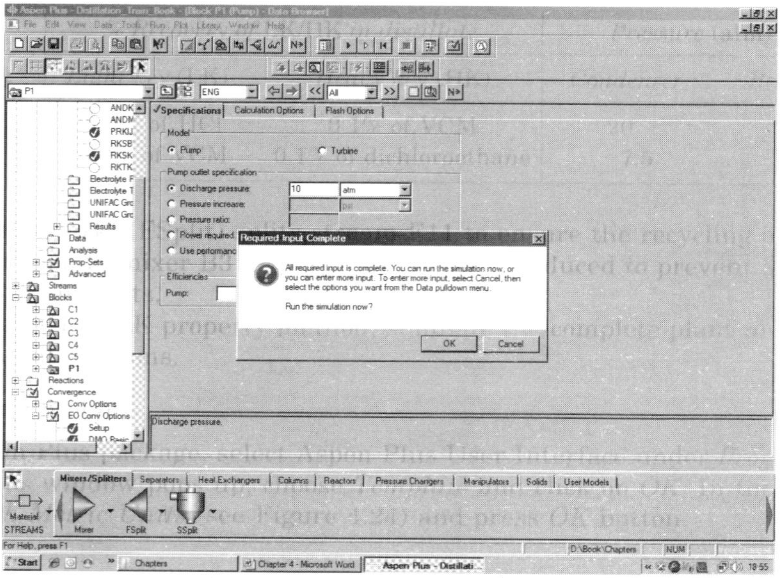

FIGURE 4.18

Running the simulation

As we approve the simulation run, the *Control Panel*, displayed in Figure 4.19, shows the progress of the flowsheet simulation in addition to a message of *Results Available*.

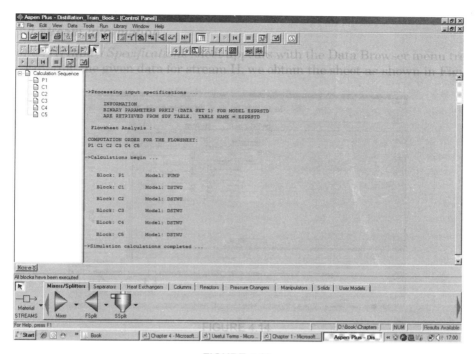

FIGURE 4.19

Viewing results

Choose *Results Summary/Streams* in the column at the left side and obtain the compositions of all distillation products as shown in Figure 4.20.

We may save the work by choosing *File/Save As/*...using the menu list on the top. We can give a name of the file whatever we like. Note that if we click on *Stream Table*, the results summary table is incorporated in the *Process Flowsheet Window,* as shown in Figure 4.21.

Viewing input summary

If we wish to have the systematic input information, press Ctrl + Alt + I on the keyboard or select *Input Summary* from the *View* pulldown menu (see Figure 4.22).

In order to create a report file (*.rep) for the present problem, we may follow the approach presented in Chapter 1. It is worthy to mention that the report file contains all necessary information on the solved Aspen Plus problem, including given process data and computed results.

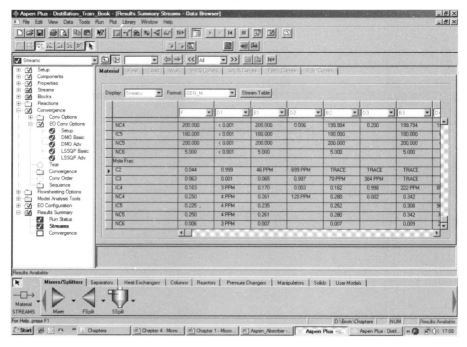

FIGURE 4.20

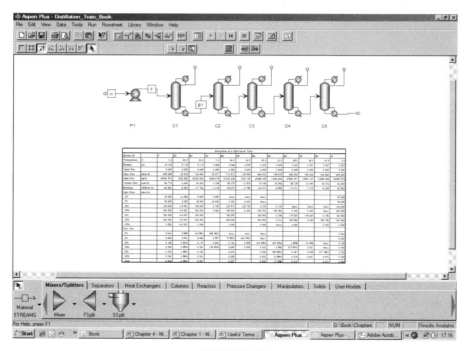

FIGURE 4.21

```
;
;Input Summary created by Aspen Plus Rel. 11.1 at 17:19:46 Sun Jul 29, 2007
;Directory D:\Book\Chapters  Filename C:\Users\akjana\AppData\Local\Temp\~apd00e.tmp
;


DYNAPLUS
    DPLUS RESULTS=ON

TITLE 'Simulation of a Distillation Train'

IN-UNITS MET VOLUME-FLOW='cum/hr' ENTHALPY-FLO='MMkcal/hr'  &
        HEAT-TRANS-C='kcal/hr-sqm-K' PRESSURE=bar TEMPERATURE=C  &
        VOLUME=cum DELTA-T=C HEAD=meter MOLE-DENSITY='kmol/cum'  &
        MASS-DENSITY='kg/cum' MOLE-ENTHALP='kcal/mol'  &
        MASS-ENTHALP='kcal/kg' HEAT=MMkcal MOLE-CONC='mol/l'  &
        PDROP=bar

DEF-STREAMS CONVEN ALL

ACCOUNT-INFO ACCOUNT=IIT-KGP PROJECT-ID=CHEMICAL  &
        PROJECT-NAME="DT" USER-NAME="AKJANA"

DESCRIPTION "
    General Simulation with Metric Units :
    C, bar, kg/hr, kmol/hr, MMKcal/hr, cum/hr.

    Property Method: None

    Flow basis for input: Mole

    Stream report composition: Mole flow
    "

DATABANKS PURE11  / AQUEOUS  / SOLIDS  / INORGANIC  /  &
        NOASPENPCD

PROP-SOURCES PURE11  / AQUEOUS  / SOLIDS  / INORGANIC

COMPONENTS
    C2 C2H6 /
    C3 C3H8 /
    IC4 C4H10-2 /
    NC4 C4H10-1 /
    IC5 C5H12-2 /
    NC5 C5H12-1 /
    NC6 C6H14-1

FLOWSHEET
    BLOCK C5 IN=B4 OUT=D5 B5
    BLOCK C4 IN=B3 OUT=D4 B4
    BLOCK C3 IN=B2 OUT=D3 B3
```

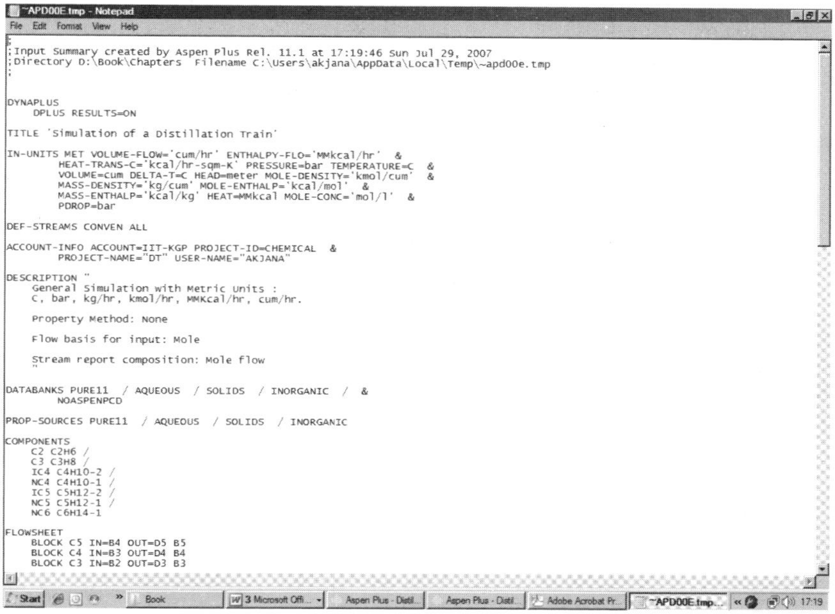

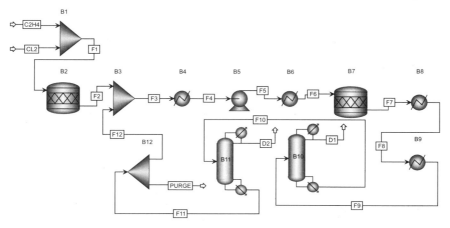

FIGURE 4.22

4.3 ASPEN PLUS SIMULATION OF A VINYL CHLORIDE MONOMER (VCM) PRODUCTION UNIT

Problem statement

The process flow diagram for Aspen Plus simulation of the vinyl chloride monomer manufacturing plant is shown in Figure 4.23. The flowsheet has been developed based on the VCM production technology reported in a book by Seider et al. (1998).

FIGURE 4.23 A flowsheet of a vinyl chloride monomer production unit.

Pure ethylene, stored as a gas at 70°F and 1000 psia, with a flow rate of 20 tons/h, and pure chlorine, stored as a liquid at 70°F and 150 psia, with a flow rate of 50 tons/h enter the mixer block B1 operated at 2 atm. The mixer outlet F1 then goes to the reactor B2 run at 363 K and 1.5 atm. In this stoichiometric reactor (RStoic), the following chlorination reaction occurs with 98% conversion of ethylene to 1, 2-dichloroethane:

$$C_2H_4 + Cl_2 \rightarrow C_2H_4Cl_2$$
ethylene   chlorine   dichloroethane

In the next, mixer B3 operated at 1.4 atm allows the mixing of the recycled stream F12 with the reactor product F2. The outlet stream F3 is then condensed fully to liquid phase in block B4 at 298 K before being pumped to an evaporator. The pump B5 has discharged the liquid at 26 atm. The evaporator B6 performs the phase change operation and then the vapor temperature is increased in the same unit to 515 K. In the subsequent step, stream F6 is introduced in the reactor B7 (RStoic) in which the following pyrolysis reaction occurs:

$$C_2H_4Cl_2 \rightarrow C_2H_3Cl + HCl$$
dichloroethane   VCM   hydrogen chloride

The dichloroethane is converted to VCM and it takes place spontaneously at 773 K and 25 atm with 65% conversion. To reduce carbon deposition in the heat exchanger, the hot vapor stream leaving the reactor is quenched in block B8 yielding a saturated vapor stream at 443 K. Quencher effluent stream F8 is condensed to liquid phase in block B9 at 279 K and then fed to a DSTWU column B10 as stream F9. In the next, Stream F10 is introduced in another DSTWU column B11. The first column mainly separates HCl from other components, while the second column purifies VCM from the rests. Both the distillation columns have 10 theoretical stages (including a total condenser and a reboiler) each along with the specifications, shown in Table 4.3.

TABLE 4.3

	% Recovery of LK/HK in distillate		Pressure (atm)	
Block	Light key (LK)	Heavy key (HK)	Condenser	Reboiler
B10	99.9% of HCl	0.1% of VCM	20	22
B11	99.9% of VCM	0.1% of dichloroethane	7.5	8

Finally, block B12 (FSplit) splits stream F11 to ensure the recycling of 99.999% of F11 as F12 stream to mixer B3. A purge stream is introduced to prevent accumulation of unreacted components.

Using the POLYSRK property method, simulate the complete plant to compute the composition of all streams.

Simulation approach

To start Aspen Plus package, select Aspen Plus User Interface under *Programs*. When the Aspen Plus window pops up, choose *Template* and click on *OK*. In the next, select *Polymers with Metric Units* (see Figure 4.24) and press *OK* button.

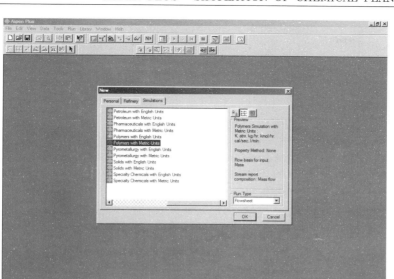

FIGURE 4.24

Click *OK* when the Aspen Plus engine window appears to obtain a blank *Process Flowsheet Window.*

Creating flowsheet

We can develop the process flow diagram (see Figure 4.25) by incorporating the following:

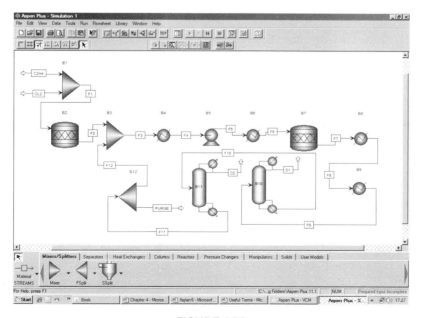

FIGURE 4.25

built-in process units available in the Aspen Plus *Model Library*:

two mixers (B1 and B3)

two 'RStoic' type reactors (B2 and B7)

four 'Heater' type heat exchangers (B4, B6, B8 and B9)

one 'Pump' type pressure changer (B5)

two 'DSTWU' type columns (B10 and B11)

one 'FSplit' type splitter (B12)

All the blocks and streams are renamed according to the problem definition.

The status message directs us to provide the input information required to run the complete Aspen Plus simulation program. In the subsequent sections, we will fill up several input forms one by one.

Configuring settings

After creating the flowsheet for the VCM manufacturing unit, hit *Next* button followed by *OK* to open the *Setup / Specifications / Global* sheet. In the following, the first screen, shown in Figure 4.26(a), includes the *Title* of the present project as 'Simulation of a VCM Production Unit' and the next screen, displayed in Figure 4.26(b), shows the Aspen Plus *accounting* information as given below.

User name: AKJANA

Account number: SAY X

Project ID: ANYTHING

Project name: AS YOU LIKE

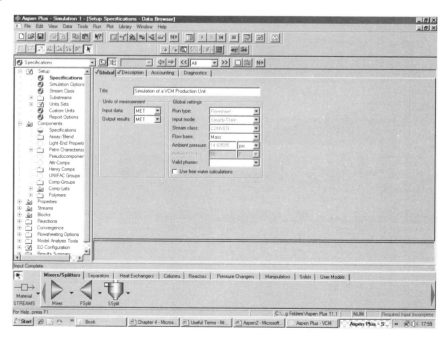

FIGURE 4.26(a)

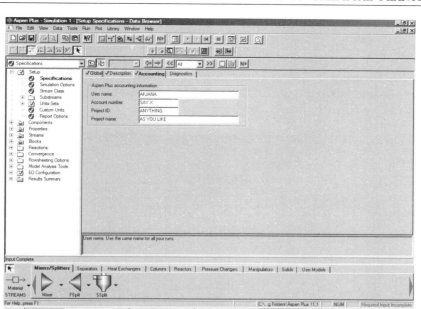

FIGURE 4.26(b)

We wish to have the streams results summarized with mass fraction basis that is not set by default. Accordingly, we choose 'Mass' fraction basis in the *Report Options/Stream* sheet under *Setup* folder (see Figure 4.27).

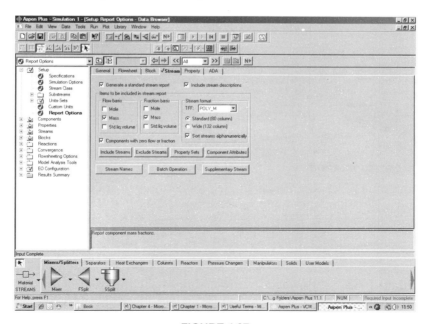

FIGURE 4.27

Specifying components

The components that are involved in the monomer manufacturing process are ethylene (C_2H_4), chlorine (Cl_2), 1,2-dichloroethane $(C_2H_4Cl_2)$, vinyl chloride (C_2H_3Cl) and hydrogen chloride (HCl). In order to get a blank component input form, choose *Components/Specifications* in the left pane of the Data Browser window. Defining all species in the *Selection* sheet, we have Figure 4.28.

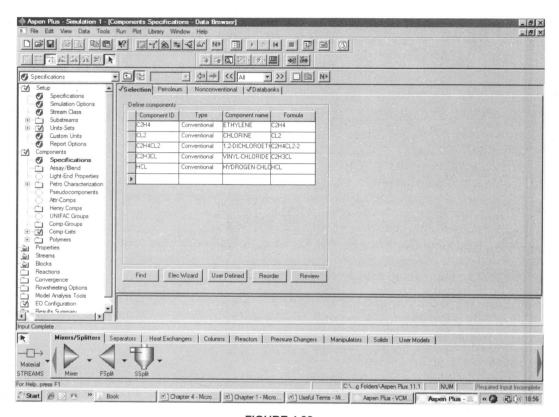

FIGURE 4.28

Specifying property method

In the subsequent step, choose *Properties/Specifications* to set the property method. As mentioned in the problem statement, accordingly select POLYSRK base method under POLYMER process type (see Figure 4.29).

Specifying stream information

From the Data Browser, choose *Streams* folder and see the name of all input, output and intermediate streams. However, we have to provide information for only two input streams, C2H4 and CL2, which are fed to the mixer block B1. Figures 4.30(a) and (b) show the filled stream input forms.

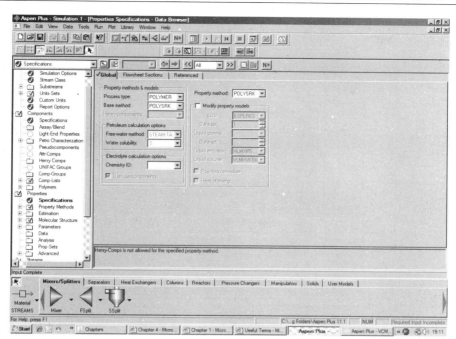

FIGURE 4.29

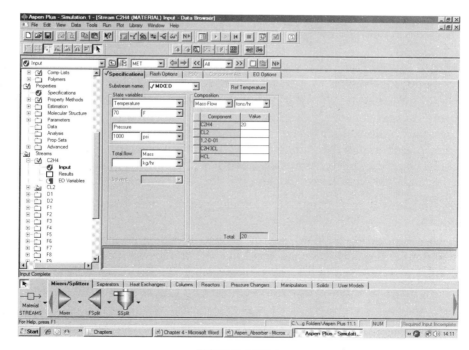

FIGURE 4.30(a)

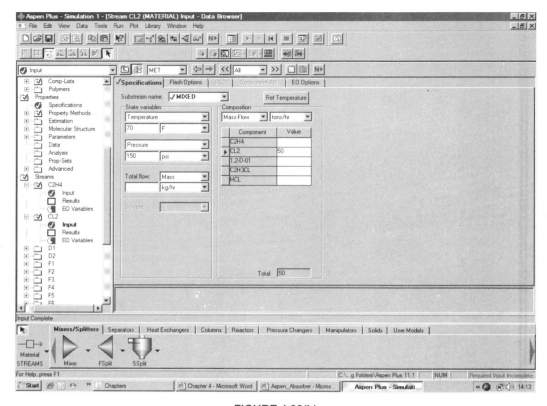

FIGURE 4.30(b)

Specifying block information

Unlike stream information, we need to input required data for all blocks of the process flow diagram. As stated earlier, the flowsheet of the VCM plant consists of two mixers, two reactors, four heat exchangers, one pump, two distillation columns and one splitter. Although discussed during the Aspen Plus simulation of several single process units in the preceding chapters, here we must remember the following points when we fill up the input forms of the concerned blocks.

- To simulate a mixer model, at least provide the pressure data and valid phases.

- In the simulation of the reactor model, coefficients should be negative for reactants and positive for products.

- In the *Vapor fraction* field of a heater model, put 0 to indicate bubble point and 1 to indicate dew point. For subcooled liquid and superheated vapor, use *Flash specifications*.

The windows, shown in Figures 4.31(a) to 4.31(n) display the block-wise information using the input forms.

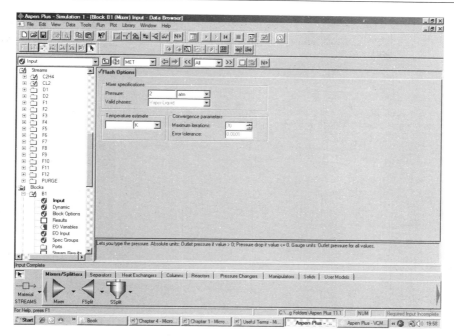

FIGURE 4.31(a)

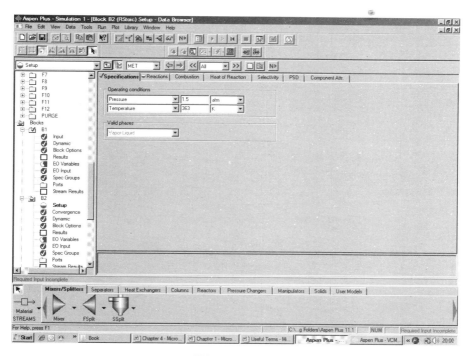

FIGURE 4.31(b)

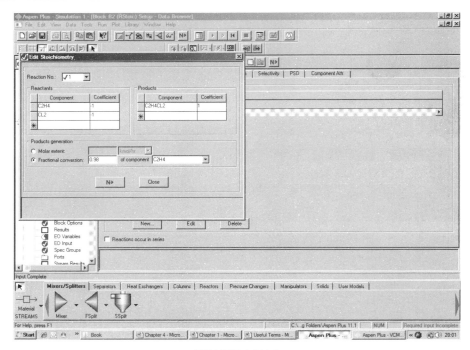

FIGURE 4.31(c)

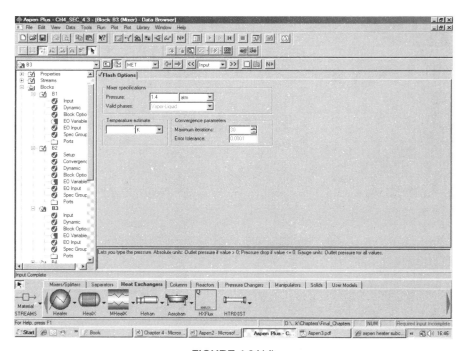

FIGURE 4.31(d)

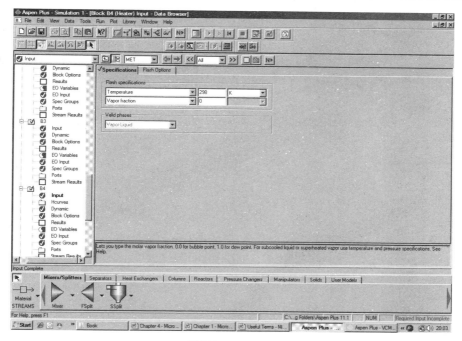

FIGURE 4.31(e)

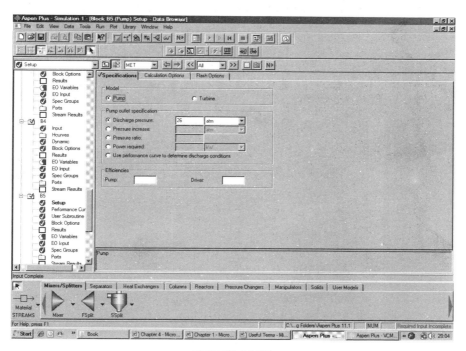

FIGURE 4.31(f)

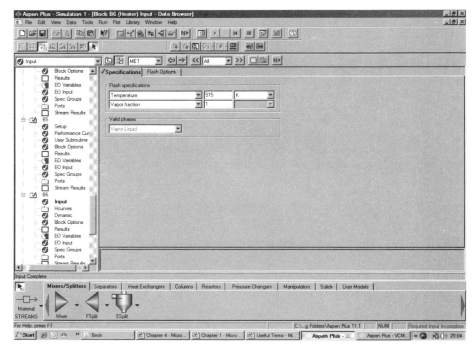

FIGURE 4.31(g)

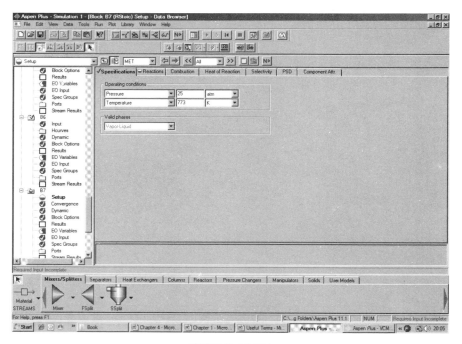

FIGURE 4.31(h)

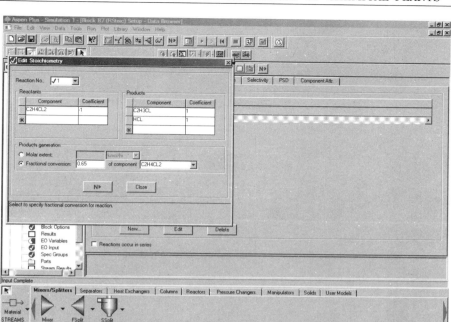

FIGURE 4.31(i)

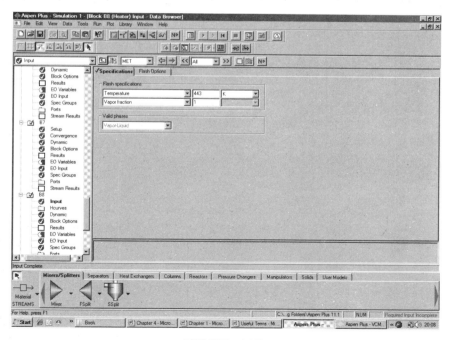

FIGURE 4.31(j)

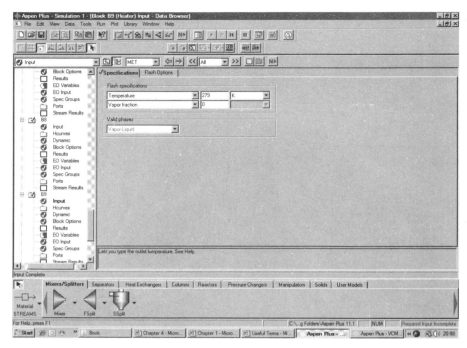

FIGURE 4.31(k)

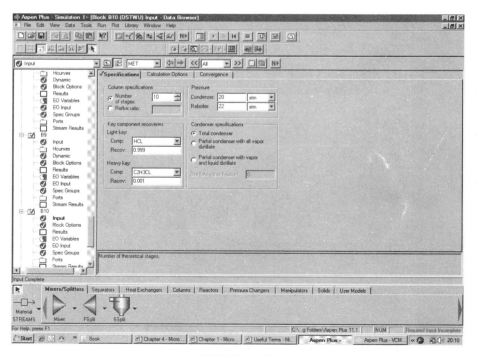

FIGURE 4.31(l)

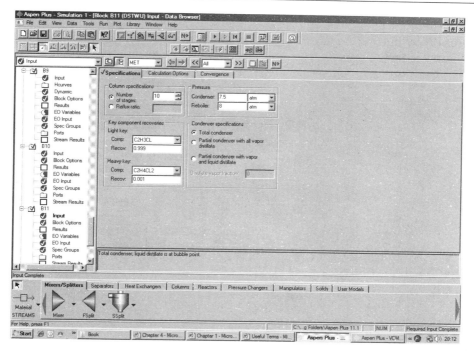

FIGURE 4.31(m)

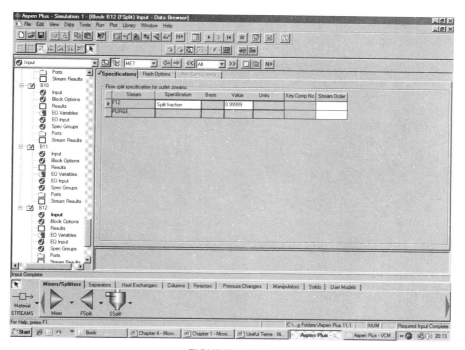

FIGURE 4.31(n)

Running the simulation

As we press *Next* button, Aspen Plus displays a message as shown in Figure 4.32. Since the data entry is fully complete, the simulator seeks user permission to run the program.

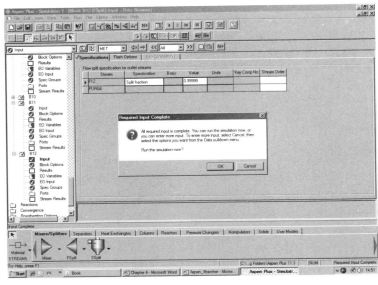

FIGURE 4.32

As we hit *OK* button on the message, the *Control Panel* window appears as displayed in Figure 4.33. It usually shows errors, warnings, convergence status, etc.

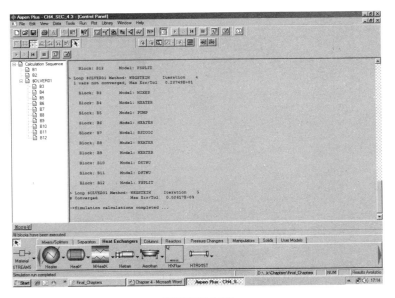

FIGURE 4.33

Viewing results

Choose *Results Summary/Streams* in the column at the left side and rearrange the table to get the results in the form as shown in Figure 4.34. Save the work positively at this moment.

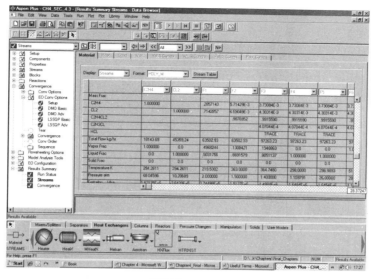

FIGURE 4.34

Viewing input summary

To obtain the input information of the present project, select *Input Summary* from the *View* dropdown menu (see Figure 4.35).

FIGURE 4.35

4.4 ASPEN PLUS SIMULATION OF A PETLYUK COLUMN WITH STREAMS RECYCLING

Problem statement

Naphtha reformate is extracted for aromatics and the aromatic compounds are fractionated into benzene, toluene and xylene in sequence. This separation is conventionally carried out using a series of binary-like columns. In order to improve the thermodynamic and economic performance, Lee et al. (2004) proposed a fully thermally coupled distillation column, which is also called the *Petlyuk column*. Originally, the first two columns of the conventional process flowsheet are suggested to replace with the Petlyuk column shown in Figure 4.36. The molar flow rate of liquid feed stream, namely FEED entering at 132.3°C and 150 kPa is provided component-wise in Table 4.4.

Table 4.5 specifies other two input streams introduced to the prefractionator (PREFRAC) before connecting recycle streams (see Figure 4.37). Two separators, namely PREFRAC and MAIN are detailed in Table 4.6. The discharge pressure of both the PUMP and COMP (isentropic model) is 120 kPa.

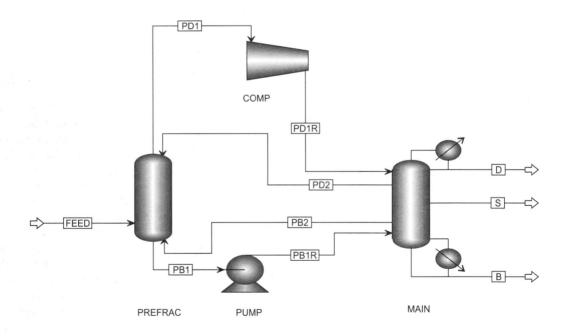

FIGURE 4.36 A Petlyuk column.

TABLE 4.4

Component	Flow rate (kmol/h)
Benzene	87.85
2,3-Dimethyl Pentane	0.0124
2-Methyl Hexane	0.0075
Toluene	338.10
n-Octane	0.049
Ethyl benzene	14.975
p-Xylene	57.798
m-Xylene	128.55
o-Xylene	60.16
n-Nonane	0.0057
n-Pentyl benzene	0.33
1-Methyl-2-ethyl benzene	26.01
1,3,5-Tri-methyl benzene	75.95
1-Methyl-2-n-propyl benzene	0.57
o-Diethyl benzene	0.33
1-Methyl-2-iso-propyl benzene	4.12
1,2,4,5-Tetra-methyl benzene	4.75
Penta-methyl benzene	2.2389

TABLE 4.5

Stream	Phase	Guessed temperature (°C)	Guessed pressure (kPa)	Guessed composition (mole fract)
PD2	–	115.0	110	Benzene: 0.107 Toluene: 0.893

PB2	vapor	155.4	177	Benzene: 0.0082
				Toluene: 0.3618
				Ethyl benzene: 0.02
				p-Xylene: 0.09
				m-Xylene: 0.19
				o-Xylene: 0.09
				1-Methyl-2-ethyl
				  benzene: 0.07
				1,3,5-Tri-methyl
				  benzene: 0.17

TABLE 4.6

Column	Specifications
PREFRAC	No. of stages (no condenser and reboiler): 22
	Entry of FEED stream: 7th stage (above stage)
	Entry of PD2 stream: 1st stage (above stage)
	Entry of PB2 stream: 23rd stage (above stage)
	Exit of PD1 stream: 1st stage (vapor)
	Exit of PB1 stream: 22nd stage (liquid)
	Column pressure: 70 kPa
MAIN	No. of stages (including total condenser and reboiler): 89
	Distillate rate: 86.8 kmol/h
	Reflux rate: 1792 kmol/h
	Entry of PD1R stream: 13th stage (above stage)
	Entry of PB1R stream: 55th stage (above stage)
	Exit of D stream: 1st stage (liquid)
	Exit of S stream: 28th stage (liquid; 337.8 kmol/h)
	Exit of B stream: 89th stage (liquid)
	Exit of PD2R stream: 15th stage (liquid; 290.1 kmol/h)
	Exit of PB2R stream: 60th stage (vapor; 492.9 kmol/h)
	Column pressure: 110 kPa

Flow rates of PD2 and PB2 are fixed at 290.1 and 492.9 kmol/h, respectively. Using the UNIQUAC thermodynamic property prediction method, simulate the flowsheet for finding the composition of three product streams.

Simulation approach

The Aspen Plus simulation of the representative Petlyuk column is divided into three sequential steps. In the following, step-wise simulation procedure is presented systematically.

Step 1: *Simulation with arbitrarily defined PD2 and PB2 streams*

The process flowsheet is developed in Figure 4.37. Subsequently, all 18 components are defined in Figure 4.38.

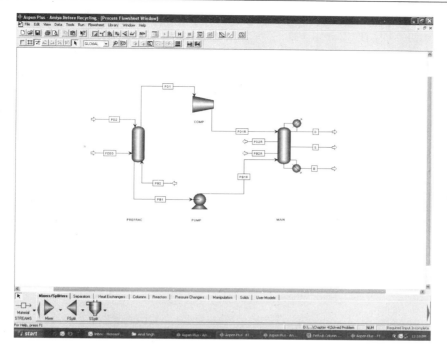

FIGURE 4.37

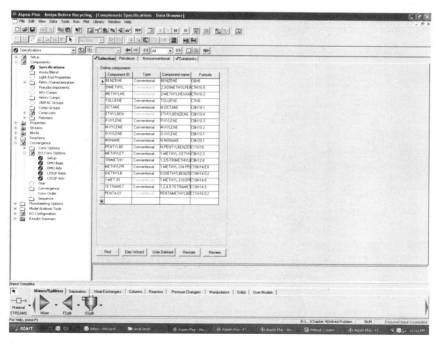

FIGURE 4.38

Use the UNIQUAC thermodynamic property package. Figures 4.39–4.44 are obtained by specifying FEED, PB2 and PD2 stream in sequence.

Feed stream

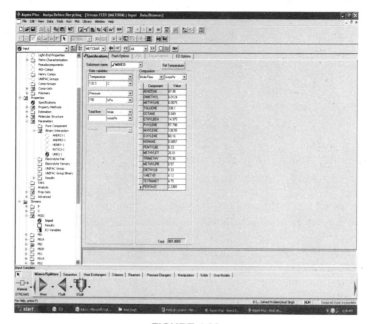

FIGURE 4.39

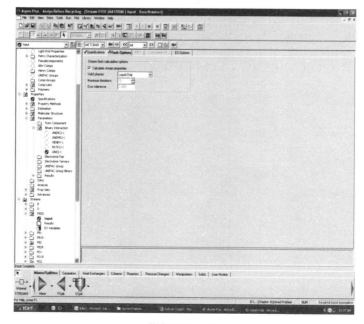

FIGURE 4.40

PB2 stream

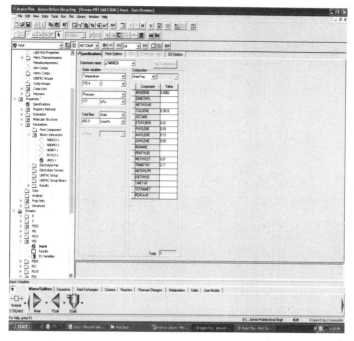

FIGURE 4.41

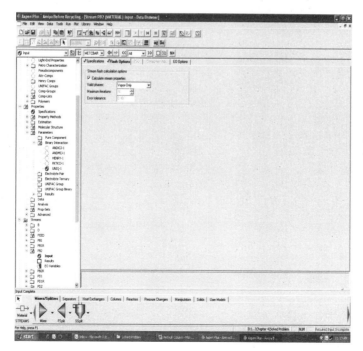

FIGURE 4.42

PD2 stream

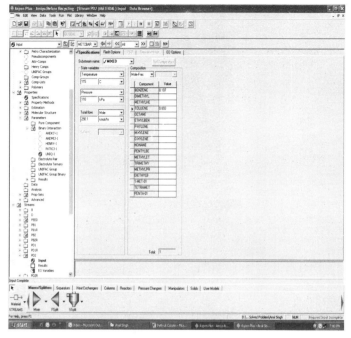

FIGURE 4.43

FIGURE 4.44

Figures 4.45 and 4.46 define the compressor (model type: Isentropic) block, namely COMP. It discharges the vapor with a pressure of 120 kPa.

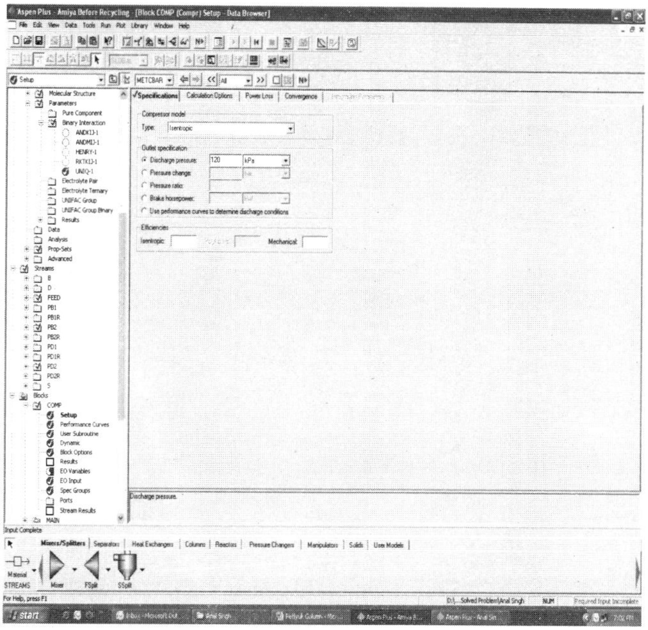

FIGURE 4.45

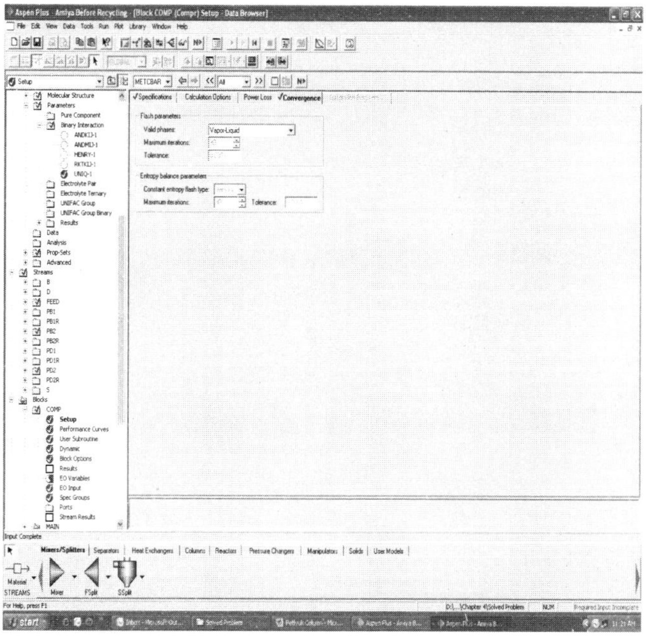

FIGURE 4.46

Using the given data, the key fractionator, MAIN is specified in Figures 4.47–4.49.

FIGURE 4.47

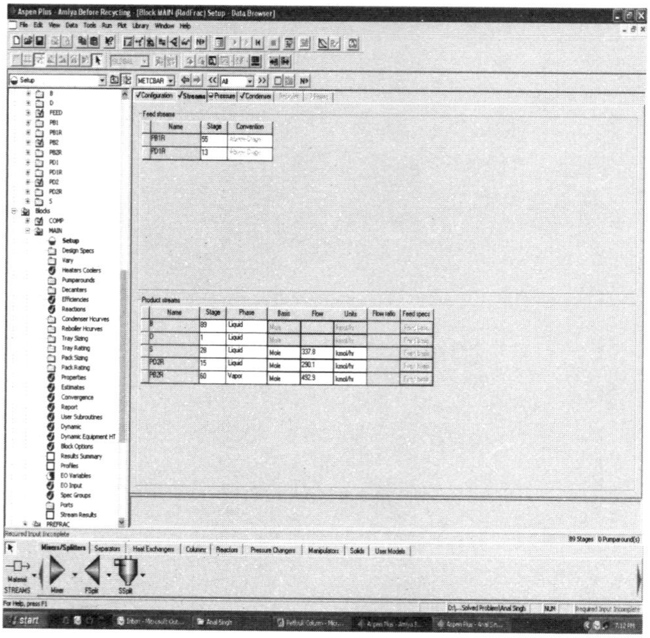

FIGURE 4.48

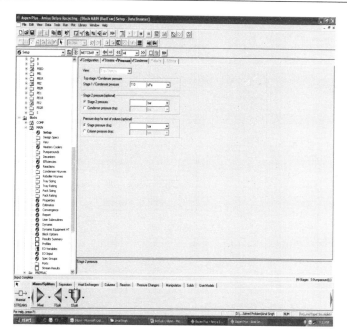

FIGURE 4.49

In the next, the prefractionator, PREFRAC is detailed in Figures 4.50–4.52.

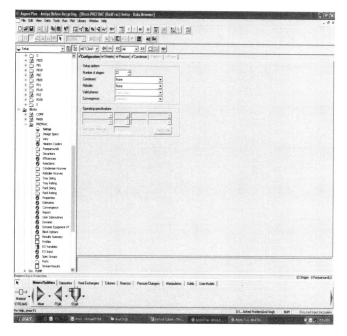

FIGURE 4.50

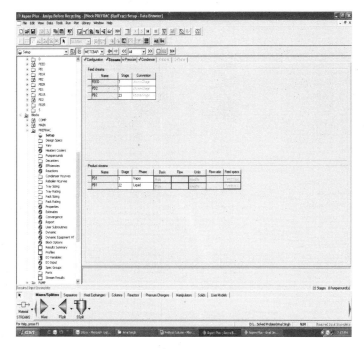

FIGURE 4.51

FIGURE 4.52

Providing the discharge pressure of pump, we have Figure 4.53.

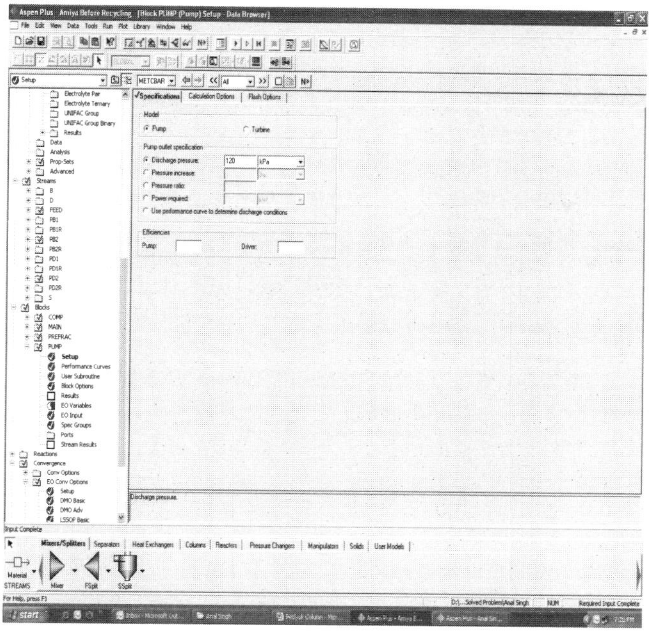

FIGURE 4.53

Running the Aspen simulator, one obtains the stream table displayed in Figure 4.54.

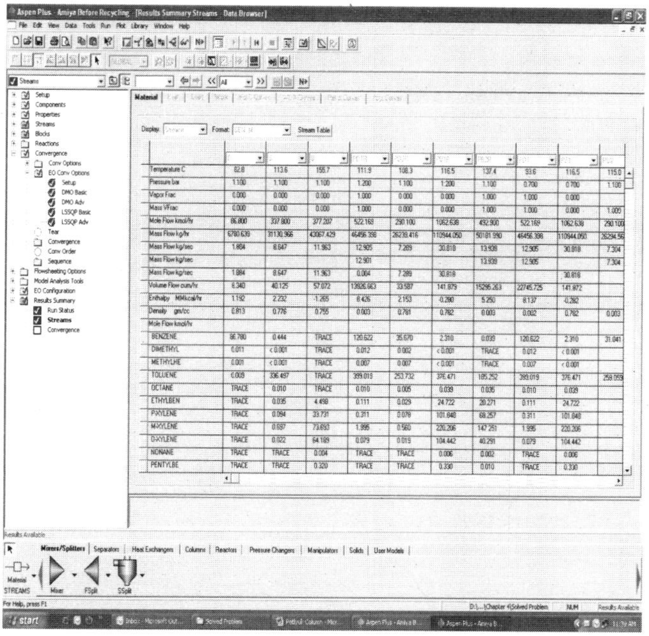

FIGURE 4.54

Step 2: *Simulation runs for close matching between PD2 and PD2R & PB2 and PB2R*

Recall that the temperature, pressure and composition of PD2 and PB2 streams are arbitrarily chosen (see Table 4.5) and used in the previous simulation (Step 1). Before closing the two loops (PD2/PD2R and PB2/PB2R), we should ensure that the variables of PD2 and PD2R, and PB2 and PB2R streams are fairly close.

For this purpose, several iterations are made. Before each run, replace the temperature, pressure and composition of PD2 stream by those of PD2R. Follow the same for the PB2/PB2R loop. It is obvious that the replacement of pressure is required for PB2/PB2R loop only once at the beginning. Do not forget to reinitialize each simulation run by pressing Shift+F5. Performing a number of iterations, we have the following stream table (Figure 4.55) and it is evident that a reasonable agreement is achieved.

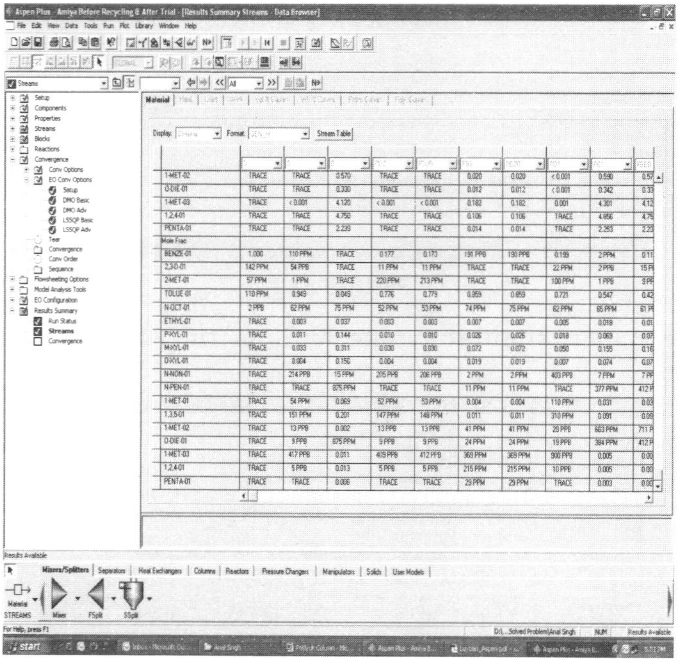

FIGURE 4.55

Step 3: *Closing of loops and final simulation of the Petlyuk column*

When the stated variables are in close matching, both the loops (PD2/PD2R and PB2/PB2R) are attempted to close by following the steps given below:

   (i) Remove the streams labeled PD2R and PB2R.
  (ii) Reconnect streams PD2 and PB2 with 15th and 60th stage, respectively, of the MAIN fractionator.
 (iii) From the Data Browser, select *Convergence/Tear*. Fixing tolerance limits for both PD2 and PB2 streams, obtain Figure 4.56.

Remember that the default tolerance value is 0.0001. For the example process, we can use that default value if more iterations was performed resulting very good matching between the variables of paired streams.

FIGURE 4.56

When the program is rerun, the simulator converges to the values shown in stream table. It should be emphasized that the 18 components are separated into three groups of components as indicated in the stream table as shown in Figure 4.57.

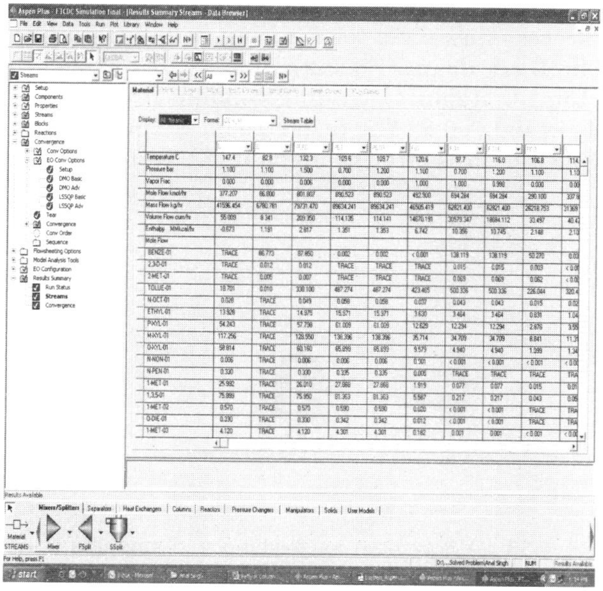

FIGURE 4.57

SUMMARY AND CONCLUSIONS

In the previous chapters, we have studied the steady state simulation of a large variety of individual process units using the Aspen Plus package. In the present chapter, several chemical processes have been assembled to develop the chemical plants and those plants have been simulated subsequently. The solved examples include a distillation train, a vinyl chloride monomer unit and a Petlyuk column with streams recycling. This chapter ends the process simulation by the use of steady state Aspen Plus package. In the next two chapters, we will study the process dynamics and closed-loop control of flow-driven as well as pressure-driven processes using the Aspen Dynamics package.

PROBLEMS

4.1 A hydrocarbon stream with component-wise flow rates, shown in Table 4.7, enters the isentropic compressor at 120°F and 1 atm. The compressor has discharged the vapor stream at 3 atm.

TABLE 4.7

Component	Flow rate (kmol/h)
C_1	15
C_2	95
C_3	150
$n\text{-}C_4$	25
$n\text{-}C_5$	10
$n\text{-}C_6$	100

The complete process flowsheet for flashing and stripping operation is shown in Figure 4.58. The flash drum (Flash2) runs at 125°F and 2.8 atm. The stripper (STRIP2) has total 6 stages (including condenser and reboiler) and bottoms to feed ratio (mole basis) is 0.8. The feed stream to the stripper is introduced above the top stage and the pressure throughout the column is 2 atm.

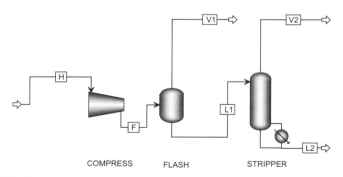

FIGURE 4.58 A flowsheet for flashing and stripping operation.

Using the UNIQUAC property method, simulate the plant to compute the product compositions and flow rates.

4.2 A ternary mixture, as shown in Table 4.8, is fed as stream H at 100°F and 290 psia to a pump P1 employed to increase 20 psi pressure.

TABLE 4.8

Component	Flow rate (lbmol/h)
C_3H_6	500
C_3H_8	300
$n-C_5H_{12}$	10

The stripper (STRIP2) has total 100 stages (including condenser and reboiler) with a reboiler duty of 10^7 Btu/h. Stream F enters above 70th stage and Stream R1 enters above 1st stage. The top stage pressure of the stripper is 280 psia with a stage pressure drop of 0.5 psi. The intercolumn pump P2 has increased 25 psi pressure. The RECT column has total 120 stages (including condenser and reboiler) with a reflux ratio (mole basis) of 10 and a bottoms to feed ratio (mole basis) of 0.6. Stream D1 enters below 120th stage. In the simulation, consider condenser pressure of 275 psia with a pressure drop of 5 psi and a stage pressure drop of 0.1 psi (see Figure 4.59).

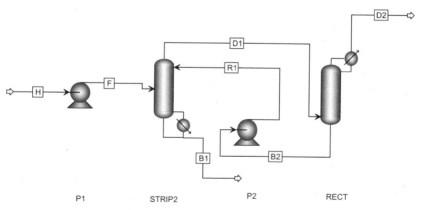

FIGURE 4.59 A flowsheet of a propylene-propane mixture separation process.

Applying the RK-Soave thermodynamic model,

(a) simulate the above propylene-propane mixer separation plant and report the product compositions, and

(b) perform the sensitivity analysis to observe the effect of the second column efficiency varied from 20% to 100% on the propylene mole fraction in the distillate.

4.3 The hydrogenation of aniline produces cyclohexylamine in a CSTR according to the following reaction:

$$C_6H_5NH_2 + 3H_2 \rightarrow C_6H_{11}NH_2$$

aniline   hydrogen   cyclohexylamine

To simulate the aniline hydrogenation process using the Aspen Plus, we develop the process flow diagram as exhibited in Figure 4.60.

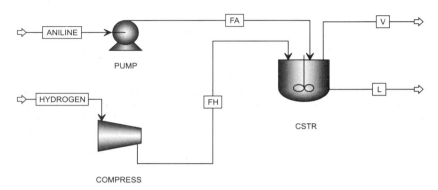

FIGURE 4.60   A flowsheet for aniline hydrogenation.

The reactor model (RCSTR) operates at 580 psia and 248°F, and its volume is 1200 ft³ (75% liquid). For the liquid-phase reaction, the inlet streams have the specifications, shown in Table 4.9.

TABLE 4.9

Stream	Temperature (°F)	Pressure (psia)	Flow rate (lbmol/h)
ANILINE (pure aniline)	95	100	150
HYDROGEN (pure hydrogen)	12	100	600

Both pump and compressor (isentropic) have discharged the fluids at 585 psia. Data for the Arrhenius law are given as:

$$\text{Pre-exponential factor} = 5 \times 10^5 \text{ m}^3/\text{kmol} \cdot \text{s}$$

$$\text{Activation energy} = 20,000 \text{ Btu/lbmol}$$

$$[C_i] \text{ basis} = \text{Molarity}$$

Use the SYSOP0 base property method in the simulation. The reaction is first-order in aniline and hydrogen. The reaction rate constant is defined with respect to aniline. Simulate the process and compute the component mole fractions in the liquid product and the vent stream.

4.4 The process flow diagram for a typical azeotropic distillation process is shown in Figure 4.61. The technique involves separating close boiling components (acetic acid and water) by adding a third component (vinyl acetate), called an entrainer, to form a minimum boiling azeotrope which carries the water overhead and leaves dry product (acetic acid) in the bottom. The overhead vapor is condensed and then separated in the decanter into two liquid phases: the organic phase and aqueous phase.

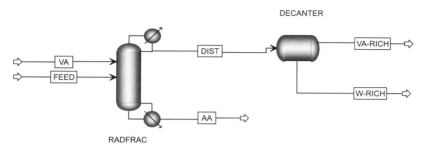

FIGURE 4.61   A flowsheet of an azeotropic distillation process.

A feed stream, namely FEED, enters above 15th stage of the azeotropic distillation column at 330°F and 90 psia in addition to the flow rates, shown in Table 4.10.

TABLE 4.10

Component	Flow rate (lbmol/h)
acetic acid	2700
water	500

The entrainer, VA (vinyl acetate), with a flow rate of 455 lbmol/h enters above 12th stage of the column at 200°F and 100 psia. The azeotropic column (RadFrac) has the following specifications:

Number of stages (including condenser and reboiler): 55

Condenser type: total

Valid phases: vapor-liquid-liquid

Reflux ratio (mole basis): 4

Bottoms rate: 2700 lbmol/h

Condenser pressure: 66 psia

Column pressure drop: 12 psi

Key component in the second liquid-phase: water

Stages to be tested for two liquid-phases: 1 to 55

The specifications for the decanter model are noted below:

Pressure: 50 psia

Temperature: 110°C

Key component in the second liquid-phase: water

Using the NRTL-RK thermodynamic model, simulate the process to compute the component-wise product flow rates.

4.5 A hydrocarbon stream H is at 5°C and 2.5 atm. The pump has discharged the liquid feed F at 5 atm. The component-wise flow rates are shown in Table 4.11 for stream H.

TABLE 4.11

Component	Flow rate (kmol/h)
C_2	35
C_3	50
$i\text{-}C_4$	130
$n\text{-}C_4$	200
$i\text{-}C_5$	180
$n\text{-}C_5$	200
$n\text{-}C_6$	5

In Figure 4.62 the schematic representation of a hydrocarbon separation process integrated with a Pump, three DSTWU columns (C1, C2 and C3) and two RadFrac (RECT) columns (CR1 and CR2) is shown.

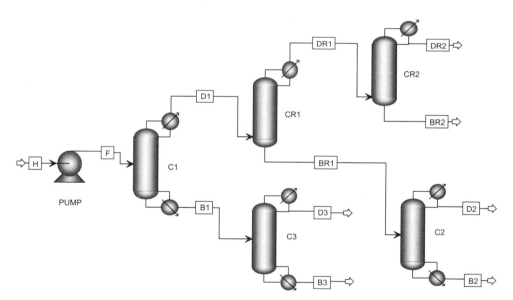

FIGURE 4.62 A flowsheet of a hydrocarbon separation process.

All DSTWU fractionators have total 20 stages each (including condenser and reboiler) and two RECT models have 10 stages each (including condenser and reboiler) with no reboiler. The specifications, shown in Tables 4.12(a) and (b), are required for simulating the process.

TABLE 4.12(a)

| Block | % Recovery of LK/HK in distillate | | Pressure (atm) | |
	Light key (LK)	Heavy key (HK)	Condenser (type)	Reboiler
C1	99% of n-C_4	1% of i-C_5	4 (partial condenser with all vapor distillate)	4
C2	99% of i-C_4	2% of n-C_4	1.5 (total condenser)	1.5
C3	99% of i-C_5	4% of n-C_5	3 (total condenser)	3

TABLE 4.12(b)

Block	Condenser (type)	Distillate to feed ratio (mole basis)	Pressure (atm)
CR1	Partial vapor	0.2	2
CR2	Total	0.5	1.5

Applying the Peng-Robinson property method, simulate the separation process to compute the flow rates and compositions of all product streams.

4.6 An inlet Stream H supplied at 80°F and 300 psia is compressed to 4000 psia by the use of an isentropic compressor B1. Stream H has component-wise flow rates, shown in Table 4.13.

TABLE 4.13

Component	Flow rate (lbmol/h)
nitrogen	100
hydrogen	300
ammonia	0
carbon dioxide	1

A flow diagram for the ammonia process (Finlayson, 2006) is shown in Figure 4.63.

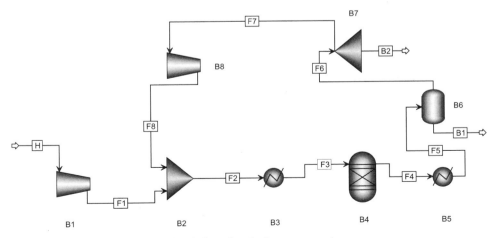

FIGURE 4.63 A flowsheet of an ammonia process.

Stream F1 is mixed with the recycle stream F8 in a mixer block B2 operated at 4000 psia. Before introducing into the reactor, the mixer effluent F2 is heated in block B3 to 900°F at 4000 psia. Note that the reactor (RGibbs) B4 runs at 900°F and 3970 psia. In the next, the reactor outlet F4 is cooled in a heat exchanger B5 operated at 80°F and 3970 psia. The flash drum (Flash2) B6 produces Streams B1 and F6 at 80°F and 3970 psia. In the subsequent step, Stream F6 enters the splitter (FSplit) B7 and 0.01% of it is used as purge. Finally, an isentropic compressor B8 has discharged Stream F8 to the mixer block B2 at 4000 psia. Using the NRTL thermodynamic model and the Newton's iteration method (from the Data Browser, choose *Convergence/Conv Options*), simulate the ammonia process to compute the component-wise flow rates and compositions of all streams.

REFERENCES

Finlayson, B.A. (2006), *Introduction to Chemical Engineering Computing*, Wiley Interscience, New Jersey.

Lee, J.Y., Y.H. Kim and K.S. Hwang (2004), "Application of a Fully Thermally Coupled Distillation Column for Fractionation Process in Naphtha Reforming Plant," *Chem. Eng. Proc.*, 43, pp. 495–501.

Seider, W.D., J.D. Seider and D.R. Lewin (1998), *Process Design Principles: Synthesis, Analysis, and Evaluation*, John Wiley & Sons, New York.

Part III

Dynamics and Control
using Aspen Dynamics™

CHAPTER 5

Dynamics and Control of Flow-driven Processes

5.1 INTRODUCTION

Dynamic simulation of a chemical process greatly helps to understand the transient behavior. Aspen Dynamics^{TM}, which is tightly integrated with Aspen Plus^{TM}, is widely used for process design and control. This powerful simulator can automatically initialize the dynamic simulation using the steady state results of the Aspen Plus simulation. Interestingly, when the file containing the flowsheet is opened in Aspen Dynamics, the default control structures are already installed on some loops. Usually, level, pressure and temperature controllers are included where appropriate. However, these default control schemes can be modified or even replaced with other suitable control loops available in Aspen Dynamics package. Note that there is a scope to include some additional controllers for the used process. Moreover, this simulation tool provides a graphical environment to show the process response.

To convert a steady state simulation into a dynamic simulation, there are several items that should be taken care of. For example, the size of all equipments must be specified and the control structures must be devised. For steady state simulation using Aspen Plus, the size of the equipment is not needed, except for reactors. On the other hand, for dynamic simulation using Aspen Dynamics, the inventories of material contained in all the pieces of equipment affect the dynamic response. Therefore, the physical dimensions of all process units must be known.

When the steady state Aspen Plus simulation is exported into Aspen Dynamics, we need to choose either simpler flow-driven dynamic simulation or more rigorous pressure-driven dynamic simulation. Pressure-driven simulations include pumps and compressors where needed to provide the required pressure drop for material flow. Control valves must be installed where needed, and their pressure drops selected. For flow-driven simulations, however, no such arrangements are required.

271

In the present chapter, we wish to study the dynamics and control of the flow-driven processes. For this intention, we choose a reactor (RCSTR) as well as a distillation column (RadFrac) example from the model library of Aspen simulator.

5.2 DYNAMICS AND CONTROL OF A CONTINUOUS STIRRED TANK REACTOR (CSTR)

Problem statement

Ethyl acetate is produced in an esterification reaction between acetic acid and ethyl alcohol.

$$\text{acetic acid} + \text{ethyl alcohol} \leftrightarrow \text{ethyl acetate} + \text{water}$$

A feed mixture, consisting of 52.5 mole% acetic acid, 45 mole% ethyl alcohol and 2.5 mole% water, enters the RCSTR model with a flow rate of 400 kmol/h at 75°C and 1.1 atm. The reactor, as shown in Figure 5.1, operates at 70°C and 1 atm.

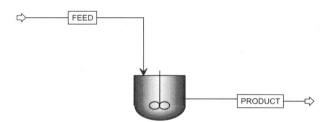

FIGURE 5.1 A flowsheet of a CSTR.

Both the reactions are first-order with respect to each of the reactants (i.e., overall second-order). For these liquid-phase reactions, the kinetic data for the Arrhenius law are given below:

$$\text{Forward reaction: } k = 2.0 \times 10^8 \text{ m}^3/\text{kmol} \cdot \text{s}$$

$$E = 6.0 \times 10^7 \text{ J/kmol}$$

$$\text{Reverse reaction: } k = 5.0 \times 10^7 \text{ m}^3/\text{kmol} \cdot \text{s}$$

$$E = 6.0 \times 10^7 \text{ J/kmol}$$

$$\text{Composition basis} = \text{Molarity}$$

Here, k is the pre-exponential factor and E represents the activation energy. The reactor geometry data are reported below.

Vessel type: vertical

Head type: flat

Diameter: 0.45711 m

Volume: 0.15 m^3

(a) Simulate the reactor model using the SYSOP0 thermodynamic model to compute the product compositions.

(b) Report the default controllers tuning parameters and control actions used, and constraints imposed on variables.

(c) Investigate the servo performance of the default liquid level and temperature control algorithms and discuss the effect of loop interaction.

(d) Show the regulatory behavior of both the controllers in presence of disturbance in feed temperature.

Simulation approach

(a) To open the *Aspen Plus Startup* dialog box, click the desktop *Start* button, then point to *Programs, AspenTech, Aspen Engineering Suite, Aspen Plus Version* and then click the *Aspen Plus User Interface*. Let's select the option with *Template* and then click *OK* (see Figure 5.2).

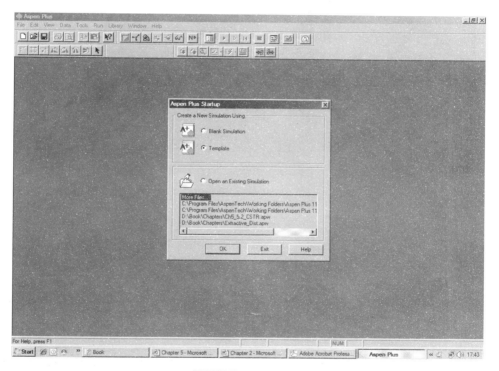

FIGURE 5.2

As the next window appears (see Figure 5.3), it is appropriate to select *General with Metric Units* and hit *OK* button.

Here we use the simulation engine at 'Local PC'. When the *Connect to Engine* dialog pops up (see Figure 5.4), press *OK*. Recall that this step is specific to the installation.

Creating flowsheet

The process flow diagram, shown in Figure 5.5, includes a reactor, namely RCSTR, with an incoming FEED stream and an outgoing PRODUCT stream.

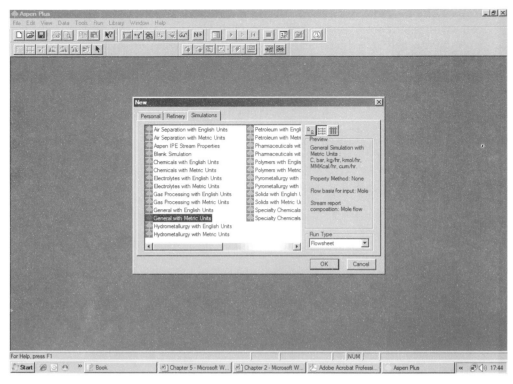

FIGURE 5.3

FIGURE 5.4

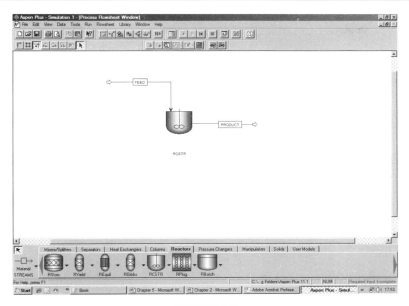

FIGURE 5.5

Configuring settings

Hitting *Next* button, we get *Global* sheet of the *Specifications* form under *Setup* folder in the left pane of the Data Browser window. Enter the *Title* of the present problem— 'Dynamic Simulation of a CSTR', change the *Input mode* from 'Steady-State' to 'Dynamic' and leave the remaining items at their defaults. The window looks like Figure 5.6.

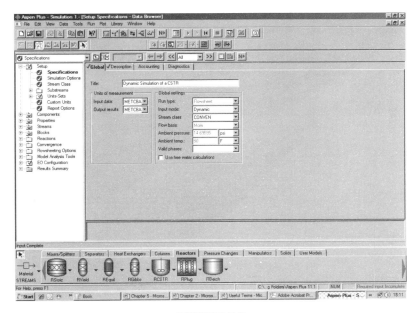

FIGURE 5.6

In the next window, as shown in Figure 5.7, the *Aspen Plus accounting information* required at some installations is provided.

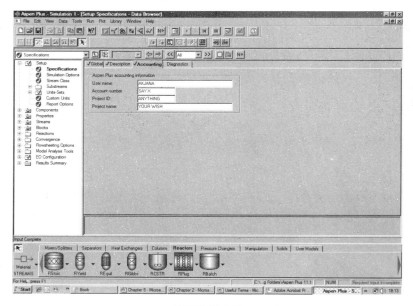

FIGURE 5.7

In the subsequent step (see Figure 5.8), select *Stream* sheet with opening the *Report Options* form under *Setup* folder and include *Mole* fraction item.

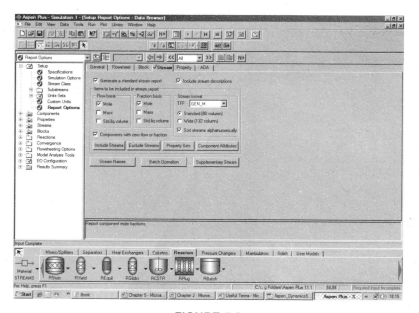

FIGURE 5.8

Specifying components

In the Data Browser window, choose *Components/Specifications* to obtain the component input form. Filling out the table with the components (acetic acid, ethanol, ethyl acetate and water) involved in the present reaction system, the screen looks like Figure 5.9.

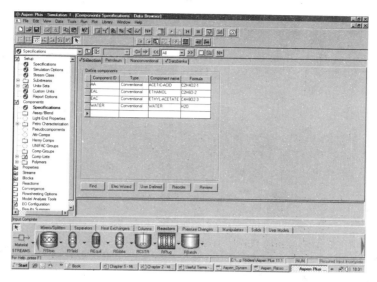

FIGURE 5.9

Specifying property method

Choosing *Properties/Specifications* in the column at the left side, one obtains the property input form. As shown in Figure 5.10, we use the SYSOP0 base property method.

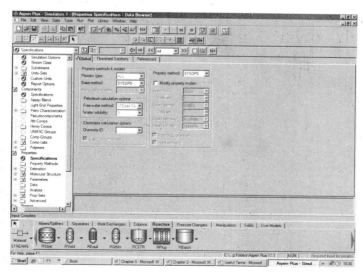

FIGURE 5.10

Specifying stream information

Use the Data Browser menu tree to navigate to the *Streams/FEED/Input/Specifications* sheet. Specifying the FEED stream by its temperature, pressure, flow rate and composition, we have this window, shown in Figure 5.11.

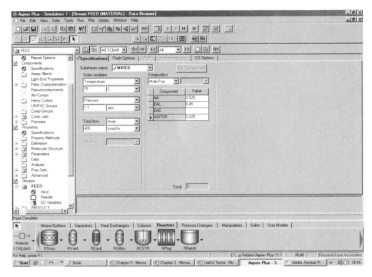

FIGURE 5.11

Specifying block information

In the list on the left, choose *Blocks/RCSTR/Specifications* to obtain the block input form. It is filled with the given data as shown in Figure 5.12.

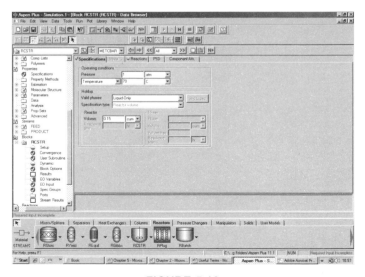

FIGURE 5.12

In the next step (Figure 5.13), select *RCSTR/Dynamic/Vessel* sheet under *Blocks* folder and enter the reactor geometry data.

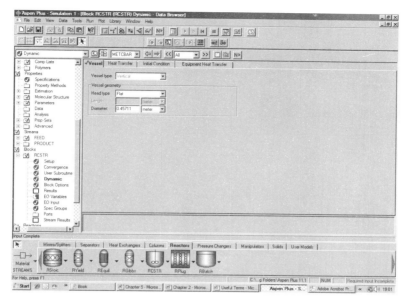

FIGURE 5.13

The forward reaction as well as the backward reaction is represented with their stoichiometric coefficients and exponents in two sheets, shown in Figures 5.14(a) and (b).

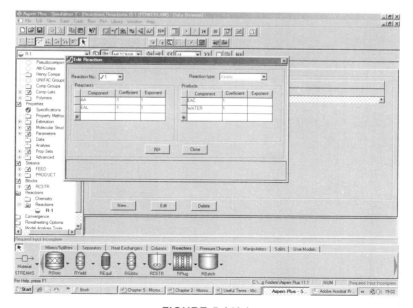

FIGURE 5.14(a)

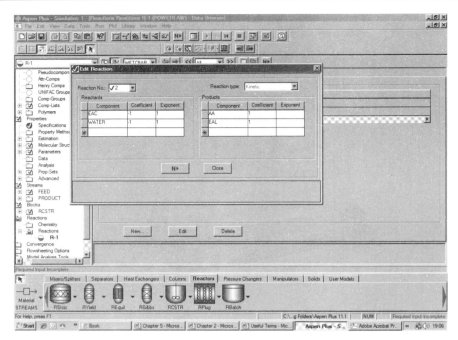

FIGURE 5.14(b)

The power law data for both the reactions provided in the problem statement are entered in the two *Kinetic* sheets shown in Figures 5.15(a) and (b).

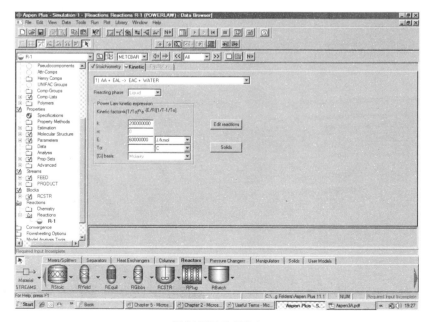

FIGURE 5.15(a)

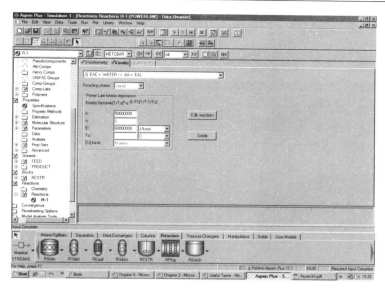

FIGURE 5.15(b)

The status indicator in the above window reveals by the message *Required Input Complete* that no more input specifications are required to run the simulation.

Running steady state simulation

As we click on *Next* button to continue the simulation, the *Required Input Complete* dialog box appears. Hitting *OK* on the message, we are displayed the *Control Panel* where the simulation messages during the run are recorded (see Figure 5.16).

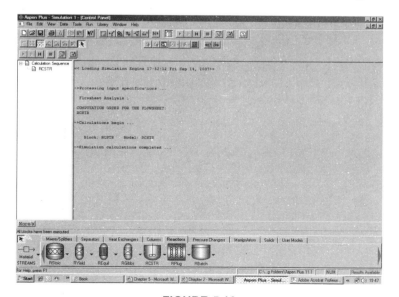

FIGURE 5.16

Viewing steady state results

In the next, select *Solver Settings*, choose *Results Summary/Streams* in the list on the left and finally get the steady state results as shown in Figure 5.17.

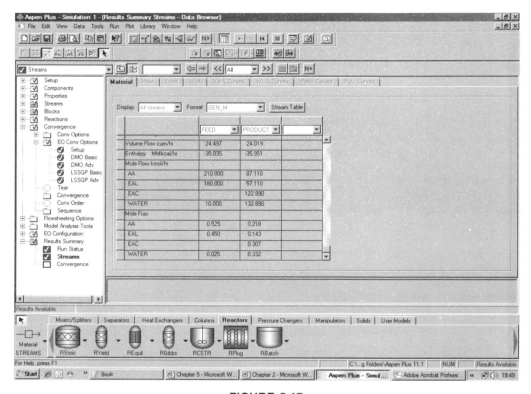

FIGURE 5.17

(b) Exporting dynamic simulation: In the subsequent stage (that starts from Figure 5.18), we wish to carry out the simulation dynamically. Accordingly, at this moment, we have to follow the sequential steps noted below:

- Click on *Export* from the pulldown *File* menu or simply press Ctrl+E on the keyboard.
- Open the Drive and then Folder where we want to save the work as a file.
- Type 'Ch5_5.2_RCSTR' in the *File name* field.
- Choose 'Flow Driven Dyn Simulation (*.dynf & *dyn.appdf)' from the options available in the *Save as type* box.
- Finally, hit *Save* button.

Also, save the work done as a backup file (e.g., Ch5_5.2_RCSTR.bkp). We may use the same folder within which the exported dynamic simulation file is saved. Originally many files are saved along with the backup or dynamic file. Anyway, we are now ready to run Aspen Dynamics and we may quit Aspen Plus.

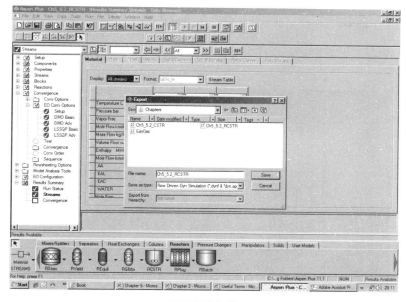

FIGURE 5.18

Starting Aspen Dynamics

As we click the *Start* button, point to *Programs*, then *AspenTech*, then *Aspen Engineering Suite*, then *Aspen Dynamics Version* and then click on *Aspen Dynamics*, a blank dynamic simulation window appears as shown in Figure 5.19.

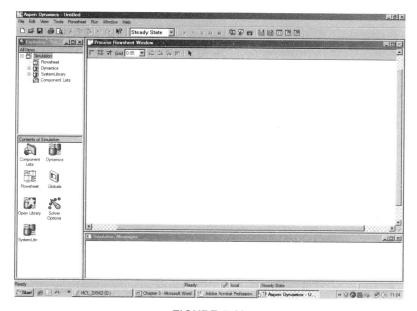

FIGURE 5.19

Opening existing simulation

To open the flow-driven dynamic file, select *Open* from the *File* dropdown menu or press Ctrl+O on the keyboard. In the *Open* dialog box, locate the drive, then folder and finally the file 'Ch5_5.2_RCSTR' (see Figure 5.20).

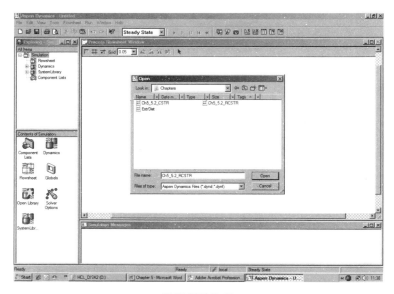

FIGURE 5.20

As we press *Open* button, the process flowsheet consisting of the automatically inserted level (LC1) and temperature (TC2) controllers appears (see Figure 5.21).

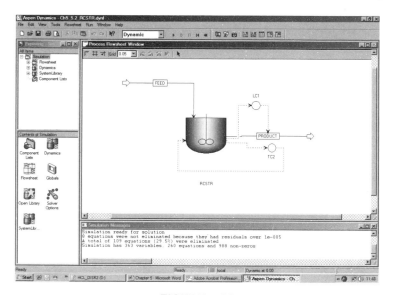

FIGURE 5.21

Details of the two control loops, to be used finally, are given below.

Loop 1

Controller: LC1

Type of controller: proportional (P) only

Controlled variable: reactor liquid level

Manipulated variable: product flow rate

Controller action: direct

Loop 2

Controller: TC2

Type of Controller: proportional integral (PI)

Controlled variable: reactor temperature

Manipulated variable: heat duty (cooling operation)

Controller action: reverse

Note that the *direct* acting control system increases the output signal as the input signal to the controller increases. On the other hand, as the input signal to the control structure increases, the output signal from the controller must decrease for the case of *reverse* acting control strategy. The direct acting control law has negative gain and increase/increase (or decrease/decrease) term is commonly used to represent it. For the reverse action, increase/decrease (or decrease/increase) term is used and controller gain has positive sign.

The reactor flowsheet includes two (LC1 and TC2) single-input/single-output (SISO) control loops. Therefore, we can say that this is a multi-input/multi-output (MIMO) closed-loop system.

In Aspen terminology, the process variable or controlled variable is denoted by PV, the set point is represented by SP and the controller output or control variable or manipulated variable is abbreviated by OP.

For the example CSTR system, level and temperature controllers are automatically implemented when the Aspen Dynamics simulation is created. The default values for SP, PV and OP are computed from the steady state simulation. To achieve better closed-loop process response, the Aspen-generated control structures can be modified or even replaced by the suitable control schemes available in the control library of Aspen software. In addition, the default values for controller tuning parameters, such as gain, integral time, derivative time, and so on, can also be changed.

Most of the control strategies are easily tuned by simply using heuristics. As suggested by Luyben (2004), all liquid levels should use P-only controllers with a gain of 2. All flow controllers should use a gain of 0.5 and an integral time of 0.3 minute (also enable filtering with a filter time of 0.1 minute). The author also mentioned that the default values in Aspen Dynamics for most pressure controllers seem to work reasonably well. But temperature controllers often need some adjustments.

Viewing default values of variables

In Aspen Dynamics, the steady state values of process variable and controller output are displayed in a table. At this stage, the set point value, displayed in table, shown in

Figure 5.22, is same with the value of process variable. To show the results table of loop 1, highlight the controller block LC1, press the right mouse button, go to *Forms* and then select *Results*.

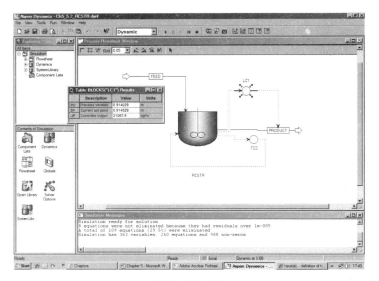

FIGURE 5.22

We can have the same information in a faceplate, shown in Figure 5.23, simply by double-clicking on the block LC1. But as a difference, the units are not mentioned here with the values of SP, PV and OP.

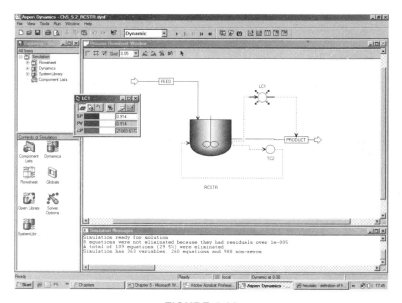

FIGURE 5.23

Similarly, we have the results table, shown in Figure 5.24, for the temperature loop 2.

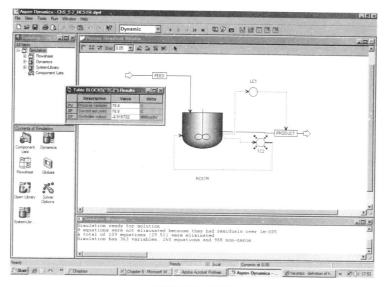

FIGURE 5.24

Modifying controller tuning properties

First we need to open the sheet that contains the controller tuning information. To do so for the level controller, highlight the controller block LC1, press the right mouse button, point to *Forms* and then select *Configure* (see Figure 5.25).

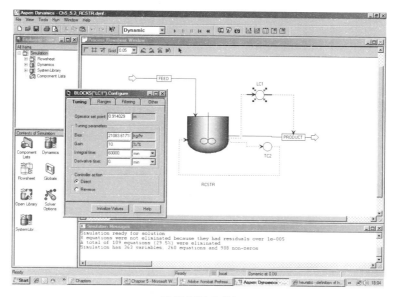

FIGURE 5.25

Alternatively, to obtain the *Configure* dialog box, first double-click on the controller block LC1 and then click on *Configure* symbol (yellow colour) in the faceplate as shown in Figure 5.26.

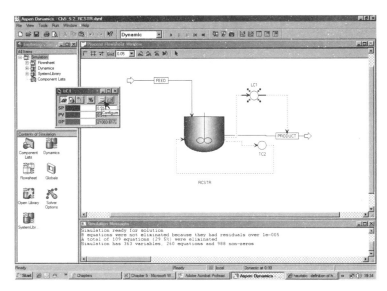

FIGURE 5.26

By the similar way, we obtain the tuning data sheet, shown in Figure 5.27, for the temperature controller TC2.

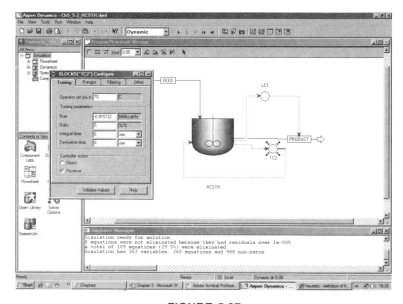

FIGURE 5.27

Note that the default *Operator set point* value is the steady state value of the process variable (PV). The reactor liquid level is the PV for loop 1 and reactor temperature for loop 2. *Bias* signal is the output from the controller when the error (= SP–PV) is zero. From the results tables shown earlier, it is obvious that the error is zero for both loops. Therefore, Aspen Dynamics has set the value of OP as the bias value.

The proportional integral (PI) control methodology is automatically installed with default values for the controller gain (= 10 %/%), integral time (= 60000 minutes) and derivative time (= 0 minute) to monitor the reactor level. However, as mentioned previously, the proportional-only controller with a gain of 2 is sufficient to effectively control the liquid level. Remember that to make the integral action inactive, we can use a very large value, for example 10^5 minutes (even the default value of 6×10^4 minutes may also be accepted), for the integral term. For loop 1, the controller action should be 'Direct' as set by default (see Figure 5.28).

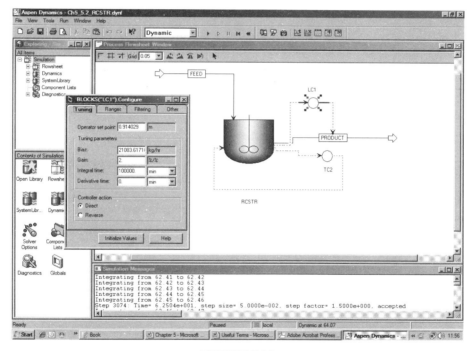

FIGURE 5.28

In loop 2, we prefer to employ the proportional integral controller to monitor the reactor temperature. In data sheet, shown in Figure 5.29, the default values are given. The TC2 is truly a reverse acting controller. However, we may adjust the values of controller tuning parameters (gain and integral time) during the closed-loop study if the control performance is not satisfactory.

Modifying ranges for process variables and controller outputs

In the *Configure* dialog box, hit the *Ranges* tab and get Figure 5.30 for level control loop.

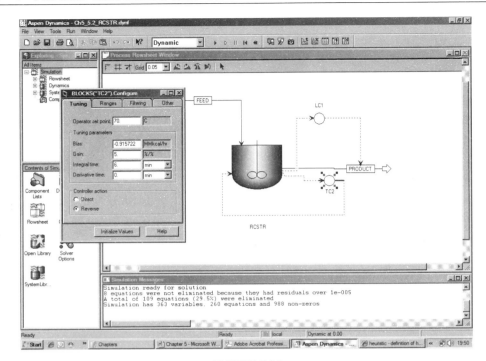

FIGURE 5.29

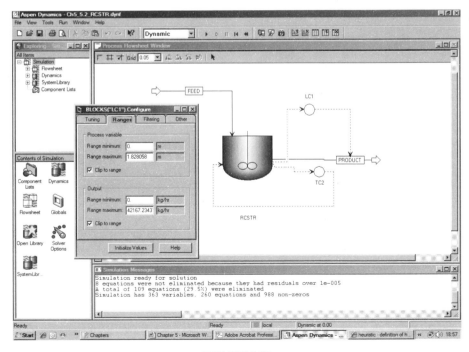

FIGURE 5.30

As shown in Figure 5.30, the default ranges for both the process and output variables are too large ($\pm 100\%$ of the steady state values). It may be practical to consider the following constraints.

Process variable

Range minimum: 0.6855 m (25% subtracted from steady state value of PV)

Range maximum: 1.1425 m (25% added with steady state value of PV)

Output

Range minimum: 15812.7 kg/h (25% subtracted from steady state value of OP)

Range maximum: 26354.5 kg/h (25% added with steady state value of OP)

Entering these upper and lower bounds, we have the window, shown in Figure 5.31, for the level controller.

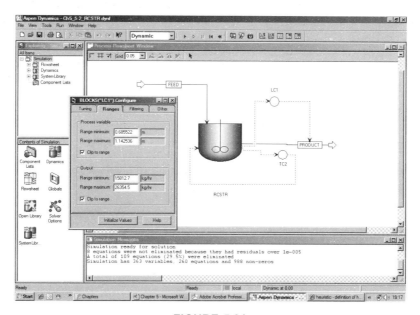

FIGURE 5.31

Again the typical ranges for the temperature control loop are noted here.

Process variable

Range minimum: 52.5°C (25% subtracted)

Range maximum: 87.5°C (25% added)

Output

Range minimum: 1.1447 MMkcal/h (25% subtracted)

Range maximum: -0.6868 MMkcal/h (25% added)

The corresponding Aspen Dynamics window is shown in Figure 5.32. It is worthy to mention that the negative value of heat duty reveals the cooling operation (heat removal).

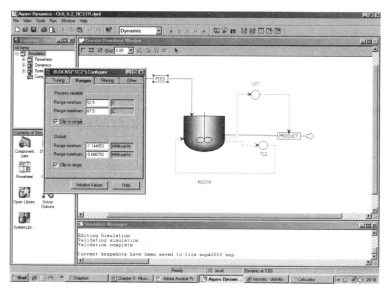

FIGURE 5.32

Both the control algorithms are completely specified above. In the next, the controller performance will be examined in terms of set point tracking (servo) and disturbance rejection (regulatory).

(c) Starting the Run: Before running the program, we must be accustomed with some frequently used items of the toolbar as described in Figure 5.33.

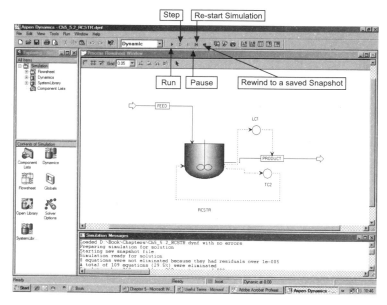

FIGURE 5.33

We wish to carry out the simulation for a certain time, say 5 hours. To fix up this time period, select *Pause At* from the *Run* pulldown menu or simply press Ctrl+F5 on the keyboard. Then select *Pause at time*, type 5 in the field or whatever we want and click on *OK* (see Figure 5.34).

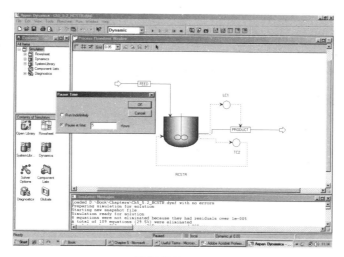

FIGURE 5.34

Viewing servo performance of LC1

As we double-click on LC1 block in the flowsheet, first the faceplate appears. In the next, press on *Configure* and *Plot* symbols in the faceplate. Alternatively, to open the faceplate, *Configure* dialog box and *ResultsPlot* dialog box, first select LC1 block, then choose *Forms* and subsequently press one-by-one on faceplate, *Configure* and *ResultsPlot*, respectively. Judiciously arrange all three items within the Aspen Dynamics window (see Figure 5.35) so that we can properly observe them together.

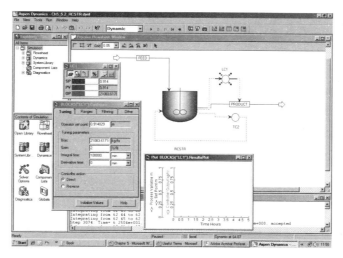

FIGURE 5.35

First make sure that all the items in the *Configure* dialog box and faceplate are correct. In order to execute the dynamic closed-loop simulation, click on *Run* button in the toolbar. During the simulation run, give a step change in the set point value of reactor liquid level from 0.914029 to 1.1 meter at time = 1.5 hours. Typing the new set point value in the faceplate, press *Enter* button on the keyboard so that the *Operator set point* value in the *Configure* dialog box also changes automatically to 1.1 meter. Note that the new set point must be within the specified ranges of PV. In Figure 5.36, the servo performance of the level controller is depicted for 5 hours as selected earlier. Obviously, the plot also includes the manipulated input profile.

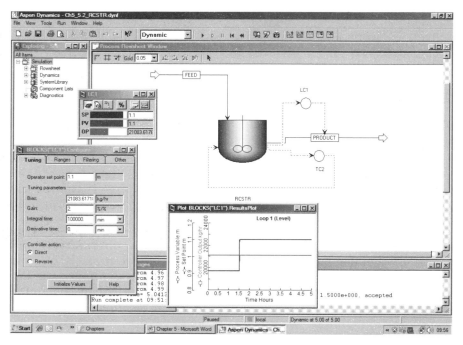

FIGURE 5.36

Figure 5.36 represents an excellent set point tracking performance of the level controller (P-only). Obviously, the LC1 provides process responses with almost no deviation from the desired set point value and with very fast approach to reach the target liquid level.

Notice that the above plot can be edited by right clicking on that plot and selecting *Properties* option or by clicking on that plot and pressing Alt+Enter on the keyboard. In the properties window, user can modify the title, axis scale, font and colour of the plot. Alternatively, double-click on the different elements of the plot and modify them as we like to improve the clarity and overall presentation.

Now, we will discuss the interaction of two control loops. When we introduce a set point step change in the reactor liquid level, the LC1 scheme attempts to compensate for the changes through the manipulation of the effluent flow rate. This, in turn, will disturb the reactor temperature and loop 2 will compensate by manipulating the heat removal of the CSTR appropriately. Thus we can say that loop 1 affects loop 2. In Figure 5.37, Aspen Dynamics window demonstrates the loop interaction under the same set point step change (0.914029 to 1.1 meter at time = 1.5 hours) as considered previously.

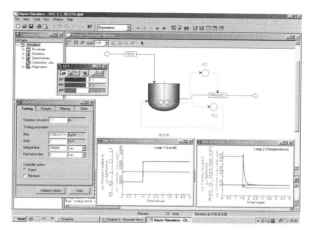

FIGURE 5.37

Viewing servo performance of TC2

As described in Figure 5.38, open the faceplate along with *Configure* dialog box and a blank plot sheet. Before starting the simulation run, carefully check all entries in the faceplate as well as *Configure* dialog box. In the next, choose *Initialization* run mode in the toolbar and then run the program once. After completion, go back to *Dynamic* mode from *Initialization* mode (see Figure 5.38).

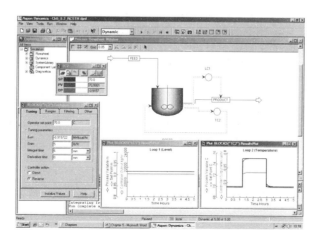

FIGURE 5.38

Now we wish to conduct the servo performance study for the TC2 controller with two consecutive set point step (pulse input) changes in reactor temperature (70 → 75°C at time = 1.2 hours and then 75 → 70°C at time = 3 hours).

Clearly, the proportional integral controller with default tuning parameters values shows a high-quality temperature tracking performance. As stated, if the performance of any controller is not satisfactory, we have the option to tune the parameters simply by trial-and-error method.

If we introduce a set point change in the reactor temperature, the TC2 controller takes necessary action with adjusting the heat duty to compensate for the changes. But interestingly, the liquid level remains undisturbed. Figure 5.38 confirms this fact. At this point we can conclude that loop 1 affects loop 2, but loop 2 does not affect loop 1. Actually here the interaction is in a single direction.

(d) Viewing regulatory performance of LC1 and TC2: To perform the regulatory study, we need to introduce at least a single change in the input disturbance. However, here we consider two subsequent step changes in the feed temperature. Initially, the feed temperature changes from 75 to 80°C at time = 2 hours and then the temperature (80°C) returns to 75°C after 1.2 hours.

To change the feed temperature twice as prescribed above, first we need to open the feed data sheet by double-clicking on the FEED block in the process flowsheet (see Figure 5.39).

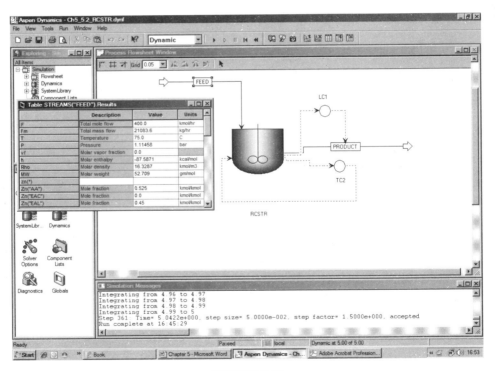

FIGURE 5.39

In the subsequent step, run the program with *Initialization* run mode. As it is finished, go back to *Dynamic* mode. Then, open the plot sheets for both the controllers. The regulatory behaviour is illustrated in Figure 5.40 giving changes in feed temperature in the feed data sheet. For brevity, the faceplate and configure dialog box are not included in the Aspen Dynamics window, shown in Figure 5.40.

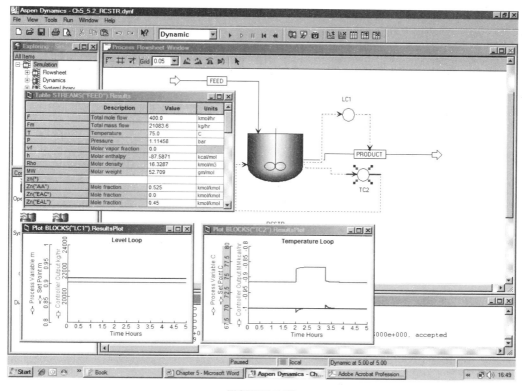

FIGURE 5.40

It is obvious that the reactor liquid level remains unchanged with a change in feed temperature since there is no interaction involved. On the other hand, the reactor temperature is disturbed. However, the TC2 controller provides satisfactory disturbance rejection performance under this situation.

So far we have studied mainly the closed-loop behavior of a reactor system coupled with Aspen-generated control schemes. We did not include any additional controller with the CSTR model. In Section 5.3, we consider a distillation example to elaborate this point.

5.3 DYNAMICS AND CONTROL OF A BINARY DISTILLATION COLUMN

Problem statement

A partially vaporized binary mixture of benzene and toluene enters a RadFrac distillation model as displayed in Figure 5.41.

The column has total 25 theoretical stages (including condenser and reboiler) and it operates at a pressure in the reflux drum of 18 psia and reboiler of 21 psia. The distillate flow rate is 285 lbmol/h and reflux ratio is 2.2 (mole basis).

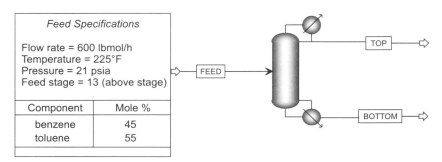

FIGURE 5.41 A flowsheet of a distillation column.

In Table 5.1, the reflux drum and the base of the column (the 'sump' in Aspen terminology) are specified. It is fair to use an aspect ratio (length to diameter ratio) of 2 (Luyben, 2004).

TABLE 5.1

Item	Vessel type	Head type	Height/Length (ft)	Diameter (ft)
Reflux drum	horizontal	elliptical	5	2.5
Sump	—	elliptical	5	2.5

The column diameter is 5 ft. Use default values for other tray hydraulic parameters (e.g., tray spacing, weir height and weir length to column diameter ratio). Consider log-mean temperature difference (LMTD) assumptions for the total condenser. Actually, the LMTD is calculated using the temperatures of process fluid and coolant. In the simulation, assume constant reboiler heat duty and apply the UNIFAC base property method.

(a) Simulate the column model to obtain the products mole fractions.
(b) Keeping the default level and pressure control algorithms unaltered, inspect the servo as well as regulatory performance of a proportional integral (PI) controller that is required to insert to control the benzene composition in the distillate by manipulating the reflux flow rate.
(c) Devising an another PI control scheme to maintain the benzene composition in the bottom product with the adjustment of heat input to the reboiler, observe the interaction effect between the top and bottom composition loops.

Simulation approach

(a) Select Aspen Plus User Interface and when the Aspen Plus window pops up, choose *Template* and press *OK*. In the subsequent step, select *General with English Units* and hit *OK* button. To open the process flowsheet window, click *OK* when the Aspen Plus engine window appears.

Creating flowsheet

From the *Model Library* toolbar, select the *Columns* tab. Place the RadFrac model on the flowsheet window and add the feed as well as two product streams. Renaming all the streams along with distillation block, we have Figure 5.42.

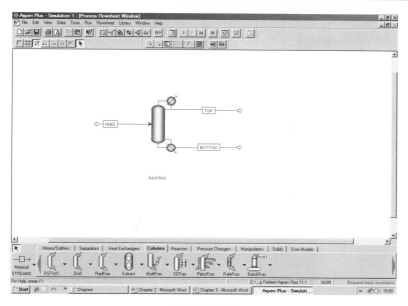

FIGURE 5.42

Configuring settings

As we hit *Next* followed by *OK* button, the setup input form appears (see Figure 5.43). The distillation problem is titled as: 'Closed-loop Performance of a Distillation Column'. Importantly, use the 'Dynamic' input mode.

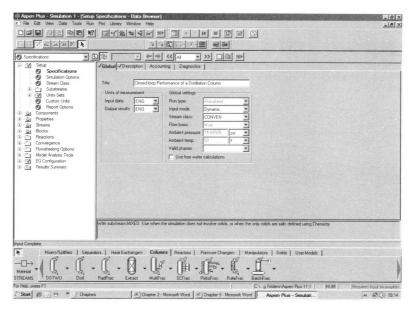

FIGURE 5.43

Figure 5.44 includes the *Aspen Plus accounting information*. We can fill up the *Accounting* sheet with any name, number and ID.

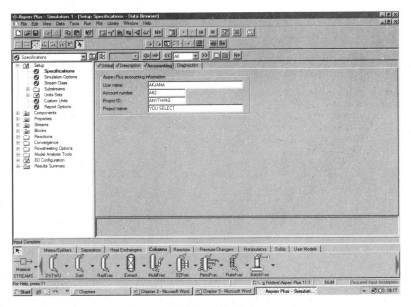

FIGURE 5.44

We like to see the composition of all incoming and outgoing streams in mole fraction basis in the final results table. Accordingly, we use *Stream* sheet under the *Report Options* of *Setup* folder (see Figure 5.45).

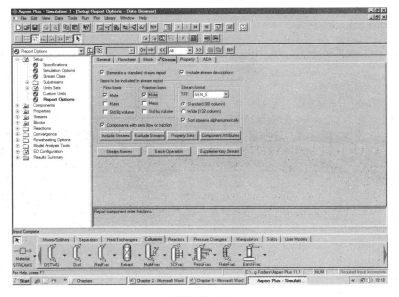

FIGURE 5.45

Specifying components

From the Data Browser, select *Components/Specifications* to open the component input form. In the table, shown in Figure 5.46, the two species are defined.

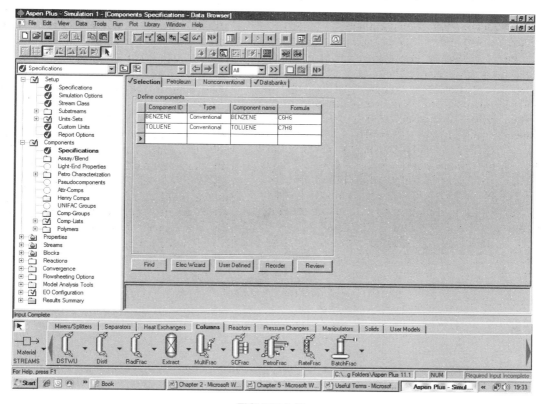

FIGURE 5.46

Specifying property method

In the list on the left, choose *Properties/Specifications* and get the property input form. In Aspen simulation, a property method originally includes several models for calculating the physical properties. For the distillation example, set the UNIFAC base method by scrolling down (see Figure 5.47).

Specifying stream information

In the next, open *Streams/FEED/Input/Specifications* sheet. Entering the given values for all state variables and feed composition, the stream input form looks like Figure 5.48.

Specifying block information

In the left pane of the Data Browser window, select *Blocks/RADFRAC/Setup* to open *Configuration* sheet and then insert the required data (see Figure 5.49).

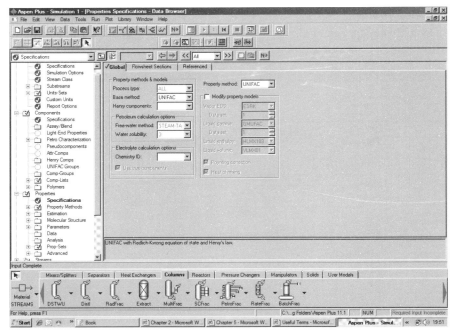

FIGURE 5.47

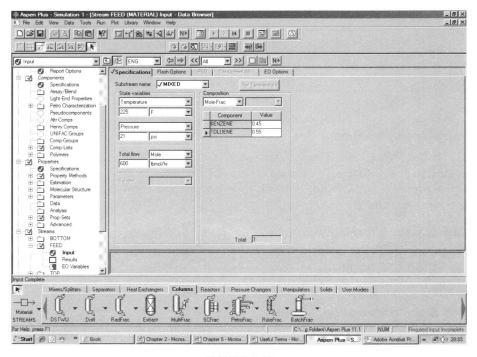

FIGURE 5.48

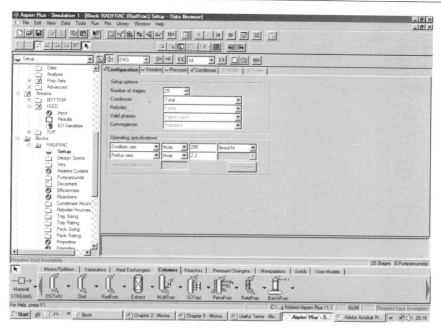

FIGURE 5.49

In the subsequent step (see Figure 5.50), fill up *Streams* sheet with informing the feed location [13th tray (above stage)].

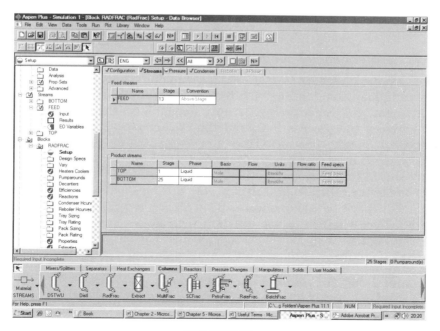

FIGURE 5.50

In Figure 5.51, the column pressure profile is defined.

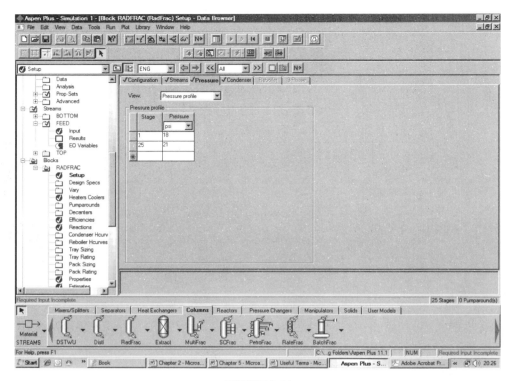

FIGURE 5.51

Entering heat transfer data for condenser and reboiler

Next select *Dynamic* under *RADFRAC* of *Blocks* folder. There are three heat transfer options: constant duty, constant medium temperature and LMTD. As mentioned in the problem statement, the condenser heat duty depends on the log-mean temperature differential between the process fluid and the coolant (see Figure 5.52). The coolant inlet temperature is set constant. Here the *temperature approach* represents the difference between the process temperature and the coolant outlet temperature at the initial steady state. Note that among the heat transfer specifications, the coolant inlet temperature and temperature approach may vary during a dynamic simulation, whereas the specific heat capacity of the coolant is fixed during a dynamic run.

For the reboiler, it is fair to use constant heat duty computed in the Aspen Plus simulation (see Figure 5.53). However, the reboiler duty may be changed at dynamic state either by manually or automatically with employing a controller.

Entering geometry data for reflux drum and sump

The reflux drum and sump are specified in Figures 5.54(a) and (b) with their given geometry data. The information on vessel orientation, head type, length (or height) and diameter are used to compute the vessel holdup.

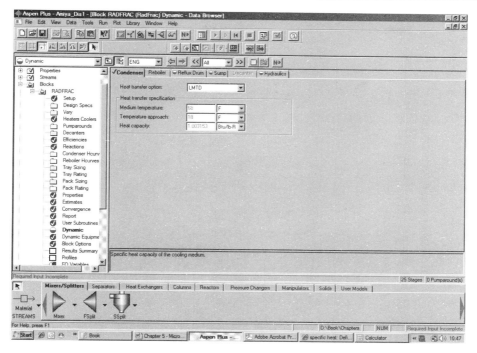

FIGURE 5.52

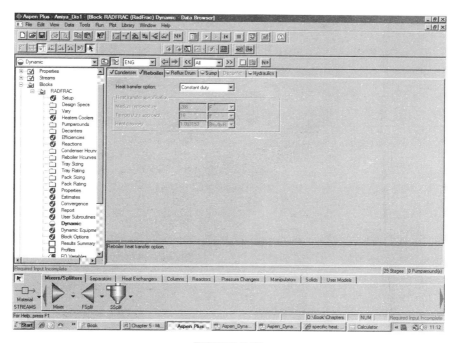

FIGURE 5.53

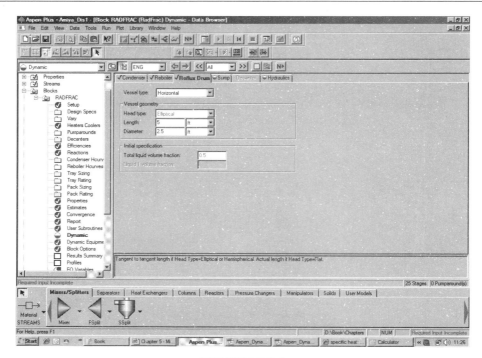

FIGURE 5.54(a)

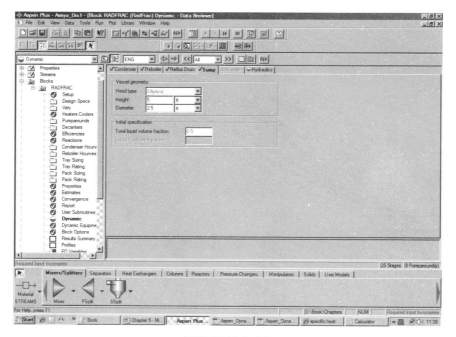

FIGURE 5.54(b)

Entering tray geometry

The example column has total 25 stages—Stage 1 being the condenser and Stage 25 the reboiler. We already have inserted the necessary information for stages 1 and 25. Now, we need to inform the simulator the tray geometry specifications for stages 2 through 24. Note that the tray holdups are computed using these geometry data (see Figure 5.55).

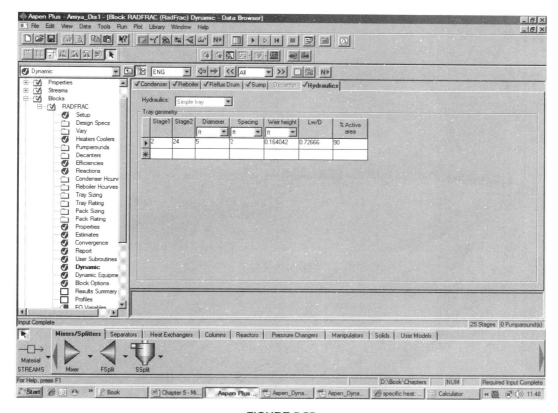

FIGURE 5.55

Running steady state simulation and viewing results

Hit *Next* button and press *OK* to run the steady state simulation. Finally, the results table, shown in Figure 5.56, is obtained. At this time, we should save the work.

   (b) **Exporting dynamic simulation:** For process dynamics study, we wish to export the steady state Aspen Plus simulation into flow-driven Aspen Dynamics simulation giving a file name of 'Ch5_5.3_RadFrac'. Then close the Aspen Plus window.

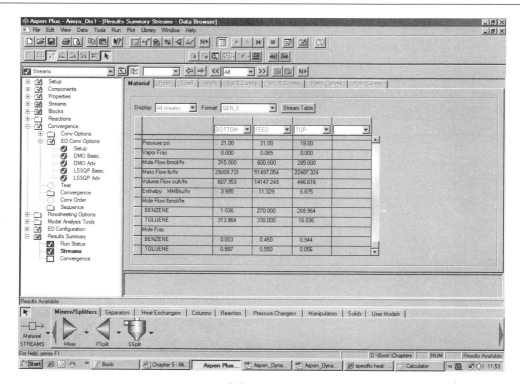

FIGURE 5.56

Starting Aspen Dynamics

Open a blank dynamic simulation window for the example column, following a similar procedure as previously shown for the CSTR problem. In the next, simply open the flow-driven dynamic file 'Ch5_5.3_RadFrac.dynf'. As a result, the Aspen Dynamics window appears (see Figure 5.57) accompanying with the closed-loop process flow diagram. The flowsheet actually includes the three default control schemes LC1, PC2 and LC3 to monitor the reflux drum liquid level, top stage pressure and column base liquid level, respectively.

In the present discussion, we do not want to change anything of the three automatically inserted control strategies. All data, including tuning parameters, ranges, bias values and controller actions, remain untouched. A little detail of these control structures is given below.

Loop 1

Controller: LC1

Type of controller: P-only (since integral time is very large (60000 minutes))

Controlled variable: liquid level in the reflux drum

Manipulated variable: distillate flow rate

Controller action: direct

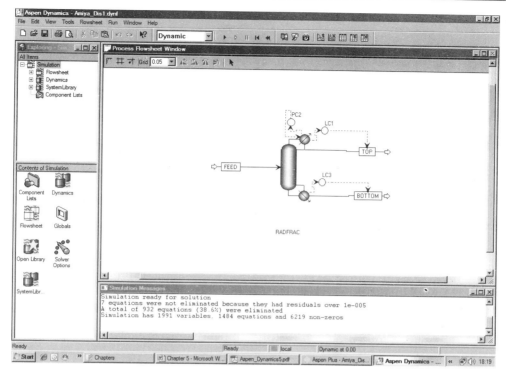

FIGURE 5.57

Loop 2

  Controller: PC2

  Type of Controller: PI

  Controlled variable: top stage pressure

  Manipulated variable: coolant inflow rate

  Controller action: direct

Loop 3

  Controller: LC3

  Type of controller: P-only

  Controlled variable: liquid level in the column base

  Manipulated variable: bottoms flow rate

  Controller action: direct

Adding a new PI controller for top composition loop

Now we wish to include a proportional integral (PI) law to control the benzene composition in the top distillation product by manipulating the reflux rate. In the top left of the window, the *Dynamics* library is included within *Simulation* folder of *All Items* pane. Click on expand (+) button of *Dynamics* subfolder. Consequently, the expand button changes to collapse (–) button as shown in Figure 5.58.

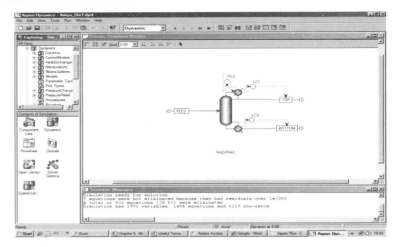

FIGURE 5.58

Again hit expand button next to the *ControlModels* icon. Then select PID controller, drag it to the flow diagram and drop the control block near to the top product stream. Renaming the top composition controller as CCT, we have Figure 5.59.

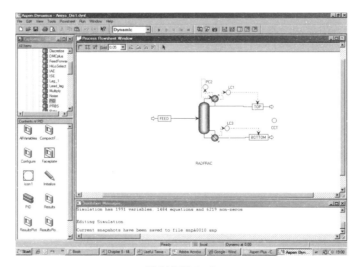

FIGURE 5.59

Connecting controller with process variable (Controlled variable)

Expand *Stream Types* under *Dynamics* subfolder and hold down the mouse button on the *ControlSignal* icon. As we drag it onto the flowsheet window, many blue arrows appear around the process diagram. Interestingly, when we move the mouse pointer with holding the *ControlSignal* icon over a port, the name of that port is displayed. Anyway, move the pointer and release the mouse button on the outgoing arrow named *OutputSignal* originated from TOP (stream) block. To select the distillate composition of benzene as controlled variable, choose 'STREAMS("TOP").Zn("BENZENE")' with 'Mole fraction' by description in the *Select the Control Variable* dialog box (see Figure 5.60).

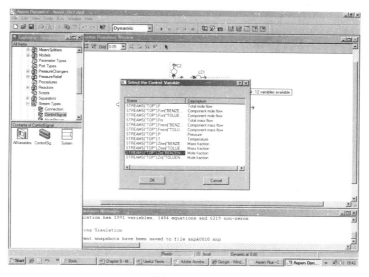

FIGURE 5.60

As we press *OK* button, the cursor becomes a solid black arrow representing the input signal to the controller. To transmit this signal to the CCT block, connect the black arrow with a port marked *InputSignal*. Since this signal conveys the process variable (PV) information to the CCT controller, select 'CCT.PV' by name with 'Process variable' by description in the *Select the Control Variable* dialog box (see Figure 5.61).

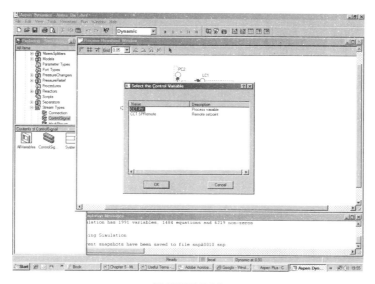

FIGURE 5.61

Hit *OK* button and obtain Figure 5.62. Obviously, the CCT controller is partially configured. To complete the top composition loop, the controller output should be connected with the manipulated variable to pass on the signal.

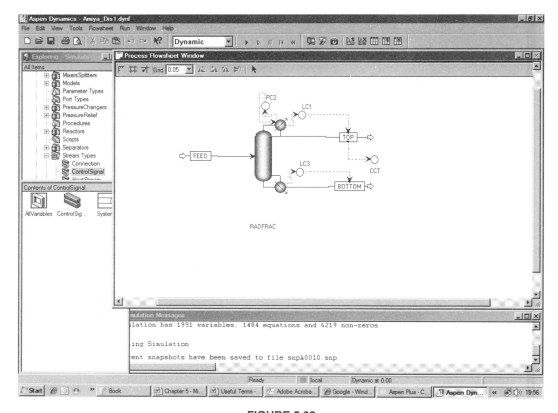

FIGURE 5.62

Connecting controller with control variable (Manipulated variable)

Again hold the *ControlSignal* icon, drag it onto the process flowsheet and drop it on the blue outgoing arrow marked *OutputSignal* from the CCT block. As *Select the Control Variable* dialog box appears (see Figure 5.63), choose 'CCT.OP' by name and press *OK*.

Immediately, a solid black arrow representing the controller output signal is automatically generated. Move the mouse pointer to reflux stream and make a connection to *InputSignal2* port. To use the reflux flow rate as control variable, select 'BLOCKS("RADFRAC"). Reflux.FmR' in the dialog box and click *OK* (see Figure 5.64).

Now the binary distillation column is coupled with four control schemes, LC1, PC2, LC3 and CCT, and the closed-loop process looks like Figure 5.65. The subsequent discussion includes the modification of different tuning properties of the CCT controller.

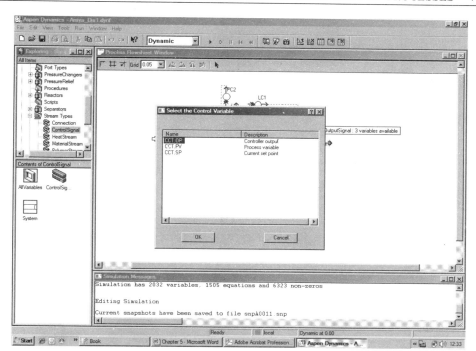

FIGURE 5.63

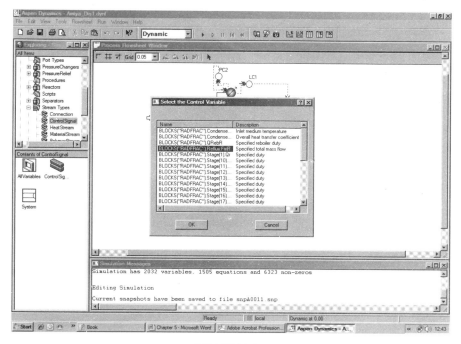

FIGURE 5.64

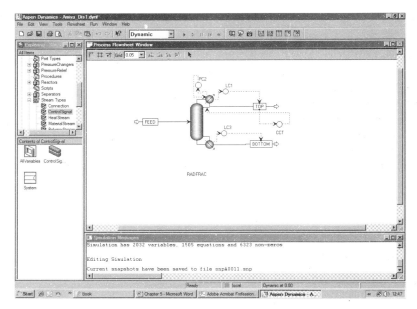

FIGURE 5.65

Modifying controller tuning properties

First we wish to see the default tuning properties. So, double-click on the CCT block and then hit *Configure* symbol in the faceplate to open the *Configure* dialog box (see Figure 5.66).

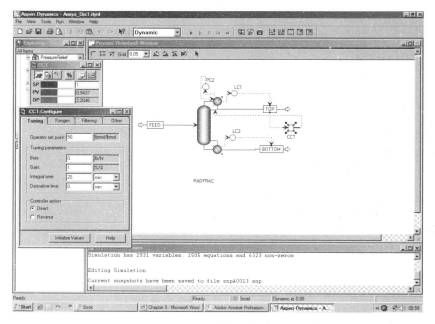

FIGURE 5.66

Obviously, some of the default values set by Aspen Dynamics are not acceptable. For example, the operator set point value of process variable (benzene composition in distillate) should not be greater than 1. Secondly, the CCT controller action must be 'Reverse'. In addition, the value of control variable (reflux flow rate) at steady state is usually used as bias value.

We have two options in our hand to correct the default values. Either manually we can do it or Aspen Dynamics can automatically initialize the values of set point, process variable, control variable, bias and ranges. Note that the controller action is changed only manually. It is wise to initialize the values by the help of Aspen Dynamics. For this, press *Initialize Values* button in the *Configure* dialog box and use 'Reverse' controller action. It is obvious in the window, shown in Figure 5.67, that the values of SP, PV and OP in the faceplate change automatically to their steady state values. If this approach fails to initialize the simulation of controller model with the steady state data, check and replace, if necessary, the values of PV and OP with their steady state values by double clicking on signal transmission lines (input to the controller and output from the controller).

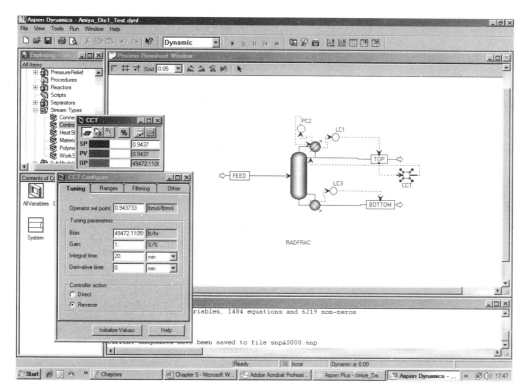

FIGURE 5.67

Modifying ranges for process variable and controller output

As we hit the *Ranges* tab, the *Configure* dialog box (see Figure 5.68) shows the default ranges imposed on process variable and controller output.

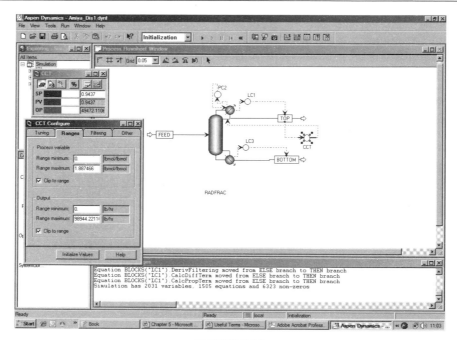

FIGURE 5.68

However, here we use the typical variable ranges, shown in Figure 5.69.

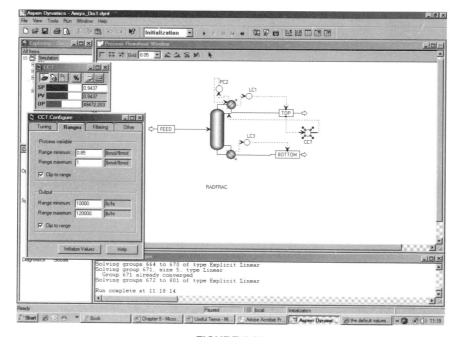

FIGURE 5.69

Process variable

> Range minimum: 0.85
>
> Range maximum: 1.0

Output

> Range minimum: 10000 lb/h
>
> Range maximum: 120000 lb/h

It is important to mention that it is a good idea to carry out *Initialization* as well as *Dynamic* run after each new change in the control scheme so that any error in controller installation can be detected individually.

We have now completed all required control specifications for the top composition loop. In the ongoing study, we prefer to conduct the simulation experiment to observe the designed controller performance continuously for 5 hours. As done for the previous CSTR problem, similarly either simply press Ctrl+F5 on the keyboard or select *Pause At* from the *Run* menu and put 5 hours as *Pause at time*.

In the next, we will inspect the CCT controller performance first dealing with the servo problem followed by the regulatory problem.

Viewing servo performance of CCT

As we double-click on the CCT controller block in the flowsheet window, the faceplate appears. Then open the *Configure* as well as *ResultsPlot* dialog box. The second one is basically a blank graph sheet that presents the variations of process variable, set point and controller output with respect to time.

Before running the program, make sure that all the items in the *Configure* dialog box and faceplate are correct. In the next, hit *Run* button to start the dynamic simulation. The plots, shown in Figure 5.70, illustrate the servo behavior of the PI control algorithm with a step increase ($0.9437 \rightarrow 0.97$ at time = 1.51 hours) followed by a step decrease ($0.97 \rightarrow 0.9$ at time = 3 hours) in the set point value of the distillate composition of benzene. To achieve an improved closed-loop performance, we have used the values of proportional gain of 10 %/% and integral time of 10 minutes. These values have been chosen based on a pulse input test in the distillate composition of benzene and using the trial-and-error approach. It should be kept in mind that the objective at this point is not to come up with the *best* control structure or the *optimum* controller tuning. We only need a control scheme and tunings that provide a reasonably good tracking performance to drive the simulation to a new steady state.

Remember that to edit the plots, shown in Figure 5.70, double-click on different elements of the plots and modify them as we like.

Viewing regulatory performance of CCT

In order to investigate the regulatory performance of the CCT controller, we give a step input change in the feed pressure ($21 \rightarrow 23$ psia) at time = 1.48 hours and that in the feed temperature ($225 \rightarrow 230°F$) at time = 3 hours. The PI controller tuning set provides good disturbance rejection performance (see Figure 5.71) although the tuning parameter values (gain and integral time) have been chosen based on a pulse set point input change.

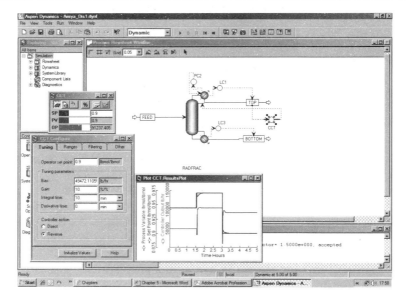

FIGURE 5.70

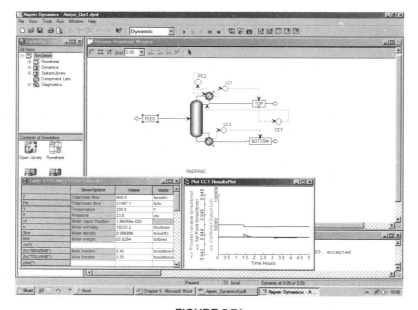

FIGURE 5.71

(c) **Adding a new PI controller for bottom composition loop:** We have to devise another PI control scheme to monitor the bottoms composition of benzene by adjusting the heat input to the reboiler. As developed, the CCT controller for the top loop, similarly we can configure the CCB controller for the bottom loop as shown in Figure 5.72.

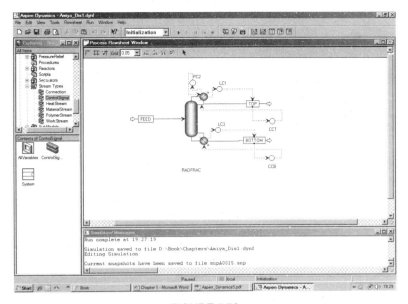

FIGURE 5.72

We have chosen the following tuning properties (see Figure 5.73):
Gain = 10%/%
Integral time = 10 minutes
Controller action: Direct

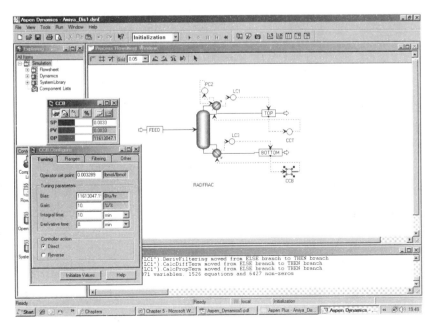

FIGURE 5.73

In addition, the used constraints are reported below:

Process variable

>Range minimum: 0.0
>Range maximum: 0.1

Output

>Range minimum: 6000000 Btu/h
>Range maximum: 18000000 Btu/h

Viewing interaction effect between two composition loops

To observe the effect of interaction between two composition loops, the set point value of bottoms composition of benzene has been changed twice. The simulation result is depicted in Figure 5.74 for a step increase (0.0033 → 0.0045 at time = 1.5 hours) followed by a step decrease (0.0045 → 0.0025 at time = 3 hours).

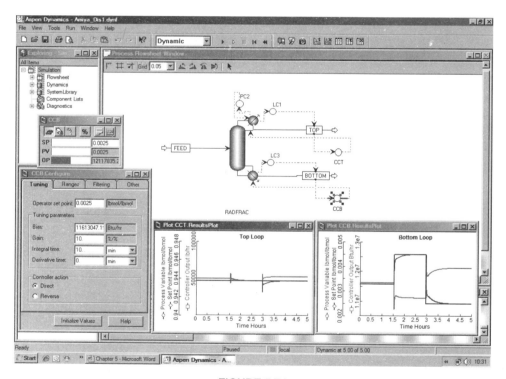

FIGURE 5.74

Clearly, the CCB controller shows satisfactory set point tracking performance against a pulse input change. It is observed from Figure 5.74 that owing to strong interaction between the two composition loops of the distillation column, the set point changes in bottom loop affect the top product composition. Similarly, when any set point change is introduced in the top composition loop, the bottom product composition will also be affected.

SUMMARY AND CONCLUSIONS

This chapter has investigated the closed-loop process dynamic characteristics using Aspen Dynamics package. To observe the controller performance in terms of set point tracking and disturbance rejection, a CSTR in addition to a distillation column have been illustrated. The default control strategies have been tested for the reactor example, whereas the two additional composition control loops have been included along with the default control laws for the distillation example. Several simulation experiments have been executed for both the processes under flow-driven dynamic simulation. Note that Chapter 6 presents the dynamic simulation and control of more rigorous pressure-driven dynamic process.

PROBLEMS

5.1 A feed mixture of benzene and toluene is fed to a flash drum (Flash2). The separator operates at 1.2 atm and 100°C. For dynamic simulation, required feed specifications are provided in Figure 5.75.

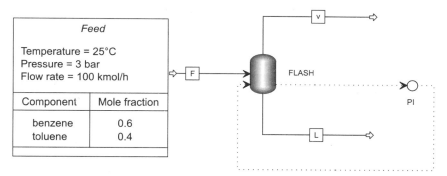

FIGURE 5.75 A flowsheet of a flash drum.

   (a) Use the SYSOP0 property method to compute the amounts of liquid and vapor products and their compositions.

   (b) As shown in Figure 5.75, employ a PI control scheme to monitor the temperature in the flash drum by manipulating the heat duty.

   (c) Show the closed-loop servo performance with +10% and then −10% step changes in the flash temperature.

   (d) Report the tuning parameters obtained by trial-and-error method, controller action and ranges imposed.

5.2 A vapor mixture of toluene, methane and hydrogen is heated using a shell and tube heat exchanger (HeatX). The superheated steam is used as a heating medium. Complete specifications required for closed-loop dynamic simulation are shown in Figure 5.76.

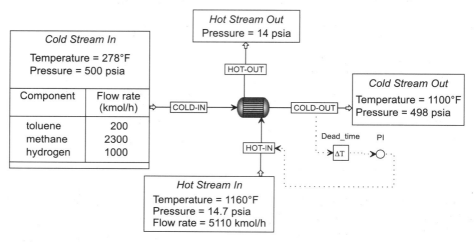

FIGURE 5.76 A flowsheet of a heat exchanger.

(a) Simulate the heat exchanger model using the shortcut method, counter-current flow direction and NRTL-RK property method.

(b) Include a PI control structure to observe the set point (cold stream outlet temperature) tracking performance and the manipulated input (steam inflow rate) profile. In the closed-loop simulation experiment, assume that the temperature sensor takes 1 minute time (dead time) to measure the controlled variable. Report the used tuning properties.

(c) Examine the regulatory performance by introducing + 10% and subsequently − 10% step changes in the inlet temperature of the cold stream.

5.3 Device a cascade control scheme for the above heat exchanger and investigate the controller performance.

5.4 A liquid mixer model with a typical ratio controller (Seborg et al. 2003) is shown in Figure 5.77. The flow rates for both the disturbance or wild stream (F_W) and the manipulated stream (F_E) are measured, and the measured ratio, $R_m = F_E/F_W$, is calculated. The output of the ratio element is sent to a ratio controller (PI) that compares the calculated ratio R_m to the desired ratio R_d (set point) and adjusts the manipulated flow rate accordingly.

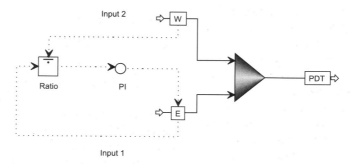

FIGURE 5.77 A flowsheet of a mixer.

The input data are shown in Table 5.2 for simulation.

TABLE 5.2

Stream	Temperature (°C)	Pressure (atm)	Flow rate (kmol/h)	Composition
E	50	1	$F_E = 100$	pure ethanol
W	60	1	$F_W = 150$	pure water

Process variable at steady state = 0.667 ($F_E/F_W = 100/150$)

Controller output at steady state = 100 kmol/h

Proportional gain = 4 %/%

Integral time = 20 minutes

Controller action = reverse

(a) Applying the SYSOP0 base property method, simulate the mixer model operated at 1 atm.

(b) Using the given controller properties and default ranges, report the ratio controller performance with two consecutive set point step changes (0.667 → 0.72 → 0.65) in the ratio.

Hint: Double-click on Input 1 transmission line and fill up Tables 5.3(a) and (b).

TABLE 5.3(a)

	Value	Spec
>STREAMS("E").Fcn("ETHANOL")	100.0	Free
<Ratio.Input1	100.0	Free

Similar table for Input 2 is obtained as:

TABLE 5.3(b)

	Value	Spec
>STREAMS("W").Fcn("WATER")	150.0	Free
<Ratio.Input2	150.0	Free

In the next, double-click on Ratio element and get Table 5.4.

TABLE 5.4

	Description	Value	Units
Input1	Input signal 1	100.0	kmol/h
Input2	Input signal 2	150.0	kmol/h
Output	Output signal, Input1/Input2	0.667	

Use *Initialize Values* button and incorporate the given tuning properties before running the program.

5.5 A reboiled stripper is used to remove mainly propane and lighter species from a feed stream, shown in Figure 5.78. It has total 6 stages (including condenser and reboiler) and no condenser.

The bottoms rate is 100 lbmol/h and the column top stage pressure is 150 psia with a column pressure drop of 8 psi. The diameter of the stripper (Stages 1 to 5) is 6.5 ft. The reboiler heat duty is assumed constant, although it changes at dynamic state. The sump has elliptical head with a height of 5 ft and diameter of 2.5 ft.

For the closed-loop simulation, use the following data:

Dead time = 2 minutes

Magnitude of noise (standard deviation) = 0.01 lbmol/lbmol

Proportional gain of PI = 1%/%

Integral time of PI = 20 minutes

Controller action = Reverse

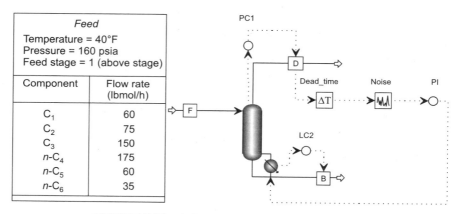

FIGURE 5.78 A flowsheet of a stripping column.

(a) Using the Peng-Robinson thermodynamic method, simulate the RadFrac (STRIP2) model and compute the product compositions.

(b) Keeping the default controllers (PC1 and LC2) unaltered, configure a composition control scheme (PI) coupling with a 'Dead_time' and 'Noise' elements to maintain the propane mole fraction in the distillate by manipulating the reboiler heat duty as shown in Figure 5.79. Use the given closed-loop data and execute the dynamic simulations to test the developed composition controller performance.

5.6 Ethylene is produced by cracking of ethane in a stoichiometric reactor. The irreversible elementary vapor-phase reaction is given as:

$$C_2H_6 \rightarrow C_2H_4 + H_2$$

ethane   ethylene   hydrogen

Pure ethane feed enters the reactor model, shown in Figure 5.79, with a flow rate of 750 kmol/h at 800°C and 5.5 atm. The reactor operates at inlet temperature and pressure with 80% conversion of ethane.

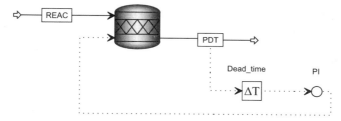

FIGURE 5.79   A flowsheet of a reactor.

(a) Using the SYSOP0 thermodynamic method, simulate the reactor model.
(b) Develop a control loop as configured in the flow diagram to maintain the desired reactor temperature by the adjustment of heat duty. Considering the measurement lag of 1 minute, inspect the servo as well as regulatory control performance. Report the tuning properties used to achieve a satisfactory closed-loop performance.

5.7 A binary feed mixture consisting of methylcyclohexane (MCH) and toluene is introduced above tray number 14 of a RadFrac distillation model, shown in Figure 5.80.

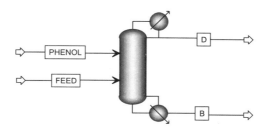

FIGURE 5.80   A flowsheet of a distillation column.

It is difficult to separate this close-boiling system (MCH-toluene) by simple binary distillation. Therefore, phenol is used as an extractant and introduced above tray number 7 of the column. The two input streams have the following specifications, shown in Table 5.5.

TABLE 5.5

Stream	Temperature (°C)	Pressure (bar)	Flow rate	Mole fraction
PHENOL	105	1.4	100 m³/h	1.0
FEED	105	1.4	181.44 kmol/h	0.5/0.5
				(MCH/toluene)

The column has 22 theoretical stages (including condenser and reboiler) with a total condenser. The distillate rate and reflux ratio are given as 90.72 kmol/h and 8 (mole basis), respectively. The pressure profile is defined with Stage 1 pressure of 1.10316 bar and Stage 22 pressure of 1.39274 bar. Use LMTD assumptions for the condenser. The reboiler heat duty is assumed constant. The reflux drum and sump are specified in Table 5.6.

TABLE 5.6

Item	Vessel type	Head type	Height/Length (m)	Diameter (m)
Reflux drum	horizontal	elliptical	1.5	0.75
Sump	—	elliptical	1.5	0.75

The column diameter and tray spacing are given as 2 m and 0.6 m, respectively.

(a) Simulate the distillation column using the UNIFAC property method to compute the composition of MCH in the distillate and that of phenol in the bottom product.

(b) In addition to the default level and pressure controllers, insert a PID structure to control the MCH composition in the top product by manipulating the flow rate of PHENOL stream.

(c) Produce the plots to show the closed-loop control responses, and report the tuning parameters, control actions and operating ranges for controlled as well as manipulated variables used.

5.8 An MTBE (methyl tertiary butyl ether) reactive distillation column with detailed input specifications is presented in Table 5.7 and Figure 5.81.

TABLE 5.7

Total number of stages: 30 (including a total condenser and a reboiler) Catalyst on stages: 15–19 Molar reflux ratio: 8 Top tray pressure: 100 psia Bottoms rate: 520 lbmol/h	Equilibrium reaction *iso*-butylene+methanol"!MTBE [Hint: select *REAC-DIST* Reaction ID and compute K_{eq} from Gibbs energies]

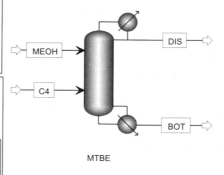

Pure methanol (MeOH) *feed*
Temperature = 70°F
Pressure = 400 psia
Flow rate = 540 lbmol/h
Feed stage = 15 (above-stage)

Mixed C4 *feed*
Temperature = 110°F
Pressure = 400 psia
Feed stage = 20 (above-stage)

Component	Flow (lbmol/h)
propane	7
iso-butane	670
iso-butylene	530
n-butane	20
1-butene	5
cis-2-butene	5
trans-2-butene	5

FIGURE 5.81 A flowsheet for MTBE reactive distillation column.

Considering the UNIFAC property method and flow-driven dynamic simulation:

(a) produce the simulation results

(b) perform the sensitivity analysis for MTBE product composition with varying the bottoms rate from 500 to 550 lbmol/h

(c) perform the controllability analysis

(d) investigate the closed-loop control performance

REFERENCES

Luyben, W.L., (2004), "Use of Dynamic Simulation to Converge Complex Process Flowsheets", *Chemical Engineering Education*, pp. 142–149.

Seborg, D.E., T.F. Edgar and D.A. Mellichamp, (2003), *Process Dynamics and Control*, 2nd ed., John Wiley & Sons, Inc.

CHAPTER 6

Dynamics and Control of Pressure-driven Processes

6.1 INTRODUCTION

To know the transient characteristics of a complicated chemical plant, we need a dynamic process simulator. It is well-recognized that Aspen Dynamics™ is such an efficient flowsheet simulator used for dynamic process simulation. As we have seen in Chapter 5, Aspen Dynamics simulator can be employed to design a process as well as its associated control strategies.

Aspen Dynamics extends an Aspen Plus™ steady-state model into a dynamic process model. If the steady state Aspen Plus simulation is exported to Aspen Dynamics, there is a necessity to choose either flow-driven dynamic simulation or pressure-driven dynamic simulation. As started earlier, in a rigorous pressure-driven simulation, pumps and compressors are inserted, where needed, to provide the required pressure drop for material flow. Control valves are installed, where needed, and their pressure drops selected. For good control, the pressure drop across a control valve should be greater than 0.1 bar. The fluid that flows through a valve should normally be liquid-only or vapor-only because the two-phase flow through a control valve is unusual.

It should be pointed out that for a pressure-driven case, we must not insert a valve in the suction of a pump or at the discharge of a compressor (compressor speed or its equivalent compressor work is manipulated). The control valves are positioned on the fluid streams such that the controllers can manipulate the valve positions.

The simple flow-driven dynamic simulations have been discussed in detail in the previous chapter. Therefore, here we are intended to study the pressure-driven simulation. A reactive or catalytic distillation column is exampled for the rigorous pressure-driven Aspen Dynamics simulation as well as control.

6.2 DYNAMICS AND CONTROL OF A REACTIVE DISTILLATION (RD) COLUMN

Problem statement

The methyl tertiary butyl ether (MTBE) column configuration (Jacobs and Krishna, 1993) chosen for the simulation is shown in Figure 6.1.

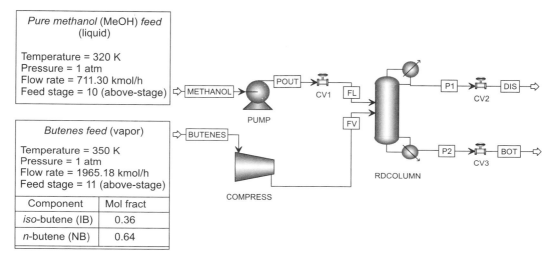

FIGURE 6.1 A flowsheet for the production of MTBE.

The RD column (RadFrac) consists of 17 theoretical stages, including a total condenser and a partial reboiler. Reactive stages are located in the middle of the column, Stage 4 down to and including Stage 11. In Aspen terminology, the numbering of the stages is top downward; the condenser is Stage 1 and the reboiler is the last stage.

MTBE is produced by reaction of IB and MeOH:

$$(CH_3)_2C = CH_2 + CH_3OH \leftrightarrow (CH_3)_3 COCH_3$$
$$\quad\quad\text{IB} \quad\quad\quad\quad \text{MeOH} \quad\quad\quad \text{MTBE}$$

The liquid-phase reaction is catalyzed by a strong acidic macroreticular ion exchange resin, for example Amberlyst 15, and n-butene does not take part in the reaction (inert). The forward and backward rate laws (Seader and Henley, 1998; Rehfinger and Hoffmann, 1990) are derived in terms of mole fractions, instead of activities (products of activity coefficient and mole fraction):

$$\text{Forward rate: } r_f = 3.67 \times 10^{12} \exp\left(\frac{-92440}{RT}\right)\left(\frac{x_{IB}}{x_{MeOH}}\right)$$

$$\text{Backward rate: } r_b = 2.67 \times 10^{17} \exp\left(\frac{-134454}{RT}\right)\left(\frac{x_{MTBE}}{x^2_{MeOH}}\right)$$

Here, x represents the liquid-phase mole fraction. The pre-exponential factors, including the activation energy (kJ/kmol), are given in SI units. The catalyst is provided only for reactive stages (8 stages total), with 204.1 kg of catalyst per stage (Seader and Henley, 1998). The used catalyst is a strong-acid ion-exchange resin with 4.9 equivalents of acid groups per kg of catalyst. So, the equivalents per stage are 1000 or 8000 for the 8 stages. In some references, the equivalents per stage are directly given.

The column, starting from Stages 2 to 16, is packed with 'MELLAPAK' (vendor: SULZER) having a size of 250Y. Use 'Simple packing' hydraulics and the height equivalent to a theoretical plate (HETP) may be considered as 1 m. The distillation column diameter is 6 m. Stage 1 (condenser) pressure is 11 atm with a column pressure drop of 0.5 atm. The reflux ratio is set to 7 (mole basis) and the bottoms flow rate is 640.8 kmol/h. In the MTBE synthesis process, it is desirable to obtain a bottom product containing high-purity MTBE and a distillate containing high-purity NB. In Table 6.1 the reflux drum and the sump (the next-to-last stage in the column) are specified.

TABLE 6.1

Item	Vessel type	Head type	Height/Length (m)	Diameter (m)
Reflux drum	horizontal	elliptical	2	1
Sump	—	elliptical	2.2	1.1

The pump delivers the liquid stream POUT at 11.7 atm. The compressor (isentropic) has discharged the vapor feed FV at 11.5 atm. The three control valves (adiabatic flash) CV1, CV2 and CV3 have the outlet pressures of 11.5 atm, 10.8 atm and 11.3 atm respectively. Using the UNIFAC base property method,

(a) simulate the process flowsheet to obtain the distillation product summary, and
(b) develop the control configurations to achieve the desired product purity under disturbance input.

Simulation approach

(a) Start the Aspen program by double-clicking the *Aspen Plus User Interface* icon on the desktop. Then select *Template* option and press *OK* (see Figure 6.2).

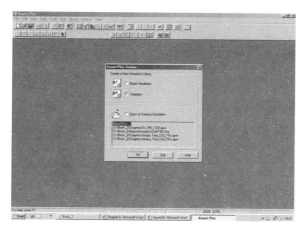

FIGURE 6.2

We choose *General with Metric Units* option and hit *OK* button (see Figure 6.3).

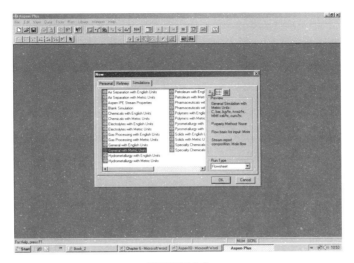

FIGURE 6.3

When the *Connect to Engine* window appears, use the default *Server type* (Local PC).

Creating flowsheet

The process flow diagram includes a feed pump, a feed compressor, a distillation column and three control valves. The complete process flowsheet drawn in an Aspen window should somewhat resemble the one shown in Figure 6.4. Recall that Aspen has a tool in the toolbar that automatically takes the user through the required data input in a stepwise fashion. The blue *Next* button does this.

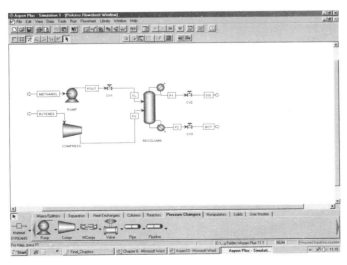

FIGURE 6.4

Configuring settings

At the beginning of data entry, fill up *Global* sheet followed by *Accounting* sheet under *Specifications* of *Setup* folder. Moreover, select 'Mole' fraction along with 'Std.liq.volume' flow basis in *Stream* sheet under *Report Options* [see Figures 6.5(a), (b) and (c)].

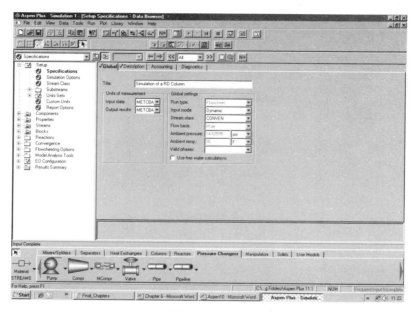

FIGURE 6.5(a)

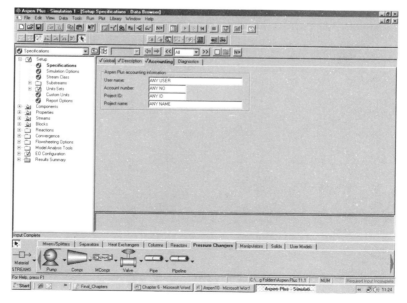

FIGURE 6.5(b)

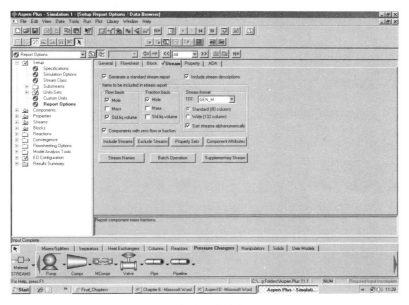

FIGURE 6.5(c)

Specifying components

The components involved in the example system are MeOH (CH4O), IB (C4H8-5), NB (C4H8-1) and MTBE (C5H12O-D2). Within the parentheses, the chemical formulas used in Aspen terminology are mentioned (see Figure 6.6).

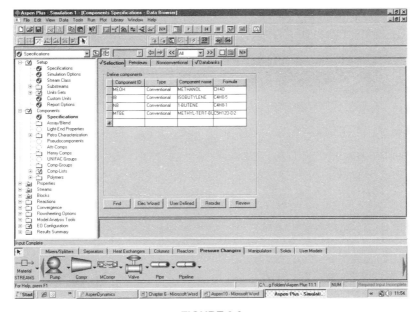

FIGURE 6.6

Specifying property method

The user input under the *Properties* tab is probably the most critical input required to run a successful simulation. This has been discussed in much greater detail in the previous chapters. This key input is the *Base method* found in *Global* sheet under *Specifications* option. Set UNIFAC for the present project (see Figure 6.7).

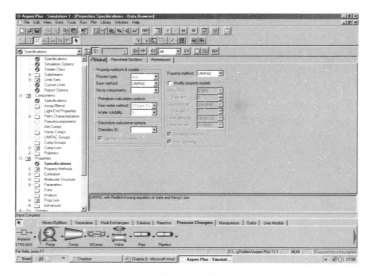

FIGURE 6.7

Specifying stream information

Under the *Streams* tab, we have used *Specifications* sheets to input the data for both the feed streams, BUTENES and METHANOL [see Figures 6.8(a) and (b)].

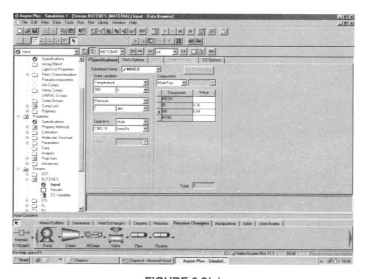

FIGURE 6.8(a)

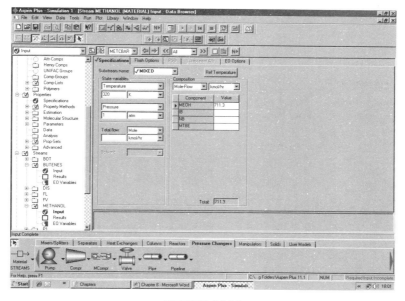

FIGURE 6.8(b)

Specifying block information

In Figures 6.9(a) to 6.9(d), first the feed compressor details are given. Subsequently, the three control valves, CV1, CV2 and CV3, are specified.

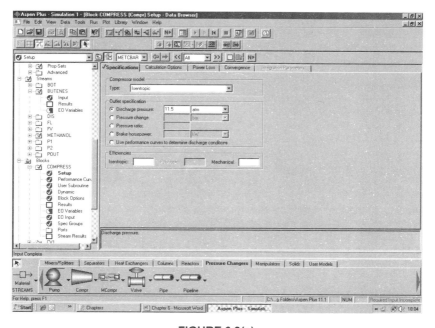

FIGURE 6.9(a)

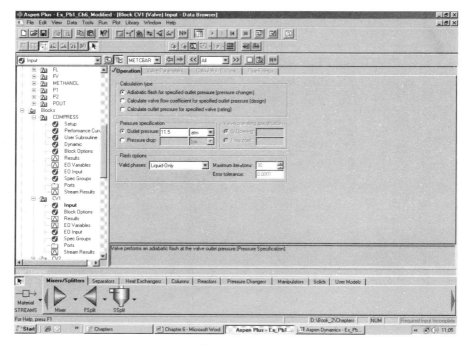

FIGURE 6.9(b)

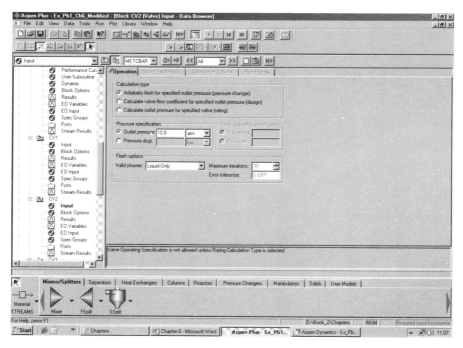

FIGURE 6.9(c)

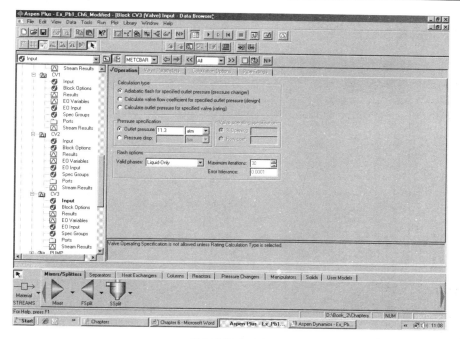

FIGURE 6.9(d)

When the data entry for the feed pump is complete, the window should look like Figure 6.10.

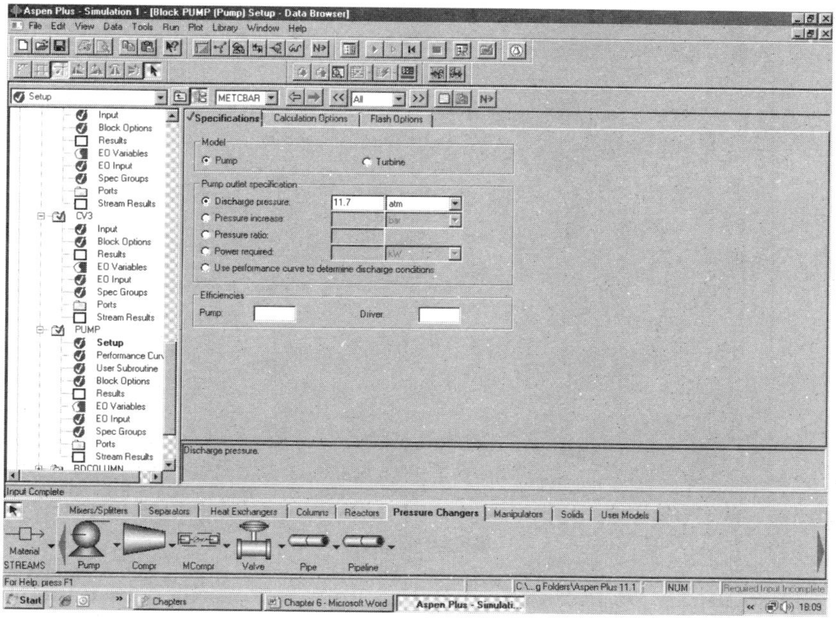

FIGURE 6.10

In the list on the left, choose *Blocks/RDCOLUMN/Setup* to fill up *Configuration* sheet (see Figure 6.11).

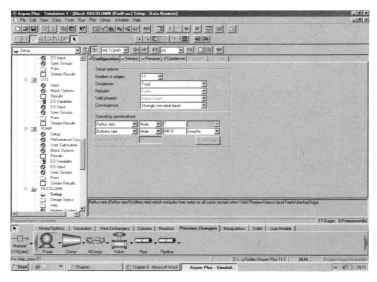

FIGURE 6.11

Streams sheet defines both the feed streams (see Figure 6.12), FL and FV, as well as product streams, P1 and P2.

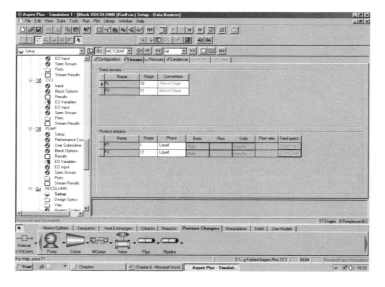

FIGURE 6.12

The pressure profile of the sample RD column is described in window shown in Figure 6.13.

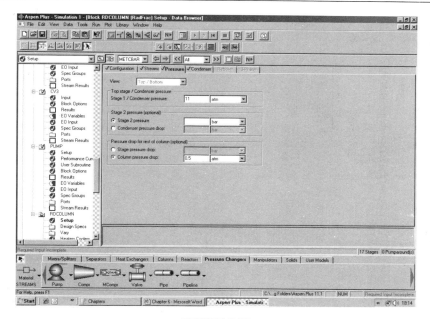

FIGURE 6.13

In the left pane of the Data Browser window, select *Blocks / RDCOLUMN / Reactions*. Filling out *Specifications* and *Holdups* sheets, we have two windows as shown in Figures 6.14(a) and (b).

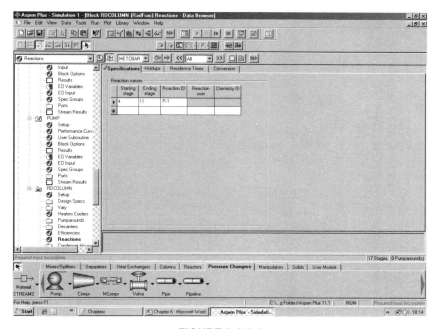

FIGURE 6.14(a)

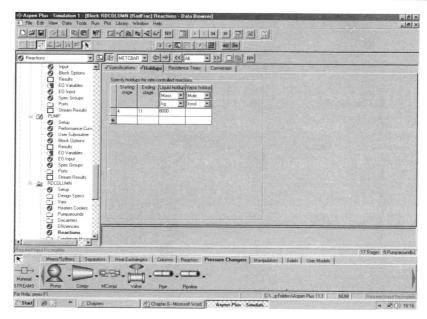

FIGURE 6.14(b)

Select *Pack Rating* under *RDCOLUMN* of *Blocks* folder. Creating a new ID, '1', and specifying the packing section as well as packing characteristics, we obtain Figure 6.15.

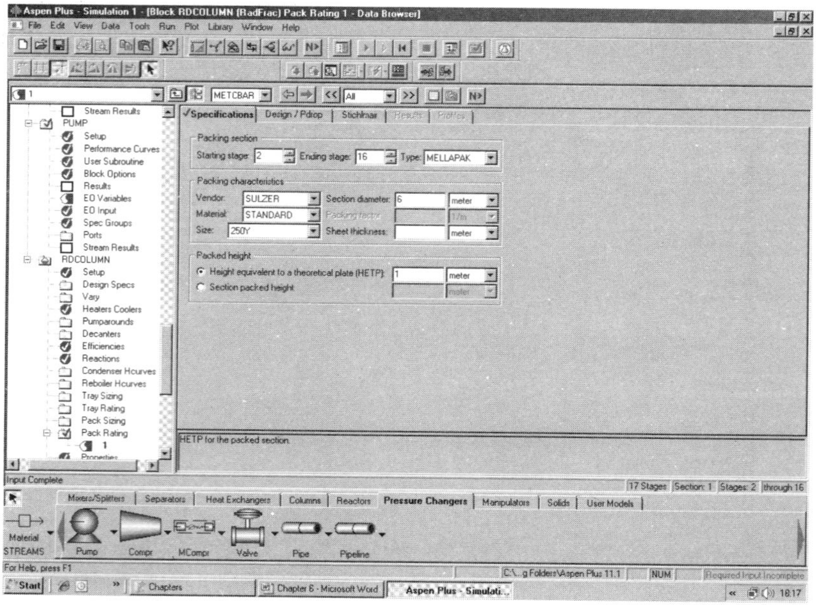

FIGURE 6.15

Choose *Blocks/RDCOLUMN/Convergence* and fix up the maximum iterations to 200 (see Figure 6.16).

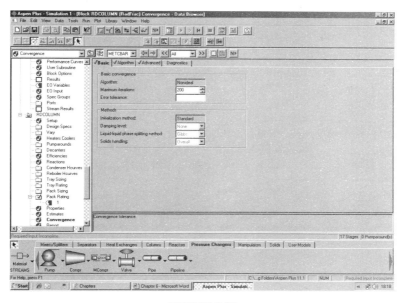

FIGURE 6.16

In the next, click *Dynamic* tab under *Blocks/RDCOLUMN*. The design specifications of the reflux drum and sump are reported in Figures 6.17(a) and (b).

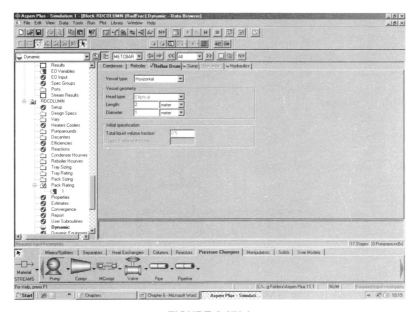

FIGURE 6.17(a)

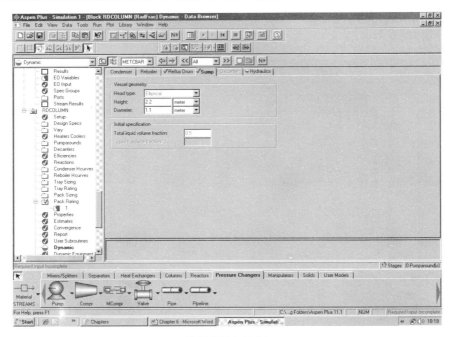

FIGURE 6.17(b)

Hydraulics sheet incorporates the information displayed in Figure 6.18.

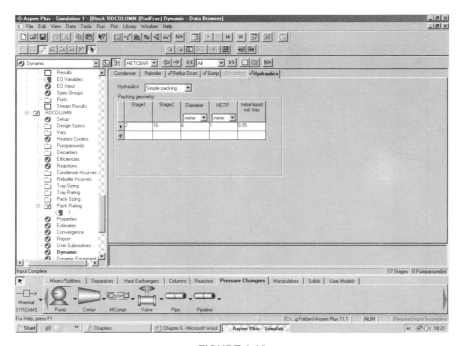

FIGURE 6.18

Hit *Next* icon to open the *Reactions* folder. For the forward reaction (Reaction No. 1) and the backward reaction (Reaction No. 2), the stoichiometric coefficients and exponents are defined under 'Kinetic' *Reaction type* in the two sheets as shown in Figures 6.19(a) and (b).

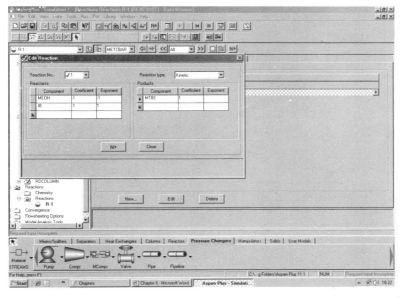

FIGURE 6.19(a)

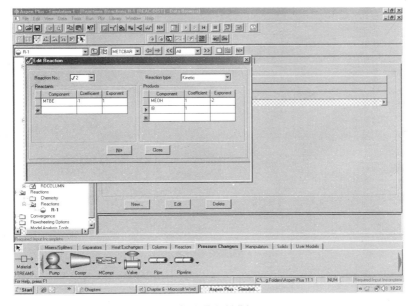

FIGURE 6.19(b)

The Power law kinetic data for both the reactions are provided in Figures 6.20(a) and (b).

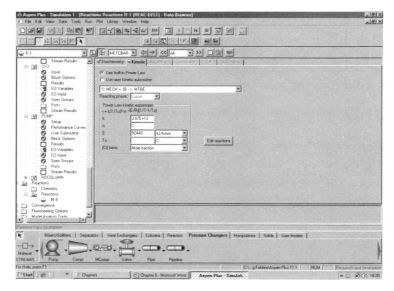

FIGURE 6.20(a)

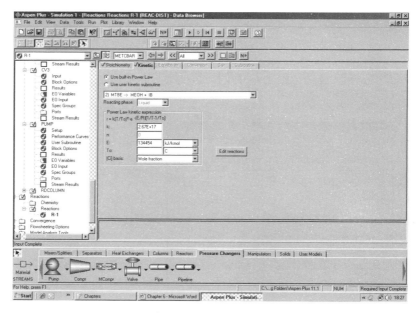

FIGURE 6.20(b)

Running steady state simulation and viewing results

As we hit *Next* knob followed by *OK*, *Control Panel* window pops up. Under *Results Summary/Streams*, the results are displayed in Figure 6.21.

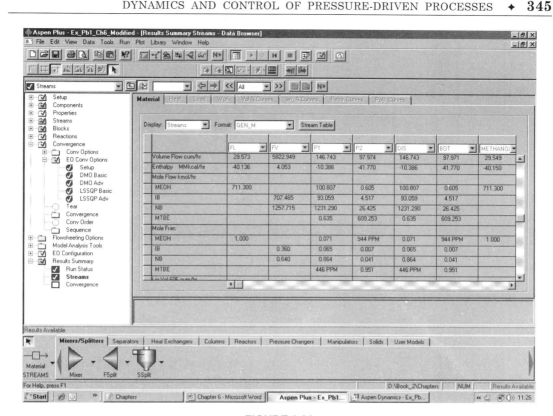

FIGURE 6.21

The mole fraction of MTBE in BOT stream is computed as 0.951.

(b) Exporting dynamic simulation: In order to conduct the dynamic process simulations, export the steady-state Aspen Plus simulation into Aspen Dynamics with saving as a pressure-driven dynamic file.

Opening existing simulation

As we press the *Start* knob, point to *Programs*, then *AspenTech*, then *Aspen Engineering Suite*, then *Aspen Dynamics Version* and then select *Aspen Dynamics*, a blank dynamic simulation window appears. In the next, open the pressure-driven dynamic file saved earlier. The screen looks like Figure 6.22.

It is obvious that the process flowsheet includes the automatically inserted two level controllers (LC1 and LC3) and one pressure controller (PC2). Each of these controllers has an operator set point (SP), a process variable (PV), also known as *controlled variable*, and a controller output (OP), also called as *manipulated variable*, whose values are obtained from the Aspen Plus simulation. These control structures also have their own tuning parameters, and so on, suggested by Aspen Dynamics. However, there is a scope to modify (or remove) the controller and its related items.

The Aspen generated control loops defined below should be used in the closed-loop study of the prescribed catalytic distillation column.

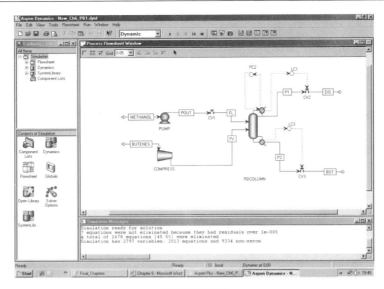

FIGURE 6.22

Loop 1

Controller: LC1

Type of controller: proportional (P)-only (since reset time is very large)

Controlled variable: liquid level in the reflux drum

Manipulated variable: distillate (DIS) flow rate (percentage opening of valve CV2)

Controller action: direct

Use all default data, except proportional gain of 2 (suggested by Luyben, 2004)

Loop 2

Controller: PC2

Type of controller: proportional integral (PI)

Controlled variable: top stage pressure

Manipulated variable: condenser heat removal

Controller action: reverse

Use all default data (suggested by Luyben, 2004)

(*Note:* It is known to us that when the proportional gain (K_C) is negative, the controller is said to be direct acting. To select the action of a distillation column pressure controller in Aspen Dynamics, we simply take an example of the proportional controller for discussion. The control law is defined as: $Q_C = Q_{CS} + K_C (P_{SP} - P)$, where Q_C is the condenser heat removal and P denotes the pressure to be controlled. Assuming direct control action, the controller equation can be rewritten for Aspen Dynamics as: $Q_C = -47.48 - K_C (P_{SP} - P)$, where 47.48 is the bias signal (Q_{CS}) and its negative sign indicates heat removal (cooling operation). If we move from steady state position, it is clear that when pressure (P) increases, the error ($P_{SP} - P$) value becomes negative. And ultimately, the negative value of Q_C decreases. Originally, the negative value should increase because if pressure increases, there is a need to increase the heat removal rate. Therefore, our assumption is wrong and it should be reverse action in Aspen Dynamics.)

Loop 3

> Controller: LC3
>
> Type of controller: P-only
>
> Controlled variable: liquid level in the column base
>
> Manipulated variable: bottoms (BOT) flow rate (percentage opening of valve CV3)
>
> Controller action: direct
>
> Use all default data, except proportional gain of 2 (suggested by Luyben, 2004)

Configuring new control loops

The primary objective of the example process is to produce a bottom MTBE product of high purity. To achieve the desired product purity in presence of disturbance and uncertainty, several control algorithms need to be employed with the reactive distillation.

It should be noted that in the control system of a RD process, the liquid level and column pressure controls constitute inventory control, maintaining the basic operation of the column. Thus, here emphasis is placed on the response of composition control methodologies to maintain product quality as well as correct stoichiometric ratio between the feed streams. In the following, different control schemes have been discussed for three distillation sections, namely feed section, top section and bottom section.

Feed section

For a chemical reaction with two reactants, the type of flowsheet depends on whether we want to operate the catalytic distillation column with *no-excess* of either reactant or *excess* reactant (Kaymak and Luyben, 2005). For a double-feed RD column, if there is any imbalance in the inflow of the two reactants ('excess reactant' case), the product purity drops. This is because one of the reactants becomes excess and exits with the product stream, and this stream would have to be further processed to purify the product and recover the reactant for recycle. Obviously, the 'excess reactant' flowsheet requires at least two separating columns and is therefore more expensive. However, it is easier to control. On the other hand, the 'no-excess reactant' flowsheet has better steady state economics but presents challenging control problems because of the need to precisely balance the stoichiometry of the reaction.

Several control structures used to maintain the correct stoichiometric ratio of the reactants have been proposed by researchers (e.g., Al-Arfaj and Luyben, 2000; 2002; Wang et al., 2003). To meet this control objective, the controller requires some type of feedback of information from within the process to indicate the accumulation or depletion of at least one of the reactants. This can simply be done by the use of an internal composition controller by manipulating the flow rate of one of the fresh feeds. There are also other efficient control techniques (e.g., cascade control, inferential control) reported for stoichiometric balancing (Wang et al., 2003). However, it is not practical to simply ratio the two feed streams, as has been proposed in some of the literature papers. Flow measurement inaccuracies and feed composition changes doom to failure any ratio controller that does not somehow incorporate information about compositions inside the reactive system and feed this information back to adjust fresh feed.

For the concerned distillation column, the methanol composition is controlled on 10th stage by the adjustment of the methanol fresh feed. The butene feed rate is flow controlled. It is worthy to mention that manipulating the methanol feed to control an internal methanol composition is preferred when the butene feed coming from the upstream units is not free to be adjusted. If this is not the case, then alternatively the *iso*-butene concentration, instead of methanol concentration, may be controlled on a reactive stage by adjusting the butene feed rate.

We are now moving on to configure the composition controller for methanol feed. To do this, click on expand symbol (+) of *Dynamics* subfolder. Then again hit expand button of *ControlModels* icon. Subsequently, select the PID object, drag it to the flow diagram, place the control block near to CV1 block and rename it as CC4. In the next, expand *Stream Types* and use *ControlSignal* icon to complete the CC4 configuration, shown in Figure 6.23. Chapter 5 presents a detail of how to configure a control structure in Aspen Dynamics.

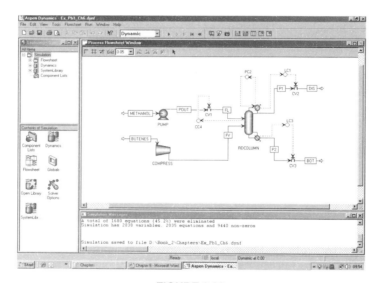

FIGURE 6.23

A little detail of the composition control loop for methanol feed is demonstrated below.

Loop 4

Controller: CC4

Type of controller: PI

Controlled variable: liquid phase mole fraction of MeOH on Stage 10

Manipulated variable: fresh methanol (FL) flow rate (percentage opening of valve CV1)

Controller action: reverse

Before executing the simulation run, it is customary to have a look on the data sheet. For this, double-click on CC4 control block and then press *Configure* knob in the faceplate to open the *Configure* dialog box. As mentioned in Chapter 5, it is wise to click on *Initialize Values* button. Still one doubt is there: is the value of process variable (PV) displayed same with the steady state liquid phase concentration of MeOH on Stage 10 obtained in the Aspen Plus simulation? Be sure about it, choose *Blocks/RDCOLUMN/Profiles* with opening the Aspen Plus simulation file. Then select 'Liquid' in the *View* field in *Compositions* sheet and obtain the table shown in Figure 6.24, with liquid mole fraction of MeOH on 10th stage of 0.04886022. This value is identical with that of PV in the *Configure* dialog box.

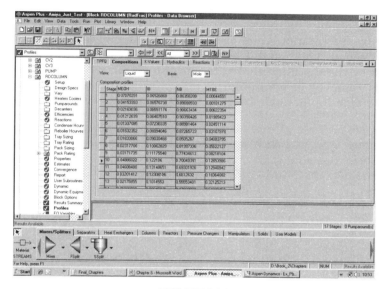

FIGURE 6.24

The controller CC4 is tuned by trial-and-error approach and the parameter values have been chosen as:

Proportional gain = 5%/%

Integral time = 5 min

Use default values for other items including bias signal, ranges, etc.

Notice that by the similar way, we can design the flow controller for butene feed of the RD column.

Top section

In addition to the LC1 and PC2 control structures, the distillate composition can be controlled by manipulating the reflux flow rate. In an alternative approach, along with the pressure control (PC2), we can control the reflux drum level by the manipulation of the reflux rate and the distillate flow rate can be adjusted by a ratio control law to give a constant reflux ratio. In the present case, the former control scheme has been incorporated for performance study.

Bottom section

In the bottom section of a distillation column, it is a common practice that either the bottom product purity or the tray temperature near the bottom of the column, which has a strong correlation with the product purity, is controlled at its desired value by the manipulation of the reboiler heat duty. For the sample process, we have implemented a composition control structure for product quality control.

As the CC4 control block has been connected, similarly we can incorporate the other control structures discussed above with the distillation flowsheet. The window, shown in Figure 6.25, includes a closed-loop scheme in which the MTBE purity is controlled in the bottoms by adjusting the reboiler heat input and the methanol impurity in the top is controlled by manipulating the reflux flow rate. As stated earlier, the concentration of methanol on the reactive stage it is being fed to (Stage 10) is measured and controlled by the manipulation of the fresh methanol feed rate. The butene flow rate is flow-controlled. The liquid levels in the reflux drum and the base of the column are maintained by the distillate flow rate and the bottoms flow rate, respectively. The condenser heat removal is manipulated to control the column pressure. All of the structures are single-input/single-output (SISO) structures with PI controllers (P-only on levels).

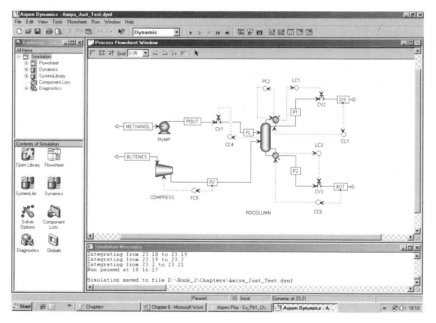

FIGURE 6.25

The details of control Loops 5, 6 and 7 are presented below.

Loop 5

      Controller: FC5

      Type of controller: PI

Controlled variable: molar flow rate of butene feed (FV)

Manipulated variable: brake power (shaft power or brake power of motor or engine required to drive a compressor)

Controller action: reverse

Proportional gain = 0.5%/%

Integral time = 0.3 min

Use default values for other terms

Loop 6

Controller: CC6

Type of controller: PI

Controlled variable: MTBE mole fraction in the bottoms

Manipulated variable: reboiler heat input

Controller action: reverse

Proportional gain = 5%/%

Integral time = 5 min

Use default values for other terms

Loop 7

Controller: CC7

Type of controller: PI

Controlled variable: MeOH mole fraction in the distillate

Manipulated variable: reflux rate (mass flow)

Controller action: reverse

Proportional gain = 5%/%

Integral time = 5 min

Use default values for other terms

Now the flowsheet is ready for closed-loop performance study. Start the program as usual. It is important to mention that to restart a dynamic simulation, click 'Restart' (F7) from the *Run* menu or press 'Re-start Simulation' button on the *Run Control* toolbar.

Performance of the closed-loop RD process

In the present study, two consecutive step changes in methanol feed temperature (46.85 → 40°C at time = 1.7 hours and then 40 → 46°C at time = 3.9 hours) have been introduced to examine the performance of the closed-loop RD process. A change in feed temperature affects the internal composition in the reactive zone. This, in turn, may deteriorate the product quality. The system responses to temperature disturbance are illustrated in Figure 6.26. It is obvious that the proposed structure is able to maintain the MTBE purity in the bottoms under the influence of disturbance variable. It can also prevent excessive losses of both methanol and iso-butene in the products.

Each Aspen Dynamics model includes different plots and tables from which we can easily access the simulation inputs as well as results. For this, first highlight a block or stream, then right-click to point *Forms* and finally select the item that we want to access.

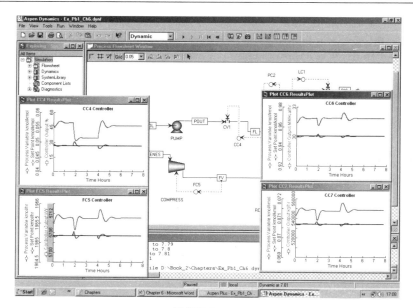

FIGURE 6.26

Performance of the closed-loop RD process with measurement lags

Aspen Dynamics screen, shown in Figure 6.27, includes three dead time blocks (DT1, DT2 and DT3) connected with three composition controllers (CC4, CC6 and CC7).

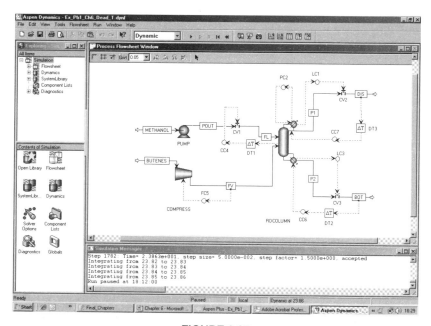

FIGURE 6.27

The measurement lag of 15 sec (0.25 min) is used in all composition loops. To incorporate a dead time for a measured variable, say methanol mole fraction on Stage 10, highlight DT1 block, right-click on the block, point to *Forms* and then select *Configure* to open the configure table. In the *Value* cell, enter 0.25 min as a sensor dead time. Follow the same approach for other two dead time blocks.

Here, we have used the proportional gain of 1%/% and integral time of 20 min for all composition controllers. The effects of disturbance in butene feed temperature have been depicted in Figure 6.28.

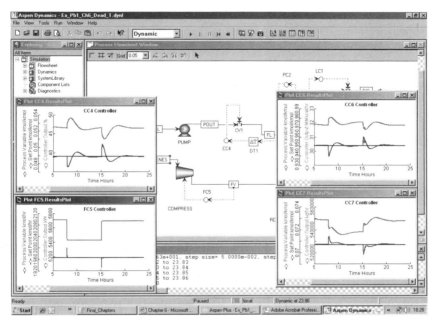

FIGURE 6.28

Initially a step decrease (76.85 → 65°C at time = 8 hours) and subsequently a step increase (65 → 76°C at time = 15 hours) have been considered in the simulation study. The developed closed-loop process flowsheet responds satisfactorily under load variable change and measurement lag.

SUMMARY AND CONCLUSIONS

In Chapter 5, we have studied the dynamics and control of the flow-driven chemical processes. Here, a case study has been conducted on a MTBE catalytic distillation column using the pressure-driven dynamics. The complete process flow diagram includes a distillation column, a feed compressor, a feed pump and three control valves. In the MTBE synthesis process, a bottom product containing high-purity MTBE and a top product enriched with *n*-butene are obtained. To maintain the MTBE purity in the bottoms stream, several control structures have been configured with the flowsheet in

Aspen Dynamics. All of the structures are SISO schemes with PI controllers (P-only on levels). The controllers have been tuned by simply using heuristics. The proposed closed-loop process provides satisfactory results under disturbance input and measurement lag.

PROBLEMS

6.1 A binary mixture of ethanol and 1-propanol enters a flash drum (Flash2). The feed specifications are shown in Figure 6.29 with the process flow diagram.

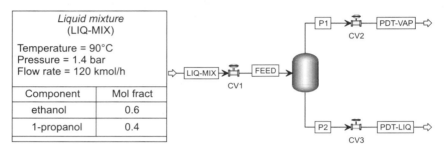

FIGURE 6.29 A flowsheet of a flash drum.

The flash chamber operates at 90°C and 1.2 bar. The vertically placed separator with a length of 2 m and diameter of 1 m has elliptical head type. All the control valves have a pressure drop of 0.2 bar. Applying the RK-Soave thermodynamic model as a base property method,

(a) simulate the flowsheet to obtain the product compositions,

(b) design the two control schemes to maintain the pressure and liquid level in the flash chamber, and

(c) examine the performance of the designed controllers.

6.2 Styrene is produced by dehydrogenation of ethylbenzene. Here we consider an irreversible reaction:

$$C_6H_5 - C_2H_5 \rightarrow C_6H_5 - CH = CH_2 + H_2$$

$$\text{ethylbenzene} \qquad \text{styrene} \qquad \text{hydrogen}$$

The process flow diagram that consists of a reactor (RSTOIC), a feed compressor (COMPRESS) and a control valve (CV) is shown in Figure 6.30.

An isentropic compressor discharges the FEED stream that enters the RStoic reactor at 2 bar. The reactor runs at 260°C and 2 bar. The control valve involves a pressure drop of 0.2 bar. Use the fractional conversion of ethylbenzene equals 0.8. Applying the Peng-Robinson thermodynamic method,

(a) simulate the flowsheet, and

(b) observe the closed-loop process response employing the flow controllers.

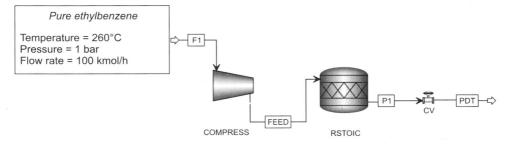

FIGURE 6.30 A flowsheet for the production of styrene.

6.3 The hydrogenation of aniline produces cyclohexylamine in a CSTR according to the following reaction:

$$C_6H_5NH_2 + 3H_2 \rightarrow C_6H_{11}NH_2$$

aniline   hydrogen   cyclohexylamine

The complete process flowsheet is provided in Figure 6.31. It includes a pump having a discharge pressure of 41.2 bar, an isentropic compressor having a discharge pressure of 41 bar, an elliptical head-type vertically placed reactor having a length of 1 m and three control valves with a pressure drop of 0.2 bar in each.

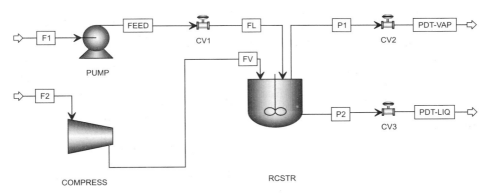

FIGURE 6.31 A flowsheet for aniline hydrogenation.

The reactor operates at 41 bar and 120°C, and its volume is 1200 ft^3 (75% liquid). For the liquid-phase reaction, the inlet streams, F1 and F2, are specified in Table 6.2.

TABLE 6.2

Reactant	Temperature (°C)	Pressure (bar)	Flow rate (kmol/h)
Pure aniline (F1)	40	7	45
Pure hydrogen (F2)	− 12	7	160

Data for the Arrhenius law:

Pre-exponential factor = 5×10^5 m^3/kmol·s

Activation energy = 20,000 Btu/lbmol

$[C_i]$ basis = Molarity

Use the SYSOP0 base property method in the simulation. The reaction is first-order in aniline and hydrogen, and the reaction rate constant is defined with respect to aniline.

(a) Simulate the flowsheet to compute the product compositions,

(b) configure the control schemes for maintaining the liquid level, pressure and temperature in the CSTR, and

(c) investigate the closed-loop process response under any disturbance input.

6.4 Repeat the above problem with adding a time lag of 0.2 min in temperature measurement and carry out the closed-loop process simulation to report the disturbance rejection performance of the developed scheme.

6.5 In addition to the level, pressure and temperature controllers, include the flow controllers with the flowsheet, shown in Problem 6.3, and inspect the closed-loop process response.

REFERENCES

Al-Arfaj, M.A. and W.L. Luyben (2000), "Comparison of Alternative Control Structures for an Ideal Two-product Reactive Distillation Column," *Ind. Eng. Chem. Res.*, 39, pp. 3298–3307.

Al-Arfaj, M.A. and W.L. Luyben (2002), "Control Study of Ethyl *tert*-Butyl Ether Reactive Distillation," *Ind. Eng. Chem. Res.*, 41, pp. 3784–3796.

Jacobs, R. and R. Krishna, (1993) "Multiple Solutions in Reactive Distillation for Methyl *tert*-Butyl Ether Synthesis," *Ind. Eng. Chem. Res.*, 32, pp. 1706–1709.

Kaymak, D.B. and W.L. Luyben (2005), "Comparison of Two Types of Two-temperature Control Structures for Reactive Distillation Columns," *Ind. Eng. Chem. Res.*, 44, pp. 4625–4640.

Luyben, W.L., (2004) "Use of Dynamic Simulation to Converge Complex Process Flowsheets," *Chemical Engineering Education*, pp. 142–149.

Rehfinger, A. and U. Hoffmann (1990), "Kinetics of Methyl Tertiary Butyl Ether Liquid Phase Synthesis Catalyzed by Ion Exchange Resin-I. Intrinsic Rate Expression in Liquid Phase Activities," *Chem. Eng. Sci.*, 45, pp. 1605–1617.

Seader, J.D. and E.J. Henley (1998), "*Separation Process Principles,*" John Wiley & Sons, Inc., New York.

Wang, S.J., D.S.H. Wong and E.K. Lee (2003), "Control of a Reactive Distillation Column in the Kinetic Regime for the Synthesis of *n*-Butyl Acetate," *Ind. Eng. Chem. Res.*, 42, pp. 5182–5194.

Index

POWER AND PRINCIPLES OF THE RUNES

POWER AND PRINCIPLES OF THE RUNES

FREYA ASWYNN

THOTH PUBLICATIONS
LOUGHBOROUGH, LEICESTERSHIRE

A CIP catalogue record for this book is available from the
British Library.

Cover Design by Avanson Design

ISBN 978-1-870450-23-2

Printed and bound in Great Britain

Published by Thoth Publications
64, Leopold Street, Loughborough, LE11 5DN
web address: www.thoth.co.uk
e-mail: enquiries@thoth.co.uk

CONTENTS

ACKNOWLEDGMENTS

A vote of thanks to:
Stormerne and Arlea Anschutz Hunt for helping out
with proofreading and computer wars.
For Kveldulfr as ever for his continued friendship and support.
For Andrew Clifton who did the proofreading
for the original edition.
And to all my students who keep inspiring me with their original
ideas, especially Julia; John; Brian; Sue and Victoria
my Alpha Crew
May the Gods shower their blessings on all the above mentioned.
Wassail

ABOUT THE AUTHOR

Freya Aswynn was born in Holland in an outwardly respectable Roman Catholic family. As a child she displayed natural psychic ablities and this, combined with her reactions to a hostile family environment, prompted the official diagnosis that she was 'maladjusted'. Entrusted to the so-called 'care' system, she was institutionalised for nine years. Her intelligence, combined with a capacity for aggressive self-assertion, allowed her to emerge from this ordeal more or less psychologically unscathed at the age of nineteen. Deprived of any secondary education, she resorted to taking on cleaning work, until she met and married her first husband, George. Somewhat older in years, George introduced her to the basics of philosophy (particularly the works of Friedrich Nietzsche) and classical music, predominantly Richard Wagner. He also was the first person who ever spoke to her about the esoteric nature of the runes. Sadly, after two years of marriage, George died of cancer; but he left Freya with a pension that, for the first time, gave her a measure of financial independence.

Determined now to make good her lack of learning, Freya taught herself English and German by means of books and tapes. At the same time, she took an interest in the development of her long-suppressed psychic gifts. She received her first paranormal training in a spiritualist environment, and from there progressed to the study of Rosicrucianism, astrology and Cabbala. At the age of 30, she felt unable to progress further in Holland and left for England, where she was soon recognised by prominent magicians and witches as a natural. With their help and training, she was initiated into the mysteries and made rapid progress as a ritual magician.

The next major turning-point for Freya was a life-changing spiritual and magical experience: an intense, spontaneous invocation of the God Woden. She took this to be a calling to open up the Northern pathway, which until then, was virtually nonexistent within the occult community. She embarked on an intensive study of the runes and began work on her first book, *Leaves of Yggdrasil* (revised and updated in 1998 under the title, *Northern Mysteries and Magick*). Within Pagan circles, this initiative was at first viewed with scepticism and suspicion as to whether she had a covert political agenda (which at that time was often the case with Odinists). Freya soon overcame these perceptions and won the respect and affection of the majority of the Pagan and occult community.

On a social level, realising that she was not suited to living in a normal nuclear family, Freya became actively involved with a community named Enclave Ex which she ran for a number of years as Managing Director, together with her partner Lionel Hornby and their friend Alison Behr, a well known Jewish feminist activist and a supporter of gay and bisexual rights. During this period, Freya decided to choose a celibate path, for her own spiritual reasons. After Lionel's death in 1994, Freya sold her share of Enclave Ex and in 1996 she bought a large farmhouse in Scotland and founded another community named 'Gladsheim' which has had its share of problems and setbacks, but as of now (2007) a real nice stable and committed community.

In 1993, Freya had started up a UK branch of the Ring of Troth, originally an American organisation founded by Edred Thorsson and James Chisholm. The UK branch soon attracted members in other European countries and, re-named the Ring of Troth Europe, rapidly became a successful and progressive Northern Tradition group. Having led this organisation as Steerswoman for a number of years, Freya stepped down from this role and other positions of power. She now concentrates on teaching and guiding by means on running a correspondence course. Freya also continues to be available for talks, workshops and readings.

FOREWORD

You look up as you reach the brow of the small green hill, and see smoke rising gently on the light breeze. A figure is there, arms raised, hands outstretched, seemingly at one with the elements. Among the stones her staff stands, scribed with sigils that seem to pull your attention. She smiles into the light of the low sun whilst behind her, arching across grey thunder clouds, breathes a perfect rainbow. Slowly, she begins her call, a chant so primal that you feel you and all the rest of the world must hear – and respond...

There is something irresistibly attractive about the magic that surrounds runes. They appeal to our intellect and to our spirit. They challenge us to find anything significant in what surely, we insist, is merely an alphabet. And they survive, despite all the abuses and forgetfulness they've suffered down the centuries.

What is it about runes that compels our attention? And more to the point, once our attention has been seized, where can we go to satisfy the need to know more about them?

When I first heard that call in the 1970s and began my own rune studies, there was scant information available. Nowadays, by contrast, there is much written on runes - perhaps too much, as my own 'black library' testifies - and a student must discriminate so as not to waste time. The best solution would be to seek out a personal teacher, but trustworthy teachers are few and hard to find. So how to distinguish one book from another?

Above the cloth, her hand moves and casts brief shadows as runes tumble. Here and there, their angular forms evoke deep feelings and memories, demanding attention. It seems as though they are laying bare parts of your life, and all the steps that have brought your life to this point. Silently she considers the patterns revealed and then, without raising her eyes, she begins to tell you of the ways your wyrd is woven...

One of the things that first attracted me to Freya's writing was the balanced approach she took. It had always seemed a very odd thing to me, and still does, that so many other people associated runes exclusively with divination. The fact that they can be used successfully for divination is undisputed, but what a waste to be limited by that! It is but part of their power, and Freya knows their wider potential for magic and personal growth.

By her own definition, Freya Aswynn would call herself an occultist. She combines the objective with the subjective, for she knows both the mythology and the fruits of academic research and yet she is quick to express her insights and inspirations. While she is utterly devoted to the god Odin and she's keen to uphold what she sees as the Heathen folk ethic, she is also a pragmatist and will not waste an available resource. These things in combination give her writing a breadth rarely found elsewhere. They create a unique style and show an approach to runes that demands to be taken into consideration. Like her or not, if you are serious about rune study you must read her work!

Later, but more confident now, you stand alone as you let the runic forces sweep within you, remaking and unlocking potential only glimpsed before. She may have ploughed the land and sown the seed for you, but it is you who have tended it and brought it to harvest. You stand transformed, and now you can choose where to apply that power and for whom...

Over the years I've learnt more and more about runes. I suspect it's a never ending journey for all who become hooked on them. I've written and lectured about runes and formed a teaching-study group called *Rune Ring*. As part of course work, I often set extracts from Freya's previous books for the group to critique, and will continue to do so. Critique? Yes...

There's no getting away from the fact that studying runes will eventually lead you to come into contact with the old Heathen gods. And whatever form that takes, you will probably come to see them as gods you must stand up to. You don't grovel before a Heathen god – they would only laugh at you with contempt. And you don't take everything they say as 'gospel'. You listen and you question and you think before acting.

In the same way I suggest you discriminate when you read this book. You won't agree with everything out of hand anyway. That's good: Freya doesn't want puppets. Even though you are benefiting from over twenty years of her hard work and experience with runes, who is to say that one day you won't stand on her shoulders and see still further? Think about what you read and question it. That way you give yourself the best opportunity to grow, and that way what we all know about runes will grow too.

You stand on the brow of a small green hill while smoke rises gently on the light breeze. Your arms are raised, your hands outstretched, and you feel at one with the elements. Your staff stands among the stones, scribed with the sigils that have changed your life – that continue to change your life and the lives of others who come to you. Climbing up the hill towards you is someone who has heard your song, someone drawn despite themselves...

On my shelves is an old copy of the *I Ching*, the Chinese Book of Changes, which was the first divination tool I ever studied back in 1973. My Richard Wilhelm translation and has a foreword by Carl Jung and in this foreword, Jung uses *I Ching* divination to enquire about the *I Ching* itself. I thought I might do a similar thing about this new edition of Freya's book, and I cast *Jera* and *Ansuz*.

What these runes tell me is that effort in studying this book will bear fruit, and that communication with the runes will be strengthened as a result. I believe this will be the case whether you're seeking to find out about runes for the first time, or whether you're more experienced with runes and want to get insights from a proven master, or whether you have students of your own and want to give them a book that will help them grow.

This book is full of practical hints and useful insights and I recommend it to anyone to feels the pull of runes and who wants an insight into the mind of a dedicated and experienced rune worker.

Stormerne Hunt-Anschütz
stormerne@here.is
Scotland, 2006

INTRODUCTION.

This is the updated version of 'Principles of Runes' previously published by Thorsons. The original text is preserved in it's totality and augmented with some newer insights and added material.

Freya Aswynn 2007

ONE

hISTORICAL ORIGINS OF THE RUNES

In Middle-Earth, the runes probably became known to humans sometime between the first century B.C.E. and the end of the first century C.E. Some Rune signs are however much older than that and have been found all over the place from Indo-European settlements to new Grange in Ireland. Their shapes - vertical staves with slanted cross-cuts - show that they were designed to be cut into wood, but because of difficulties in preservation, the oldest inscriptions that have survived are in metal and stone (with the exception of some Frisian staves which were preserved in peat bogs). Some elements of runic writing are almost certainly derived from Mediterranean alphabets (Roman, Greek, and North Italic have all been discussed as hypothetical models); others from the Bronze Age stone-signs of the Germanic peoples. Their use and understanding, however, are thoroughly Germanic; and as tools of magic, suitable for memorial and operative inscriptions, they were naturally known as the gift of the god Wodhanaz to his people. According to the Norse poem "Hávamál" Odin won the runes through a shamanic initiation of fasting and death upon the world-tree, through which he reached down to the deepest roots of the origins of the cosmos to draw up their holy pattern. The pattern of Wyrd.

The oldest runes known to us are those of the Elder Futhark ("futhark" is the runic equivalent to "alphabet", derived from the first six runes - F, U, Th, A, R, and K/C), a 24-rune pattern divided into three aetts ("eights" or "families"). The impotance of the runic order and the division into aetts was so strong that even when the Scandinavian futhark (Younger Futhark) had been reduced to sixteen runes, not only the order but the division of the runes into three aetts survived, even though they no longer filled three "eights". The worst offence any modern practitioner has ever perpetrated against runic lore is to confuse the holy rune-row and change the ancient order of the futhark according to whim, destroying the pattern which held such great meaning to our forebears and generating a huge burst of destructively chaotic energy in those unfortunate and gullible enough to accept this warped version without bothering to refer to any other works on the runes. It is the purpose of this book to bridge the gap between advanced books such as *Northern Mysteries*, by myself, *Teutonic Magic,*[1] by Kveldufr Gundarsson, *Futhark* and *Runelore*, by Edred Thorsson, and the mediocre rubbish which is so common place today. In this book I strive to make the runes accessible to those people who are sensitive to their energies and meanings without necessarily having to wade through the scholastic or deeply esoteric material available today, whilst maintaining the integrity of the runes and their applications.

As the original language of the North (Proto-Germanic or Common Germanic) developed in several different directions, the futhark also changed to reflect the shifting sounds of different dialects, though the basic order was always preserved. In North Germany, the Angles and Frisians added several runes to the futhark, a development which ultimately ended in the thirty-six rune Anglo-Saxon futhork (the old a-rune taking on the sound-value of o). In Scandinavia, as mentioned above, the system was simplified to sixteen runes. Like the futhark order and the division into aetts, the names of the individual runes were generally preserved, though the Anglo-Saxon futhork, written down in the Christian period, is thought to have substituted similar words for the rune-names Oss (Ás, "god") and Tiw (Týr, the god's name) to de-heathenise the original meanings. As the twentyfirst century opens, the runes enjoy a resurgence and are reaching a wider audience than at any time in the past thousand years. But that audience is different than that for which they were originally intended, and this has resulted in many different perspectives emerging and converging upon them, and consequently, originating a whole new set of interpretations based on the Old Lore but re-evaluated and updated to modern usage. There is nothing wrong with this as long as respect is paid to some of the basic principles which, although originating from our ancestral past, have stood the test of time, in particular with reference to the order of the futhark, the shapes of the runes, and their names (albeit in different languages).

ORIGINS OF THE RUNES MYSTICAL

Runes have been around outside time and space since the origins of the universe itself. The signs used in the various futhark or futhorcs are the Germanic embodiment of these mysteries, as is borne out by the primary meanings of the word Rune itself i.e. secret, mystery and song. The song or sound of the coming into existence of the material universe in so far as we may know and understand this.

The runes are the physical embodiment of the Mysteries of the origin of consciousness itself, in the form of Odin/Woden, the god mostly associated with the runes poetically. One could say that the runes are the DNA of Odin, all this within the context of Germanic understanding. It goes without saying that other tribal peoples have their own often similar systems of interpretation and understanding of these mysteries.

For the purpose of this book whenever I say 'runes' I mean the pictographic sigils as they are known today. Runes are a tool for communication, therefore they are a language, albeit a language of symbols rather than a language of letters. However they also function in the latter manner as they can be used adequately to write with as is know from the carvings and writings of the past. The term 'Runes' is often descriptive of obscure, cryptic, unidentifiable signs referred to in various science fiction, horror and fantasy literature. Most of the time these runes have nothing to do with the futhark runes (with the exception of course of Tolkien). These writers instinctively feel the hidden power of 'Runes' futhark or otherwise. Runes are a language of the most simple kind (used for mundane and even profane sayings), at the same time they are a language of the deepest collective unconscious, submerged in mystery and fraught with hidden dangers.

There is a dark Lovecraftian feel to 'Runes'. Some magicians may choose to work on these lines, I do not. I attempt to work with the runes for the advancement of understanding, personal spiritual growth, and communication with subtle realms of knowledge and power or might. Runes can be equally well used to create a message to your kinsman, asking him to 'feed the cat' as they can be used to create a message to your unconscious or the 'deep mind', which, after a period of gestation, manifests itself through the higher mind either as the result of an objective or an insight or a teaching. The three selves we can correlate with the Germanic three fold system of the soul (actually it's nine fold but that is a further division of three). I will be more precise in later chapters.

HISTORY OF THE RUNES, THEIR USE AND ABUSE FROM ANCIENT TO MODERN

We don't know much about the use of runes historically speaking. There are some artefacts preserved which date from second century e.v (era vulgaris) with inscriptions including archaic names of Germanic gods. These may have been hallowed items of magic. This would indicate that they were used in a religious and/or magickal context from at least the second century onwards (if not before) as a hidden tradition not accessible to everyone. By the time they emerge in the outer world in public, most magickal traditions have had already a hidden existence for some time. We'll never know for sure, but it is something to bear in mind. After all the runes appear almost overnight in a neatly outlined 24 row! Just like that? Not very likely. Someone or other has constructed the rune row in this format probably for a specific reason.

We know from the Roman Tacitus that runes or 'notea' were used for divination in a specific manner. When the Roman Christians enslaved and subverted our nations (with the help of some elements of the aristocracy, I must admit) the runes went underground, and much later in the Middle ages re-emerged as Masonic marks. In Iceland in the middle ages the possession of runes actually carried a death penalty similar to the death penalty for 'witchcraft' in the rest of Europe.

We still have similar silly laws. Possession of certain natural herbs and plants still carry penalties theoretically. Evolution of consciousness and freedom takes along time indeed. Perhaps 500 years from now we too will be viewed as a 'dark age'!

Back to the runes however. So, the runes were 'used' for Magick, religion, divination and writing. The runes were also abused, by the Nazis. They misappropriated them and attempted to pervert their meaning to fit in with their despicable political objectives, and, as is inevitable, they paid the price for their stupidity, proving that the runes are not to be messed with. Odinists and Asatruar (for the most part born after the war, and only being familiar with what went on by second hand education and information) often are immediately lumped in with the Nazi spectre, and often our integrity is pulled in to question more than is the case with other forms of Paganism.

SETTING THE RECORD STRAIGHT!

The runes are the most holy, mightiest and highest emanations of the consciousness of the Gods, whatever one perceives these to be. They are there for those who feel inclined as a mystery part of empowerment and evolution and no one human being may say No to any other, using the runes! 'Using the runes' I said. Misuse will carry its own retribution. Race, or sex bear absolutely no relevance to the use of runes! Culture may play a part, but certain cultures may interface and correspond with each other. So who can use Runes? Well, basically, everyone who feels the calling. It is very difficult to draw a hard fast line and even more difficult to enforce it. There are various opinions as to who may use runes and /or practice Asatru, the religion which often goes with it. Ultimately, if it's good enough for the gods, who am I to deny anyone?

I've heard or read all the arguments for a number of years now, both for and against 'Folkish' Asatru. It's assumed that this automatically implies some form of Fascism and/or White Supremacy and in some very few cases this is correct. As stated in the previous sentence, in most cases this is incorrect, the vast majority of Folkish Asatru are just people who wish to practice their religion within a group of like minded people who share a genetic and/or cultural back ground. Fortunately the real white supremacists are very few in numbers.

This small minority are very insecure of their racial identity (hence their need for an exclusive approach) and therefore don't need to be taken seriously. Some of those will, however, get very upset if it is pointed out to them that the Asatru is essentially part of the Native European Mysteries and their knowledge is ultimately derived from secondary sources if they were not born and bred in the Northlands, there's only so much which can be taught, the real connection one is born with and gain access through being brought up in a Germanic speaking environment. A few of my assertions relating to the meaning of runes, can not be objectively verified as they may not be in keeping with standard etymological meanings, but relate very well from a associative perspective working with a chain of words in various Germanic languages, utilising folk lore; sayings; proverbs derived from Dutch, my native language.

I remember a Black voodoo High Priest giving a talk in London at one of the Pagan Gatherings, afterwards we met up for a drink, and he told me how he made a statement about Voodoo more or less to the effect that it was for black people only and white folk had no business here. Well I know enough white folk who are very accomplished Voodoo or Santeria practitioners for that matter, but the thing that Colin and I agreed upon was that he could make a statement like this whereas I or any other Asatru/Odinist would not be able to do so, peacefully. The fact that he was perfectly safe to make this statement but if I, representing the Northern 'voodoo' so to speak, would have made a similar remark, I would have been fucked over by the press like nobody's business.

Yes, these are some dodgy groups with the wrong racial attitude, but they are very, very few, and they are so busy politicking they never find the time and discipline to use magick. So really there isn't a lot to worry about.

However, we must not forget that Asatru, Vanatru, the Northern Mysteries are essential an Northern European tradition, and what's right for one country is not necessarily right for another. Apart from the Native Americans everybody in the United States is an recent immigrant, some of their own choosing. Some poor buggers such as the Blacks had no choice at all. Therefore as the Runes are not native to America every American should be able to use them, certainly to learn them and incorporate them into what ever system they work. America is the ideal country to develop the concept I have coined: 'synergistic magick'. The 'Racial soul', to use a term borrowed from Dion Fortune, of the United States of America has actually very little to do with racial identity. It's evolving into something more, hopefully something better. After all the Americans more or less represent this planet as the hyper-power and have the responsibilities which go with that position. Perhaps the greatest gift of the Germanic Folk Consciousness as a whole, past, present and future, is that North, West, East and South Germanic tribes all collectively bequeathed the English Language as the planetary language. You can bet that when first contact occurs it will be in English!

In Europe this amalgamation of different racial souls has not yet occurred and may be avoided, and as an occultist I can see good reasons to keep it that way. I am not convinced that what is best for America would be necessarily best for Europe as the majority of people living in Europe are still native born, and therefore have more easier and deeper access to their own native mysteries than a non European. Perhaps the European directive is preservation and conservation and the American directive is change and integration. It's an interesting point of speculation but ultimately it's up to the Gods. Odin never was one to put all his eggs in one basket.

TWO

TRADITIONAL, ACCEPTED AND ESOTERIC MEANINGS OF THE RUNES

The oldest version of the runes is known as the Common Germanic (or Elder) Futhark. The word 'Futhark' derives from the names of the first six rune names. The 24 runes of the Elder Futhark are divided into three groups of eight known as Aettir (singular 'Aett'). 'Aettir' translates roughly as 'family' and this indicates that the eight runes in each aett form a group and have something in common.

The first Aett is named after Frey. The runes of this aett represent the forces of creation and establishing of the Gods, Giants, and all beings humans included.

FREY'S AETT

Germanic name: FEHU ᚠ
Anglo-Saxon name: FEOH
Old Norse name: FE
Phonetic value: F
Traditional meaning: Cattle; wealth

The classical meaning of Fehu as handed down through various sources is cattle or wealth, in particular, movable wealth, and therefore by implication, perishable wealth. In a so-called primitive society such as existed in Northern Europe at the time when the runes were first developed, cattle represented wealth.

The status of the chieftain was usually measured by the number of cattle which he or she owned. In those days, cattle provided both a livelihood and a barter medium used in much the same way as money is used nowadays. The present-day English word "fee," is a reminder of this name of this rune.

Today, the meaning of money itself has undergone many mutations. From coins made of metal and notes made of paper we are now moving into the era of digital money: plastic 'smartcards' and meaningless figures stored somewhere in cyberspace. Only time will tell whether this is an improvement. This brings to mind an overlooked but pertinent aspect of the Fehu rune, namely, it's relevance to the food chain. Cattle eat grain, as do we; and in spite of fears about BSE, most of us still eat beef - or at least milk, cheese and other dairy products derived from cattle. So, food is energy recycling in many forms, and, in fact, food is the real bottom line with the Fehu rune, on the material plane. Money is ultimately useless in itself: you can't eat paper or metal, let alone figures on some computer! I urge everyone who works with Fehu to emphasise food rather than money, for physical as well as spiritual well-being. Synchronistically, the shape of the Fehu rune resembles an ear of wheat.

Fehu on the basic level as we have seen represents money or wealth. Money is only the densest level meaning of this rune. Money does not come out of nothing, it's either earned or inherited. It's mobile. It moves around. The esoteric aspect of money is energy, energy that can be harnessed and used. It's higher meaning is the vital force of life, the initial spark, creative and mobile. The Fehu rune can be the key to accessing the personal powerbase from which money can be obtained. The greater power of which money is only a pale shadow is the 'hamingja' or personal power. Strong hamingja enables one to achieve whatever the objective is, including making money.

Fehu can endow you with this ability if that is within your 'örløg'. Hamingja as well as örløg, which very roughly corresponds to 'karma', can be passed on from one generation to the next. From there is it either added to or subtracted from by acts of honour or lack thereof. The hamingja is a storehouse of 'Luck' on the inner plane. There are people who inherit good luck and still manage to screw up and others who inherit bad luck and succeed. In everyday life the greatest influence on you hamingja is the esteem in which you are held by your peers, i.e. your reputation, and of course that includes self esteem which is often negatively programmed by upbringing. Thoughts too are energy and negative thought therefore will create a negative thought-form. The ultimate exercise of power using the hamingja is to be an ever vigilant and conscious participant in creating your own reality in the service of your gods and not allow others (usually with their own agendas) to 'program' this for you. Your Hamingja shapes your thought, shapes your words, shapes your deeds and ultimately shapes your personal wyrd. The Fehu rune is the first and foremost rune to call upon for help in any given situation, it functions a bit like a rescue remedy, it infuses you with power (psychic or magical) to deal with any problems on the spot.

The Fehu rune can be used to increase your hamingja and access this creative power source. Fehu can give the impetus to any intent followed by an act of will. It can be used in conjunction with other runes to forcefully break down negative programming and 'turn' your 'luck'. It won't necessarily help you win the lottery though!

The essential power of the Fehu Rune is the power of the first ripple in space or the out breath of the universe. This power is a continuing stream of creative force. Stepped down through the various planes to the accessible levels of the Astral plane it can be transferred from person to person or from person to object as in charging magical items.

Spiritual aspect: Life force.
Emotional or psychological aspect: self respect, personal power

Negative aspect: fear of failure, lack of self esteem.

Germanic name: URUZ ᚢ
Anglo-Saxon name: UR
Old Norse name: UR
Phonetic value: U
Traditional meaning: Aurochs

The Aurochs was a ferocious native species of wild ox similar to a European bison which is now extinct. In olden days, in Continental Germania, the young warriors were subjected to a test of strength, an ordeal in which they had to go out armed with only the most basic weapons and single-handedly slay one of these beasties, bringing back the horns as proof that they had succeeded. (The horns were highly prized as drinking vessels.) The hunting and slaying of the Aurochs with a spear or knife was a risky business, and probably the chances of success were not very great. Having succeeded, one presumably had passed a test of man making and was accepted into the tribe as a adult warrior. The task of hunting, tracking and slaying an Aurochs demanded strength, endurance, patience and above all courage, and these are the values embodied and encoded in the rune. Uruz is a direct representation of the life force pertaining to physical health. Characteristic values belonging to this rune are stamina, independence, endurance, the ability to take responsibility and the ability to control aggression.

Perhaps there is a shamanic connection with this rune such as exists in Native American buffalo lore. Before the religion of the Aesir in Germany, back in Neolithic times when the Aurochs was still roaming free in the forests, this animal may have been seen as a manifestation of the horned god of nature or the Wild Man of the Woods; a very early form, perhaps, of Wodan, or Frey.

The magical power of Uruz is health, vitality, strength on all levels. Whereas Fehu can be invoked to energise and initiate an operation, Uruz will enable you to shape it into physical form. Fehu is force, Uruz is form. First and foremost, much more important than money or wealth in whatever form, is health. Health can generate wealth, if you are healthy you can work all hours of the day to create wealth, but, no amount of wealth can create health.

As Fehu is the rune of the power of the hamingja or potential ability to manifest your will, or intent, Uruz is the power to manifest in the real world of matter. Metaphysically speaking, Fehu is largely a rune of intent and Uruz more of materialisation. For example, Fehu properly invoked will give you an inspiration to improve your financial circumstances. Uruz will, if properly invoked, empower you with the determination and persistence to carry the idea out in the world. So many original and inspirational ideas keep floating in the air until their power is dissipated. The idea loses its original attraction and is dismissed, i.e. forgotten. Uruz will keep the initiative going. Uruz, therefore, is the primary rune to be invoked for the physical energy to keep going at times when this is necessary. However the energy of Uruz is raw, primitive and has to be moderated by the mind's understanding. It cannot be invoked and let loose. The correct application of this force, i.e. when controlled by personal discipline, makes us both resourceful and persistent. Uruz is the will to live, the primal impulse to be and to become.

In German and Dutch, ur means "primal" or "ancient"; this is contained in the name of one of the Norns, Urd the Eldest Norn, who rules 'the past' or 'that what has been' and this is an example of how I work, my proof reader Arlea Anschutz bless her, pointed out that Ur and Urd are not etymologically related according to linguistics which is a discipline based on linear logic, I think lateral and see connections which are so obvious and may be explained as coincidence?

Uruz is very much associated with growth and, through conflict and challenge, the overcoming of obstacles, the force to assert oneself and to assert one's right to one's own space. This interpretation is also applicable on the psychological level. Uruz is the healing rune, in combination with other runes which we will discuss later. Uruz can bestow rejuvenation and regeneration of physical health.

A simple and effective way to enhance one's strength at a moment when it is needed is to take a glass of clean water and, with the index and middle fingers, trace an Uruz rune over the surface of the water. Visualise the rune in brilliant red, and chant the rune name in a deep voice, letting the power of the rune charge the water. Maintain the visualisation for several seconds and then drink the water. Of course this would also work for a bath, chant Uruz over a bath nine times and feel your exhaustion being washed away. I think the higher form of the power of Uruz would be akin to what is known in the east as Chi. Invoke the power of Uruz into your hands until they feel as if the glow, with this force-field you can heal people, animals and charge objects of power especially in combination with other runes.

> Spiritual aspect: Healing and rejuvenation
> Emotional, psychological aspect: moral strength
> Negative aspect: uncontrolled rage
>
> Germanic name: THURISAZ ▶
> Anglo-Saxon name: THORN
> Old Norse name: THURS
> Phonetic value: TH
> Traditional meaning: Giant

Thurisaz is an old Germanic word for a 'Giant' - also known as a 'Jotun' in Old Norse or 'Eoten' (Middle-English 'Ettin') in Anglo-Saxon. In *Leaves of Yggdrasil* I speculated that the rune represents a 'Giant' god or god of the Giants; thus Thurisaz is mostly connected with Thor. Whilst writing this it occurs to me that it equally well may have connection with Utgard Loki the giant.

Certainly the aggressive form of the fire element, mostly in destructive form, would support this possible interpretation. Giants are representative of the forces of chaos as opposed to order in Norse cosmology. They are however not to be viewed as 'the enemy' in a strict black/white dualistic system, but rather should be viewed as necessary participating forces of evolution, chaos and order, gods and giants, form an interwoven pattern in perpetual flux, as seen in nature itself. Thurisaz is the most destructive rune, representing natural forces, from erupting volcanoes to hurricanes and floods. These all may be symbolised by giants on the rampage. However, it seems to me that in this day and age, these giant forces are not necessarily opposing the gods, but rather that they might be attempting damage control to off-set the incredible destruction wrought by ..yes, humans! Perhaps the gods and the giants have formed an uneasy alliance to save as much of the planet as possible!

This is a two sided rune, it can destroy as well as protect. The rune's powers are to assert oneself, to defend oneself, and to protect oneself. It's a battle rune, a rune of war and conflict. Ideally we would live in a world of unconditional love, positive regard and universal peace. Unfortunately that's going to take some time. In the meantime one has to look after oneself and one's kin. This rune is as active as a power cable. It's power is enormous. It is not lightly used negatively. Believe me, the only time I've used was when my life was in danger. It was simply him or me. Thurisaz is the rune of ultimate physical fighting force as such it has been attributed to both Thor and the giants. The power contained in this rune is the power of Thor's hammer Mjoilnir, and mythologically Thor never slays innocent humans, only certain evil forces represented as giants. Mjoilnir's prime function is to ward or protect and it is here we find the best side of the rune Thurisaz. It is equally strong to protect and defend as it is to attack; a circle of eight fiery red Thurisaz runes on the eight quarters will keep everything at bay. More than that, it acts as a automatic missile and chucks everything back at sender, with accelerated force, so there is no need to attack. Never start a fight if you can avoid it, however, finish it when it is needful! A healing aspect of this rune is it can be invoked for men, to have more sexual prowess. The shape of the rune is phallic. Thurisaz as the third rune in the futhark can be seen as the result of the interaction energetically of Fehu and Uruz synthesising these energies and generating a combination of great force which can be directed and even 'thrown' like Thor's hammer, and yes, it could get back to the sender.

Spiritual aspect: Courage, self empowerment
Emotional, psychological aspect: assertiveness, male
    sexuality
Negative aspect: disease, explosive violence

Germanic name: ANSUZ ᚠ
Anglo-Saxon name: OS
Old Norse name: ASS
Phonetic value: A
Traditional meaning: Ancestor/ god

The fourth rune of the futhark is complementary in its meaning to the previous one, represents order and the gods, especially Odin. The Ansuz rune represents consciousness, intelligence, communication and reason. The element of Odin, the god-force behind this rune, is Air. Air is a penetrating medium necessary to all life-forms. It is through air that sound becomes audible. In an airless environment, sound cannot be heard, and sound is the other main feature of the Ansuz rune, including the origins of sound as a medium of communication between people, and the sounds of Nature. The magical use of sound to expand one's consciousness and the chanting of runes in a manner in which sound is closely linked with breath can give spectacular results.

Ansuz is derived from the proto-Germanic word 'Ansis' which means both 'god' and 'ancestor' and in the Northern Mysteries the God most frequently seen as a divine ancestor is Odin. Our gods may have originated as ancestral spirits which gradually acquired divine attributes. The gods themselves have changed and evolved as human understanding of them has changed and developed over the centuries. Even in one lifetime, such change takes place, and after my experience of working with Odin, I find myself continually redefining him in my own understanding. Beyond my or anybody else's understanding of him is the greater mystery of his being with cannot be fully comprehended by incarnate spirits such as humans. The Anglo-Saxon traditions calls this rune 'mouth' i.e., the origin of speech. Speech is one of Odin's attributes, as are poetry and the aforementioned magical uses of sound. It is no coincidence that Odin, the god of words and communication, is credited with giving mankind knowledge of the runes. The extra dimension of knowledge gained when discussing the runes with a like-minded individual illustrates the communicative aspect of Odin as the power of consciousness.

This rune is more in tune with the higher mental abilities, thinking, connecting, communicating. The name Ansuz can and does mean both 'god' and/or 'ancestors'.

In Asatru this God is Odin, but it could mean any deity or divine intelligence. It is specifically the rune of intellectual learning, understanding, abstract thought processes and the ability to communicate these in concrete terms. Also language skills, the ability to write and speak in words as well as the language of symbols. Ansuz is one of the runes of leadership connected with spiritual leaders, teachers or Elders, these older, wiser people from whom advice is sought. The chanting of Ansuz or meditation on Ansuz can open channels of inspiration previously inaccessible. Out of all the runes it the most meaningful for me, as to me it represent the force of Odin himself, the highest source of consciousness and awareness. Odin has three main aspects expressed by Ansuz, Wunjo and Othala. The Aspect of Odin emanating from Ansuz is the aspect of Warrior. This is not to be interpreted on a lower scale. We are referring to the concept of Warrior in it's highest aspirations. The Warrior as assertive Consciousness against Ignorance.

Ansuz can open hidden 'doorways' to obtain occult knowledge about the ancestors who are 'dead' and whatever residue of ancestral knowledge remains on the Inner planes. The modern practice of channeling which in reality is a practice as old as humankind can be developed safely with the invoking help of the Ansuz rune, especially so called conscious channeling, where one does not surrender consciousness but merges with a greater mind to transmit knowledge.

Spiritual aspect: Higher consciousness, mental stability
Emotional psychological: knowledge of self and discipline
Negative aspect: neurotic compulsive liar

Germanic name: RAIDHO ᚱ
Anglo-Saxon name: RAD
Old Norse name: REID
Phonetic value: R
Traditional meaning: Riding

The traditional and accepted meaning of this, the fifth in the futhark, is 'Ride', originally referring to long journeys on horse back or in a chariot such as some of our gods and goddesses were known to use. Odin, for example, travels on Sleipnir his eight legged horse; Thor travels in a chariot drawn by goats and Freyja likewise travels in a chariot drawn by cats. Raido thus represents both means of transport and the activity of journeying. Mythically, it is also associated with 'Asgard's Ride' in Scandinavia, which was known in German and British folklore as the 'Wild Hunt': the wintry ride of Odin and his retinue of battle-slain warriors, manifested in the Winds and storms of Yule-tide.

Riding has also connotations of creating movement, generating motion, taking charge of situations, being in control, taking the initiative, starting a new venture, decision-making or directing a course of action. Rad, the Anglo-Saxon name of the rune, may be cognate with the German Rat and Dutch raad, or rede; meaning "counsel" or "advice," which is one of the meanings of Raidho.

It gives sound advice in a reading. There happens to be another word of Gothic origin, 'raiht', which in my opinion is closely related to rad or Raidho.

This word means "right," or in Dutch 'recht', the related verb being rechten, meaning in English "to do right, to dispense judgment, to litigate." The Dutch word for "judge" is rechter. Raiht is linked to the institution of kingship. In the old days, kings had the task of dispensing justice. The Latin word for king, 'rex', derives from the same root source as raiht as does Heimdallr's alternative name 'Rig' in the Eddic poem Rigisthula, in which this enigmatic god is seen as the originator of the three main divisions of society: thralls, farmers and rulers. All these associations explain why the Raidho rune is a rune that speaks to us of leadership and nobility. In olden times, nobility was not something which was automatically acquired by virtue of birth. Rather, it was a position which was earned or held by merit, and it implied moral responsibility and integrity. Raido refers to the correct balance between respecting the rights of others versus one's own individual rights.

The Raido rune represents the journey of life, the direction in which one travels and the amount of control one may exert to determine the direction. This holds true for both real-life physical progress, as in growth from child to adult, and metaphorical progress, the initiatory path of life, whether it's consciously undertaken in co-operation with, and in the service of, higher forces, or unconsciously subject to the decree of the

Norns, the Nordic embodiment of 'wyrd'. The magical and psychological significance of this rune relates to the ability to move within one's natural limits and, consequently, to become aware of what these limits are — the realistic assessment of parameters given in life. The rune also refers to the power of making a conscious decision and the discipline to carry it out. This rune can also be used in a magical working to establish control over one's own circumstances, and to put things in order and make them subject to one's will, in accordance with one's 'wyrd'.

Inherent in Raidho are the ideas of freedom and moral responsibility to the Self (even if this conflicts with the norms prevailing in the present-day social structure), as well as the knowledge of right and wrong in the personal developed conscience and the courage to act accordingly. Thus, working with the force of Raidho means to be in charge of your own path in life; to keep your own counsel; to ride, not to be ridden; to be master of your own circumstances as far as is possible; to extend gradually the degree of control that you exercise over these circumstances; and finally to make a conscious choice of the direction you wish to follow. How much of yourself is being ridden, so to speak, depends upon your magical control, i.e., the conscious controlling the ego as a rider controls his horse, or a driver his vehicle.

Spiritual aspect: Ethical conduct; personal discipline
Emotional psychological: control issues
Negative aspect: control freak, dictator, hypocrite.

Germanic name: KENAZ ᚲ
Anglo-Saxon name: CEN
Old Norse name: KAUN
Phonetic value: K or hard C
Traditional meaning: torch, light

Most rune-workers interpret the meaning of the Kenaz rune as "torch", "which has always been regarded as a symbol of knowledge, consciousness and intellect. Kenaz means "to know"; the Scottish dialect verb 'ken' and the Dutch and German 'kennen' all mean "to know, to be familiar with", but also connote skills and abilities, i.e., "to be able to." The English word "cunning" is etymologically derived from the same root as Kenaz. Kenaz therefore implies the acquisition and application of knowledge. The terms 'cunning man' or 'cunning woman' were often used as euphemisms for 'witch', denoting someone skilled in magic, herb-craft, or other 'specialist' knowledge more or less occult, especially when it concerned the hidden knowledge of a female nature. There is a certain English word, often used as a crude expletive and derived from the same Indo European root-word as Kenaz, which demonstrates both the femininity of this kind of intuitive knowledge and also the extent to which it has been denigrated and de-valued in our culture. While Raidho is the rune which directs us towards the right path, Kenaz sheds lights on this path so that we may know where we are going. Together, they can guide us in the proper direction. The rune's element is fire in its contained form, that is to say the fire of enthusiasm and inspiration.

One of the gods who can be successfully invoked with a Kenaz working is Heimdal, who is known as the "Shining God". He is the god who, in the Eddic poem Rigisthula, teaches the runic mysteries to the child Kon - whose name is related to all those words meaning 'know' and whose descendants become kings and rulers.

Thus, Kenaz is associated with the transmission of knowledge – the passing on of the torch to the next generation of kin or cyn. Both of these words are of Anglo-Saxon origin and are related to Kenaz. As such it indicates the learning and teaching process. The Old English word 'cyning' is also related to Kenaz. Cyning meant king. The king, according to tradition, had to be a descendant of Woden. He acted as the focal point of the collective folk-soul, for the king was the torch-bearer of the folk's consciousness and conscience.

Kenaz mostly known as 'Torch' is the rune which lights your path, it exposes, it enlightens, it informs. This rune is also associated with intellectual abilities, especially the ability to recognise, to learn. Kenaz will make the invisible, visible, it dispels darkness and ignorance. Invoking the power of the Kenaz rune by for instance visualising and repetitive chanting of the rune and visualising it as a imprint in brilliant white light on ones 'third eye' space before falling asleep, can illuminate an issue which was difficult to understand at first, it can confer a bright idea, a solution. An AHA Kenaz also will often be helpful when enquiring into others possible hidden or ulterior motives. Kenaz represents the light within; confidence in the knowledge that one is descended from the gods, kings, and tribal chieftains; and with this, the responsibility to hand over the torch to the next generation, or else to share its light by teaching.

On a psychological level the attributes of Kenaz are clarity of thought, insight, consciousness of the self, inborn or hereditary knowledge, confidence and trust in one's own intuition, and finally concentration and determined effort. Kenaz and Raidho together can give you recognition of recurring patterns in your life and relationships.

Spiritual aspect: Knowledge from within, intuition, enlightenment.
Emotional psychological: Self knowledge, self awareness
Negative aspect: pig ignorance, arrogance.

Germanic name: GEBO **X**
Anglo-Saxon name: GIFU
Old Norse name: none
Phonetic value: G
Traditional meaning: Gift

This rune, the seventh of the futhark, is named Gifu in Anglo-Saxon, and means a gift, or giving. All energy systems are exchange mechanisms; we take, we give, we trade. The life force itself ebbs and flows in us in this way as we interact with our environment, even at the most basic level of breathing in oxygen and breathing out carbon dioxide. Plants complement our interaction with the atmosphere by assimilating carbon dioxide and exchanging this for oxygen; thus a balance is maintained. Similar processes of exchange are constantly active in us, sustaining life and consciousness. When we enter this life, we are endowed with 'gifts': talents, abilities etc. These gifts are passed on from our particular ancestors, and in general from the collective genetic pool of which the origins lay with the gods themselves. Metaphorically, certainly we are descended of our gods. Perhaps this is true genetically as well; it all depends on's one's point of view and whether one can envisage the possibility of the gods at some point in history having had a physical existence.

At any rate, we are endowed with gifts, from the start, and these gifts have to be passed on in one form or another. Having received such gifts, we are obliged to give in our turn, both to the gods in the form of sacrifice, which can mean anything from a flower or a glass of drink before a statue to a commitment to a project and seeing it through, and to our fellow men for the greater good of all. The Gebo rune teaches us to become conscious of the process of exchange in all its forms. It teaches us not to take more than we are willing to give, for when we create an imbalance, we shall have to compensate for that some time or other. Sometimes the compensation required can take a different form and be extracted on a different level. If we eat more than our bodies need, sooner or later (sometimes much later) our bodies will object to the abuse and give us trouble in the form of illness. This is a simple example of the natural working of the Gebo rune.

The shape of the rune is perfect balance on all four directions, it is symmetrical and bilateral, above, below, left and right all look exactly the same. The rune is neutral, it's neither a 'good' rune or a 'bad' one, simply indicating equilibrium and balance. Giving and receiving in equal measure. Giving does not exist in a vacuum, even if it does not look for overt payment or recompense, because the giver needs to receive something back - even if this is only an emotional benefit. In this realistic view, altruism in the sense of absolute selflessness cannot exist and would be out of balance. Everything in nature, everything in the soul and everything in mind seeks its equal and opposite for complementary equilibrium.

This is the rune of giving in all its forms. Its the rune of fair exchange. The higher aspect of the Gebo rune is the unconditional giving of self to Self i.e to the gods from whence we came and to whom everything ultimately will return. The soul on its journey through life, gaining knowledge, always homeward bound to return the gift, those gifts of life and consciousness which are the gifts of the gods to man. On the higher level, the Gebo rune represents the gifts given by the gods to man. In the Mythology the story is told of how the three gods Odin, Hoenir and Lodur encounter two trees, Ask (male tree) and Embla (female tree). Each of the gods gives a gift of life, by which Ask becomes the first man and Embla the first woman. The Gebo rune also implies the gifts that man gives to the gods in return. Man's gifts to the gods include service, loyalty and dedication to whatever is conceived to be "the gods." In this sense Gebo may imply the voluntary sacrifice of one's resources, time and energy to whatever one holds sacred, without an expectation of reward other than the development of one's own potential. In extreme cases, this gift is the supreme sacrifice. On the highest level, the gift transcends both giver and receiver, thus implying the synthesis of gift, giver and receiver of the gift in a unity; the result is the dissolving of barriers between all, mystical union through complete giving, and the surrender of one's ego to the divine consciousness, which is also known as the higher Self, and has a connection to the Gods.

> Spiritual aspect: Integration of complementary opposites
> Emotional psychological aspect: balancing opposite aspects in the self
> Negative aspect: unbalanced behaviour, dishonesty

Germanic name: WUNJO ᛈ
Anglo-Saxon name: WYNN
Old Norse name: none
Phonetic value: W and sometimes V
Traditional meaning: Joy

The name of this rune is usually translated by other rune-workers as "joy" or "pasture."

This association with joy is evidently derived from the rune-name's similarity to the modern German word 'Wonne', which may indeed be derived from the proto-Germanic word Wunjo. Although this interpretation is not altogether wrong, the original meaning of the word Wunjo in the oldest Germanic language known to us the word is "perfection," according to the philologist Jacob Grimm. A variant of this word is 'Wunsch' which means 'wish'. From the combination of these words one can deduct that Joy is attained when fulfillment of Wish has occurred and thus both meaning ancient and modern converge neatly. The key to joy then is the ability to wish, the correct application of the will or 'intent'. The reason most people are unhappy is that they are discontent and wish for the impossible, and then rage in frustration at Wyrd when they are disappointed. I think the lesson in Wunjo is to wish wisely, and be content with what little joys and comfort Wyrd may bestow on you.

The Anglo-Saxon name of the rune; 'wynn' may have been wenn, which supports the above as 'wenn' also means 'hope' and, one of Odin's bynames is Oski, which means fulfiller of wishes. This corresponds to the German tradition expressed by the word Wunsch, which, like the English "wish," and the Dutch 'wens', derives from the proto Germanic word Wunjo, meaning "perfection." or 'fulfillment'. Odin certainly has to power to grant 'wishes', but as the saying goes be careful what you wish for - you might just get it! What most people wish for more than any material wealth, beyond the everyday necessities of course, is companionship, either in a committed one to one relationship, or family bonding, or trusted friends and Kinfolk with shared ideals and objectives. Wunjo therefore also has been associated with fellowship and bonding of clan. Often groups affiliated for religious or spiritual reasons or even something as common as football have a banner, a symbolic display which conveys their shared identity and unison of will.

This is the rune of Joy and fulfillment of desires, optimism and contentment. Wunjo contains the power of the will. Odin, who controls the power of this rune, has three aspects: Odin, Vili and Ve. Odin is Ansuz, Willi is Wunjo and We or Ve is Othala as we shall see. The name Vili, in Germanic Wili, fits the concept of the magical application of the Wunjo rune and the art of correct wishing, or in other words, the use of the power of the magical will. This identifies the second major aspect of Odin as Shaman, or Magician if you prefer. Since all magic is fundamentally an act of will, we see that this rune can be extremely useful in realising one's goals, whatever these may be, combines very well with Raido in that Raido operates as a means of controlling and directing the will symbolised by Wunjo. Wunjo is excellent for group bonding and sharing of ideals, it generates harmony and cooperative effort, a togetherness of like minded people. Wunjo especially is the rune for genuine friendship and comradeship.

From an esoteric and spiritual point of view, Wunjo can mean joy when it is combined with Gebo, which symbolises the gift of one's own will to the gods, and in particular the alignment of one's own individual will with the will of the god or goddess one serves.

Spiritual aspect: complete harmony with everything.
Emotional psychological aspects: happiness, optimism
Negative aspects: loser, miserable, loneliness

SECOND AETTIR

This the second aettir introduces forces outside the human control such as, the Norns, Time, Wyrd/Orlog, multidimensional space, Higher spiritual issues, initiation. This is the aettir of development of consciousness and magic, psychological growth and personal development.

Germanic name: HAGALAZ N
Anglo-Saxon name: HAEGL
Old Norse name: HAGALL
Phonetic value: H
Traditional meaning: Hail

This is the first rune of the second aett: the aett of Hagalaz. This aett is named after an element, whereas the others are named after gods Frey and Freyja for the first aett and Tyr for the third. The literal meaning of the name has often been accepted by modern sources as 'hail' or 'hagel' as in hailstones. The meaning of this rune is therefore quite clear: destructive elements of nature; severe weather. No one can control the weather. The military in various countries are said to experiment with technologies trying to do exactly that, much to the detriment of the rest of the planet, as we are currently subjected to the most severe weather phenomena ever recorded. Hagalaz is the rune of unavoidable unpleasantness outside individual control. Whatever is symbolised by Hagalaz, when combined in a reading with other runes, basically is inevitable. Hence this rune is strongly associated with Wyrd, even Orlog. The natural elemental destructive power of this rune cannot be harnessed by human beings. It is completely impersonal. It is the ultimate power of nature unveiled. The best one can do is seek shelter and ride it out, and of course this can also be applied metaphorically when confronted with adverse happenings one cannot control, be it death, disease or the tax man! The lesson of Hagalaz is acceptance of that which cannot be changed, fortitude in the face of adversity and patience in the knowledge that the wheel soon turns again.

This rune is strongly associated with early winter, in particular November, when often the weather is worse than, for instance, in January. But as the Anglo-Saxon rune poem says, Hail turns to water, and we all know water is the most vital substance of life itself. Hail will melt at a slight temperature change and becomes water which is nutrition for the earth and all that live on her. Conversely hail can turn to Ice when the temperature changes the other way. Hagalaz therefore is a rune that embodies a force which is in itself potentially destructive but will change of its own accord give time. It either changes to Ice or more often to Water and fertilises the earth. The change embodied in Hagalaz can go either way.

This is the rune of the hidden realms of the past and the dead. Goddesses associated with the very dark female rune are Hella, Holda and Urd. This rune can be used as a doorway to the past, the personal past in this lifetime where one may inquire about early environmental conditions and also the transpersonal past to obtain information about past previous lives. This is also one of the runes to call and communicate with he dead. This rune gives access to the 'underworld' in the Northern tradition Hella's Halls. The realm of Hel is the equivalent of the shamanic underworld and can be reached by shamanic means through travelling while in altered states of consciousness.

Hagalaz contains a lot of dark feminine power, and has a strong connection with negative witchcraft, such as destructive female magic. The first witch known in our mythology is a lady called Gullveig. Like Hella, she represents the dark aspect of the

Goddess. Gullveig introduced to Valhalla the lust for gold and corrupted the inhabitants of Asgard so that Odin himself had to put a stop to it. She was burned three times and from this the Norns came forth. Although the Norns come forth in Voluspa verse 20, and Gullveig is burned in verse 21, I think it not too much of a liberty to suggests that these verses are expressed in linear time but actually do not take part thereof.

There is an alternative form of this rune which is very protective and is especially used to provide protection against bad weather. For this purpose it used to be displayed, over doorways or on top of the house or farm. The protective aspect of Hagalaz, or Hail, is mostly referred to as "heil," which can be used as both a blessing and a greeting. This form of the rune was particularly used as a marriage blessing and, as one can see, the shape of this rune looks like the two forms of the Algiz rune intertwined, indicating the union between the male and the female form of Algiz. This also resembles a snow flake. Snow is one of the prime elements of the Goddess Holda. In winter she shakes her bed and it snows. If one connects the six points together with a line it forms a hexagon, the molecular shape of carbon the stuff of life itself. Perhaps this rune is representative of the creation of the material world Midgard. As Uruz is representative of the etheric matrix patterns.

Spiritual aspect: confers ability to learn from the past.
Emotional psychological aspects: acceptance of past mistakes of self and others, letting go.
Negative aspects: victim consciousness blaming others, stuck in the past.

Germanic name: NAUTHIZ ᚾ
Anglo-Saxon name: NEED
Old Norse name: NAUDR
Phonetic value: N
Traditional meaning: Need

Again no ambiguity here, the Rune name speaks for itself. This is the rune of Need or necessity, again there is the feeling of inevitability similar to the above Hagalaz rune. But, no, need can be turned into an omen of help, if attended to early enough according to the Anglo-Saxon rune poem. This is a rune which will warn and advice when recognised as such, when one encounters it, either in a reading or in dreamscapes. The first lesson of this rune is in practical down to earth terms; look at what is needed now! There is always the feeling of urgency about this rune. One of the most atavistic needs of humans was fire, without fire one died, period. The shape of Nauthiz represents two pieces of wood being rubbed together to generate fire. We in modern days perhaps cannot appreciate the early man's need for fire. Fire as well as water is another vital element of life itself. The "need-fire" was originally a ritual kindling of fire associated with various festivals such as Beltane, Midsummer, Yule and Samhain. It originates from a cult of fire-worship and evidence shows that this cult is older than the religion of the Aesir. It is possible that here is a overlap with the Celtic tradition as around that time the tribes were sharing the same territory. Fire rituals were used for a variety of purposes, including sacrifices, healing and cleansing. In the need-fire ritual two fires were kindled a distance apart from each other by two chaste youngsters, one male and one female. Each made a fire by rubbing dry wood together until it caught light. Cattle were then driven between the two fires to ensure their fertility and health. (The custom of engaged couples leaping over a bonfire is a survival of this rite.) The need-fire was usually kindled in times of need such as dearth, drought and disease.

It is easily understandable why fire was held to be sacred, for the climate in which the Northern folk live is prone to extremes of cold, our lands in the past were covered in dense forests and fire could easily become a destructive and deadly foe. The need for fire was balanced by the fear of fire. The Nauthiz rune also then has associations with an element, as has Isa the next rune in the Futhark. Being the rune of Need, this is the rune of the force friction and resistance generating creative energy and progress or evolution. Need fire used to be made by the friction between two pieces of wood. Likewise progress and growth is often generated by friction, for example the friction between a growing adolescent and his elders. The need to achieve, as the saying goes "Need is the mother of invention". Need is usually triggered by distress. Need therefore often has a connection positive or negative with Fehu, for whatever the cause is of the distress, the necessity to act is ultimately a force of growth albeit often unpleasant. The Need rune kicks hard! It generates action and by inference change. Without the counter force of Nauthiz, life would be a lot more static and growth would be at a much slower rate. Nauthiz, like Hagalaz, is a force of nature outside the individual's control. One can work with it, one can work around it, one can magically use it, but, one cannot ignore it!

As we have seen Hagalaz represents, among other things, the forces of the past. Nauthiz can teach us how these forces can still exert their effect on us in the present, as the present is a continuation of the past and the past really an incorporated part of the present. There is no sharp definition between the two, they overlap. Nauthiz then can endow us with insight and motivation to heal from our past, to extract the lessons from our and others' errors, to get out of the blame-game and create something constructive using the power of the Nauthiz rune to built strength of character. In the words of Nietzsche "that which does not destroy me makes me stronger"! Nauthiz helps us to grow in spite of opposition, resistance and constraint, by simply becoming aware of what is needed and doing it.

Spiritual aspect: a generating force for the greater good and the long term.
Emotional and psychological aspect: confers the ability to recognise one's needs and cater for these.
Negative aspect: guilt ridden, constrained by fear, moral cowardice

Germanic name: ISA**I**
Anglo-Saxon name: IS
Old Norse name: IS
Phonetic value: I
Traditional meaning: Ice

And here is the third rune of the second Aettir. Isa, or Ice, another elemental rune. Ice, like hail, is a variant of water. Ice is water in its densest frozen state. It's cold and slippery but it has its uses. It preserves. Where would be without our freezers? Water is fluid; ice is solid and static. Isa, therefore, represents the principle of preservation and resistance to change. Isa is a necessary force whose controlling effect is essential in order to prevent random growth. Nevertheless, our ancestors saw it as a challenge, and not only succeeded in the struggle for survival in unfavourable conditions, but were also spurred on by Isa to evolve at a faster rate — and tough. Isa is static. It "is." Its function is to keep things as they are, to maintain whatsoever "is." On its own, it is an inert rune which merely preserves and conserves. In combination with a fire rune it can raise energy for various magickal objectives. Ice as an element is inimical to life and hostile to the environment.

It maintains its integrity. It is immobile, does not change by itself but only when the opposite element of fire in brought into play. They both generate 'steam'. Fire and Ice are in the Northern tradition in exactly the same intertwined relationship as Yang and Yin in the more well known Taoist tradition. They are indispensable to each other and between them create and maintain all existence. They balance each other perfectly. So far we have seen that Hagalaz is a so-called 'elemental' rune, hail being a state of water. Nauthiz, we see, clearly contains the element of fire, and Isa being ice, is also a state of water.

The lesson of Isa is stability and maintaining one's integrity and focus in a crisis situation. Psychologically, Isa represents the "I" in the most mundane sense, and the capacity for personal survival through concentrated effort. Isa can be a great help in concentrating the will in a single-minded action. It is the rune of self-preservation and self-containment, the positive aspect of which is our individualism and the ability to survive against all odds; the negative aspect can be self-centredness.

This is one of the more enigmatic runes in the futhark: Isa, Ice. This rune is unyielding like the element after which it is named, Ice, is! The Isa rune has no glamour, although it is smooth and slippery according to the Anglo-Saxon rune poem, it shows itself as it is. Likewise the Isa rune is the rune of the Ego and the harsh reality of life. Ice is water in its densest frozen state. Ice is the complementary opposite of Fire, these two elements create and maintain all life. Isa often is a force of inertia, a block psychological or otherwise. Working with one of the gentler runes of fire such as Kenaz or Sowulo can gradually dissolve blockages of this nature. Sometimes Isa makes one slow down and re-evaluate one's position, as such Isa can confer a clarity of sight and an objectivity devoid of emotion and unable to be manipulated by these. Isa therefore has a recuperative ability, slow down take stock and slowly proceed. Rest and regeneration are the positive aspects of Isa.

Spiritual aspect: calming, reflective influence.
Emotional psychological aspect: stable, conservative, upholding
Negative aspects: gross materialism, greed

Germanic name: JERA ᛃ
Anglo-Saxon name: GER
Old Norse name: AR
Phonetic value: J or Y
Traditional meaning: year, harvest

That this rune is the fourth in the aett of Hagalaz and the twelfth in the Futhark is a good example of synchronicity, since there are twelve months in the year and twice twelve hours in a day. Also four is a third of twelve, as eight is a third of twenty-four. Is there anybody who would dispute that the order and numbering of the Futhark is a 'magical' act, and well one of the highest order. We'll get to magic later in the book though. For now we examine the meaning of the Jera Rune, the literal translation of Jera is 'year' and this is the first indication that this rune has an important bearing on time and the divisions of time. This rune represents time itself. The shape of the rune is divided in two halves, clearly portraying two halves of a year circling around each other in perpetual swirling motion from light to darkness and back again. The Jera rune is associated, in particular, with the turning of the year at Yuletide, when the Sun returns, Yule was about the biggest festival of the North. The return of the sun, the days lengthening, gently turning the wheel of the year till Midsummer's day, and so on in perpetual motion spiraling upwards to the realm of the gods themselves. Jera is the rune of time and fruition.

What comes around goes around as the saying goes. No where is that better portrayed than in the lessons of the Jera rune. Jera gives growth, there are mysteries only time can teach, only experience, no hearsay will do, this is the rune of personal experience through time. This rune expresses the mystery of time and cyclical recurrences, the natural cycle of the year the turning of the wheel. The Jera Rune furthermore is representative of the Greater Cycles of Cosmic origin. For instance, the processional cycles of the equinox. All these cycles are in fact spirals. Although they seem to be recurring in exactly the same format over longer periods of time, a progress can be detected suggesting that these cycles are spirals. A repetitive shape of the Jera rune can resemble the spiral arm of this galaxy of which our solar system is a part. All recurring cycles of time, such as seasonal, astrological, astronomical, are contained within the mysteries of the Jera Rune, as is the turning of the Earth itself. I would therefore suggest that this rune is very much connected with the Earth mysteries, the weather and the ecology of the planet. Others as well as myself have pointed out that the Jera rune has the secondary meaning of 'harvest' and this (apart from referring to the yearly harvesting of food) also bears connotations with 'wyrd' as in what goes around comes around, but after another manner.

In a reading the Jera rune often indicates a turning point for the better, like the return of spring after a dark and cold winter, as such it is more positive than the three preceding runes in this aettir. In the Runic calendar the Jera rune is situated at Yule, the festival of the return of the sun. That time of year (or perhaps on the more conventional accepted date of 31 of December) is the optimum time to do a yearly forecast with the runes, I will give an example of this in the chapter on runic divination.

Spiritual aspect: growth and progress
Emotional and Psychological aspect: patience and healing.
Negative aspect: harking back to the past and repeating negative patterns.

Germanic name: EIHWAZ ᛇ
Anglo-Saxon name: EOH
Old Norse name: none
Phonetic value: E
Traditional meaning: Yew

This the thirteenth rune in the Futhark is named after a Tree, and not just any old tree, the mother of all trees: Yggrasil itself. This rune is associated with the world of nature, specifically trees. The Germanic tribes like their Celtic neighbours held trees in high regard. They in fact worshipped trees, or rather they conducted services of worship to the gods to whom groves of trees were dedicated. Only in later Scandinavia do we hear about temples or more properly named Hofs being used as places to worship the gods.

There is archeological and literary evidence that Heathen Anglo-Saxons had temples too.

The older Germanic tribes like the Druids with whom they had far more in common than is known today preferred a natural environment to conduct their spiritual and magical practices. The evergreen Yew tree was perceived as very sacred. Its wood was used for bows in hunting, and also for the crafting of rune staves. The wood is hard —the hardest wood to carve I have come across upon. The Yew tree is associated with both life and death, as it is an evergreen and a hardy tree it symbolises longevity. At the same time, the berries and the firs are toxic. A well known Druid initiate chose to end his life by eating a bowl full of yew leaves. The same toxicity when prepared properly by a qualified herbalist or chemist can be used as an hallucinogenic, for shamanic initiationary purposes. Homeopathically, Taxus Baccata is a remedy to eliminate toxins from the body, especially residues from inoculations received in childhood.

Yggdrasil is the tree most holy in the Northern Tradition, the tree on which Odin was said to have hung for nine days and nine nights to received the Runes. 'Runes' to be understood in this context as the 'Greater Mysteries' of life, death, evolution, ascension and transformation. These mysteries are contained in coded form in the Futhark runes as they have been handed down to us humans. Eihwaz teaches us the mysteries of life, death and that which is in between. Eihwaz can meld and transcend opposite extremes. This is the rune of endurance and initiation as is born out in the mythology of Odin's ordeal on Yggdrasil when he hung suspended between life and death for nine days, transcended the barriers and obtained knowledge of the runes i.e the mysteries and laws underpinning working of the universe and evolution.

If Jera represents time and the turning and returning of motion, Eihwaz is the pivotal point on which everything turns: the Cosmic axis. It is the point in space which holds that which is turning (for example, a planet such as earth) in place.

Yggdrasil itself gives access to all the other worlds. Here all the forces are in balance. Here is the point of stability. A beneficial exercise is to visualise Jera spinning around Eihwaz, preferably against a backdrop of a starry night. Eihwaz is the point of equilibrium of life and death, Eihwaz is suspended between both outside the circles of time. As Jera represents time, Eihwaz take us beyond time. The three Runes Jera, Eihwaz and the next one Perthro collectively embody the mysteries of Cosmic might. Working with these three runes together can give access to all knowledge which may be known by us humans in our present state of evolution. The Eihwaz rune is also the shamanic tree. One can travel to other realities in the Upper Worlds as well as to the Netherworlds and obtain altered states of consciousness, for example to confer with those who have gone before, in combination with the Hagalaz rune. Lastly the Eihwaz is also a strong protective device when made into a wand and carried around with you.

Spiritual aspect: the possibility to obtain at least
    temporarily cosmic consciousness.
Emotional and psychological aspects:Inner power of
    stability despite external circumstance, endurance.
Negative aspects: Confusion, lack of attention, hysteria.

Germanic name: PERTHRO ⟦
Anglo-Saxon name: PEORTH
Old Norse name: none
Phonetic value: P
Traditional meaning: unknown

This is the rune everyone speculates about. The word Perthro has not been clearly identified with any of the known words in Germanic languages, both ancient and modern. There is no traditional and accepted meaning for this rune. I therefore follow everyone else and rely on my intuition to offer an provisional interpretation. To me this rune contains the mysteries of birth and rebirth. It seems a logical progression that in view that the forgoing rune Eihwaz is connected with the mysteries of death this one would be related to birth. The shape of the rune resembles a birth giving position when turned sideways. Furthermore I also associate this rune with the collective memory, the well of Mimir. Memories of previous existences and ancestral memories can be retrieved from the Well of Mimir (or the Akasha Records) as this is also known in the esoteric body of lore. Destiny or 'Wyrd' and that which one brings with one when entering life can also be perceived through this rune. Perthro can connect us with the 'unseen', the hidden knowledge. Through Perthro one may perceive one's Wyrd and be able to alter certain probabilities within that Wyrd.

Perthro is the most mysterious rune of all. As I said before, not much is known about meaning other than speculations. When I said that I associate Perthro with birth I meant that in the most extensive meaning: the birth of the universe, the externalisation of consciousness, the realisation and actualisation of the hidden potential from the Ginnugagap, the Nordic version of the unborn universe. All knowledge is contained within Perthro, hidden in various layers of Orlog and Wyrd. It is here wyrd is shaped, it is here records are kept. Orlog is the larger 'destiny' of fate, the primal laws of the universe itself, Wyrd is more individual and can be tampered with using magic. For example the occurrence of Ragnarok is Orlog. It's going to happen, whether it be by the human race blowing itself up or the sun turning into a nova or an asteroid hitting the earth. Whatever has a beginning has also an end, at least on one level of existence. Orlog is laid down, an implicit condition which cannot be altered. Wyrd is the individual layers of apparent cause and effect within the individual's life. However, it is also intertwined with other people's wyrd. From a family to the larger community all wyrd is interwoven and interdependent. Insight into the workings of wyrd can be obtained through divination with runes and/or any other method. With the knowledge obtained wyrd can be manipulated to a certain extent depending on the magical might or hamingja available to the individual and the degree of consciousness of the individual. The Perthro rune is the doorway to the knowledge of wyrd and for this reason is mostly associated with the Norns.

Spiritual aspect: Well of knowledge for growth.
Emotional and psychological aspects: Access to the individual and collective unconscious
Negative aspects: Delusions, fantasies, megalomania

Germanic name: ALGIZ **Y**
Anglo-Saxon name: EOLH
Old Norse name: YR
Phonetic value: Z
Traditional meaning: protection

The tradition and accepted meaning of protection has been derived from the word Algiz which seems to have been related to 'alcis'. Alcis, according to Tacitus, was the name of the divine twins worshipped by Germanic people. Another form of the name of this rune is the Old Germanic word 'alhs' meaning temple or sanctuary. Because all the magical accouterments and ritual equipment were kept in the alhs, these sacred places of worship had to be strongly defended and protected magically.

This rune is the best for defence and protection of the individual as well as the group. The sign itself looks like a splayed hand, and is reminiscent of Tyr's hand, which he sacrificed in order to bind Fenris the wolf. This rune can be seen to have two forms, upright and inverted. It has been a historical custom in Germany to use Algiz runes on gravestones in the following manner:

Date of birth **ᛏ**   Date of death **ᛘ**

The upright form is a classical pose of an invocation or supplication to the gods, especially to ask for protection and guidance. It is the most prominent rune to be used in devotional worship, as it both protects the human from undesirable astral or mundane interference and acts as a channel to download the god force, allowing safe communication to occur. This is the rune of hallowing, warding, and the rune of access to higher spiritual awareness. As we have seen, the Algiz rune is often used in a stance to invoke the Gods. In a way it is the holiest of runes. It is used to create sacred space and draw down the power of the Upper Worlds of the gods, the realm of the Aesir, Asgard; the realm of the Vanir, Vanaheim; and the realm of the Lightalfar. The Lightalfar are most akin to Devic beings and also may include the hallowed dead. Turn the rune upside down and it can be used to access the Underworld, Hella's realm, the place of the mundane dead. The Algiz rune symbolises Bifrost, the Rainbow bridge. The three elements: Fire, water, and air comprise the actual rainbow. It is over the rainbow bridge one travels to meet with the gods.

Algiz can be used to empower the spiritual body or rainbow body. When one stand in the Algiz position one can draw fire from the right and Ice from the left hands, these two merge in the middle extending upwards and down wards creating a rainbow pathway. The god Heimdallr is the guardian of the rainbow bridge. He has to grant his permission to anyone crossing the bridge. As in other occult traditions this may involve a testing of worthiness. Algiz furthermore is strongly connected with the Valkyries, female emanations of Odin mythologically portrayed as his daughters. Valkyries are mostly known for the function to choose the dead and carry them off to Valhalla, but they are also involved in shielding, warding and protecting. In this function they are better known as 'shield maidens'.

Spiritual aspect: Invokes the highest protection.

Emotional and psychological aspects: communication with
other realms and other parts of the self.

Negative aspects: none

Germanic name: SOWULO ⚡

Anglo-Saxon name: SIGIL

Old Norse name: SOL

Phonetic value: S

Traditional meaning: Sun

The sun, the life-giving force, in the human body the heart, the seat of the soul, the higher potential. The guiding path of solar initiation.

Our ancestors depended very much of the power of the sun, hence it is so important as to have been preserved as a rune name, whereas there is no equivalent name for the moon preserved in the Futhark. However one important difference with present day pagan traditions is that in Asatru the sun is regarded as being of the female polarity (in contrast with popular paganism derived from the Greek tradition and expressed in modern Wicca as well as the Western Mysteries) where the sun is seen as masculine. In modern German the grammatical gender of die Sonne is feminine. Both interpretations of course are as valid an each other. It's all a matter of perspective and relative to the aeon. Male or female we know the sun gives growth and healing. Our ancestors navigated their ships by the position of the sun in day time, and by the Pole Star at night. Sowulo then represents the clear light of day.

There is nevertheless a destructive element in this rune. It is shaped like a lightning flash. Like lightning, Sowulo strikes suddenly, sweeping aside everything in its way, usually in order to prepare for something better (just as when the atmosphere exudes negative ions following a thunderstorm). The last rune in the second aettir, the Sun, is the solar force of life, invigoration, dedication, optimism and healing. This is another rune of power: the power of the magical will, the power to win, and the power of confidence in oneself undeterred by other's opinions. This rune can empower the individual and confers greatly the ability to persist in any endeavour. The Sowulo rune represents the Higher Self, the guiding light of the principles one holds. Sowulo will always inspire one to do 'the right thing'. Whenever in a quandary about an difficult ethical question, meditating on and invoking the Sowulo rune will often be of great help. As such Sowulo brings out the highest values and potential of the individual working with it. In adverse circumstances, focus on the Sowulo rune, as it is the rune of invincibility and final triumph. The sun was valued greatly by our forebears. In Northern mythology the deity associated with the Sun is the goddess Sunna, who is the conductor or chariot-driver of the Sun's life-force. Chariots on which Sun-disks have been mounted have been found in Scandinavia and mostly date from the Bronze Age. The combination of chariot and Sun-disk hints at a complementary relationship between the runes Raido and Sowulo. Raido is the principle of control, or the act of controlling whereas Sowulo can be the higher spiritual force which is in control, and/or the final goal. Sowulo can reveal the true purpose in life.

Spiritual Aspect: enlightenment
Emotional and psychological aspects; confidence, optimism
Negative aspect: impulsiveness, burn out.

Third Aettir

This I see as the aettir of transformation, the first aettir was largely dealing with the gods, the second aettir, with the Norns and Wyrd, time and conditions. The third aettir involves more the human condition, social aspects and spiritual transformation.

Germanic name: TEIWAZ ↑
Anglo-Saxon name: TYR
Old Norse name: TYR
Phonetic value: T
Traditional meaning: the god Tyr

The third aettir is called the Aettir of Tyr as is the name of the first rune of this Aettir. Tyr is one of the oldest gods, known to have been 'top dog' before Odin took over. This we know from various historical writings, however the Norse pantheon actually was a lot more 'democratic' to use a modern word. The stratification of the gods is largely due to Snorri Sturluson a classical scholar, who, thank the gods, recorded most of the material and modelled the Norse pantheon on the Greek. Tyr nevertheless is one of the most important gods. Whereas the Ansuz rune refers to Odin, Tyr is actually named. The Tyr rune is a Warrior's rune. It is invoked and inscribed on weapons, for example a sword to ensure victory. Tyr represents the right course of action in a military conflict. Most of us however never will be confronted with actual battle in the military meaning of the word. Most of us though have small battles all the way through life, usually with other people and cheating companies. For most of us a shortcut to victory is a most welcome concept, it is however not that easy. This rune is ruled by Tyr, the god of justice! Not the god of lawyers and politicians, no, common justice of the people. Tyr was never exclusively a war god or battle god. He also is strongly involved with arbitration and the 'Thing' a folk meet where disputes were resolved by agreements and peace was kept. Tyr however is uncompromising in his judgment. He is impeccably fair and even handed. Tyr is a warrior motivated by a sense of justice. He is more of a god of law and order, governing social values, legal contracts and oaths.

The Teiwaz, or more popularly, Tyr rune can only be used to win a dispute if one is morally and ethically in the right. The power of this rune cannot be perverted for one's own ends.

Attempts to do this have already lead to disastrous consequences in some people's lives. This is the rune of the 'spiritual warrior' the values of it are closer to what Carlos Castaneda' modern mythical sorcerer don Juan describes as impeccability. The shape of the rune is an arrow or a spear and as such it is a weapon and it can be used to smite opposition. Teiwaz, combined with Wunjo can resolve disputes in an even handed manner to the benefit of the community rather than the individual.

It cannot be denied that, however much undesirable, warfare and conflicts are part of human evolution. Tyr is the god of natural justice, hence the Tyr rune is often of great help assuring a favourable outcome in litigations, legal warfare. Týr also is the god of the 'Thing' a folk assembly where disputes could be arbitrated. Tyr therefore would be an excellent god for judges, whereas Odin's rune Ansuz would be a good rune for a solicitor! The Tyr Rune stands for bravery, honesty and even handedness. The name Tir also is the Old Persian name for the Pole Star and the Pole star was that by which our ancestors navigated by at night. Like the Sowulo rune, the Tyr rune also is a guiding principle. Tyr was the original Sky father. The shape of the Teiwaz rune resembles Irminsul, which is a symbolic representation of Yggdrasil as the cosmic axis.

Spiritual aspects: Valour, altruism
Emotional and psychological aspects: strength and honour
Negative aspects: defeat.

Germanic name: BERKANA ᛒ
Anglo-Saxon name: BEORC
Old Norse name: BJARKAN
Phonetic value: B
Traditional meaning: birch

In the majority of the Germanic languages this rune name translates as "birch." However, the Anglo-Saxons translated it as "poplar."

The birch tree is a very sacred tree. Due to its resilient nature it was the first tree to reforest our lands after the Ice Age. As such this rune represent all trees and plant life. First and foremost this is a goddess orientated rune. Pictographically, it resembles a pair of breasts when viewed sideways from on top. This rune especially relates to the goddess Holda, in southern Germany known as Berchta, who is the patron of children and (as I found out since) domestic animals, especially dogs. She resides in the underworld; there she has a beautiful garden where the young children who die in infancy are said to go, as do dogs. The Berkana rune, in particular, refers to the processes of gestation and birth. Perthro looks like an opened-up Berkana, which suggests that what remains a hidden promise in Berkana will be brought into the open by Perthro. For this reason Berkana is also associated with secrecy and containment, another goddess who co-rules Berkana is Frigg, the goddess of silence.

Berkana is a rune of healing, recuperation, rejuvenation and purification. In Scandinavia birch twigs are used for scourging after a sauna to stimulate the circulation and de-toxify the blood. It stands to reason that there was a spiritual element involved in this practice.

The Berkana rune is very feminine and in this respect is a great healing rune for female problems. It can be used for healing and alleviating both menstrual and menopausal symptoms. Apart from all the life giving aspects, there is a connection with death as well. The goddess Berchta is known in Germany as 'the White Lady' and, according to folk lore, used to appear as a ghost whenever a person of royal birth was to die. This, the most goddess orientated of all runes, is mostly connected with the Vanir deities and Frigga. It is the rune of motherhood and nurture. This is also a rune of protection especially for children and infants. Berkana is a life giving rune, in contrast with Teiwaz the previous rune which, as we have seen, has a death dealing aspect. In the mythology the Goddess Idun is in charge of the golden apples, which the gods have to eat to stave off old age and death. Even the gods are not absolutely immortal, they are relative immortals.

Berkana is one of the birth runes, it helps bringing forth, this applied to everything from actual birthing to magical workings. Berkana is magically useful for hiding and concealing things until the right time for fruition. Berkana is the rune of growth and rites of passage from one stage of life to the next, for post menopausal women this is a most powerful rune to work with. The creative energies of growth and bringing forth can be used for magical workings or any other creative projects. As the rune of life, Berkana has, nevertheless, relevance to death and initiation and rebirth, for one cannot be without the other. Similarly Teiwaz has a life aspect to it, as one man's death is another man's bread.

Both runes complement each other as Sky father and Earth mother. Both Teiwaz and Berkana represent aspects of the Tree of Life, Teiwaz as the Irminsul and Berkana as its female counterpart.

For children, and for young girls in particular, the Berkana rune is a lovely, protective symbol. I often advise parents to obtain a silver name-plate or bracelet inscribed with their daughter's name on the outside and with three Berkana runes on the inside. This amulet, which invokes the protection of the Goddess in all her aspects, has produced favourable results.

Spiritual aspects: Growth, nurturing
Emotional and psychological aspects: mature wisdom
Negative aspect: immaturity, clinging, addictions

Germanic name: EHWAZ M
Anglo-Saxon name: EOH
Old Norse name: none
Phonetic value: E
Traditional meaning: horse

The Ehwaz Rune means 'Horse'. If the previous rune represents trees and plants this one represent animals in particular and foremost the Horse.

Horses have always been sacred. They are second only to humans in value of sacrifice. I think that for the slaying of a horse 'wer gild' was due as was for humans. Man's life often depended on the Horse. My first impression of this rune was that it represented a mare. Unlike most warriors, who rode stallions, priests of the old faith of the Aesir rode a mare. Metaphorically Ehwaz also represents the physical vehicle and may be incorporated in a healing working. On a deeper level, the Ehwaz rune represents the vehicle, or in psychological terms the persona, which is used to relate to the external world through one's own emotional attitudes. The horse is also a power animal and shamanically is often of assistance to travel to other realms. In dreams the horse is representative of the astral body. Ehwaz, moreover, emphatically relates to Sleipnir, Odin's eight-legged horse. In the continental Germanic tradition, Wodan, especially was worshipped as the god of horses, in Scandinavia this was more attributed to Freyr.

There is still a taboo about eating horse flesh, especially in England.

Even in recent times rumors have been circulated about the existence of a society of "horse whisperers" who were a magical fraternity. This may well be a vestige of one of the horse-cults about in olden days.

Horses were used for magic, in a variety of ways, one of these was to use a horse's head on a 'nithing pole', a pole with insults and curses written upon it. The night-mare is another variant of malicious magic involving horses, it consists of shaping the astral body into a horse and hunting and attacking your enemy. Personally I prefer the wolf shape, it comes more natural to me.

Lastly, Eihwaz is also the rune most associated with marriages or other forms of partnerships, and of course this includes the partnership between a horse and its rider. Horses are stronger than us. They cannot be forced to do our will, only persuaded. Therefore this rune is an important asset to any form of co-operation.

Eihwaz is, as we have seen, the rune of the horse. In human terms I relate this rune to relationships, partnerships, and friendships. Horses are highly intelligent and sensitive animals, they have their likes and dislikes much like people. This I know through personal observation, having had horses on my land. They are capable of forming a deep friendship with another horse, a friendship going further than mating matters. When the owner removed a horse, it's 'friend' became very disturbed and upset, clearly missing its companion. In any relationship horsey or human there is always the question of control: who is the boss? We can learn a great deal from horses. They argue and fight but the also will help each other out and co-operate. Partnerships are based on co-operation. This, of course, is also true in a committed relationship such as marriage, and this is also part of the Ehwaz package of meanings. A negative influenced Ehwaz rune often mean false friends and betrayal. The Eihwaz rune also has some bearing on the practice of divination as Tacitus recorded some divinatory practices of the Germanic tribes involving horses. For example, the behaviour of a horse was held to predict the outcome of a battle. Horses have always been thought of as highly sacred among the Germanic peoples. Psychologically, the Ehwaz rune can teach us the art of adjusting. Whereas Raido, a rune strongly in sympathy with Ehwaz, enables us to take control of a situation, Ehwaz enables us to adjust to a situation and make the best of it. Ehwaz therefore teaches one to be flexible, adaptable, pragmatic and most importantly to follow your instinct, your gut feeling. This is not a rune of intellectual analysis of a situation, the next one is though!

Spiritual aspects: cooperation with Higher Powers
Emotional and psychological aspects: instinct and intuition
Negative aspects: indecisiveness, treacherous

Germanic name: MANNAZ ᛗ
Anglo-Saxon name: MAN
Old Norse name: MADR
Phonetic value: M
Traditional meaning: man

Whereas the previous two referred to plants and animal kingdoms respectively, this the third rune of the third Aettir refers to 'man'.

Another example of the logic and reason behind the sequence of the runes in the futhark. The previous rune, Ehwaz, was mainly related to animals, in particular to horses. The Mannaz rune is similar in shape to the Eihwaz rune, which makes sense. Man is after all an animal with above-average intelligence. It goes without saying that 'man' means homo sapiens. In the Anglo-Saxon branch of the Germanic language, 'man' denoted not just the male section of the folk.

This rune not only means "man" or "mankind", it is also the name of mankind's ancestor and progenitor. In the Continental Germanic tradition Mannaz, who is mentioned by Tacitus as the progenitor of the Germanic people, was the son of Tuisco, who was Earth-born. Tuisco is an older form of Tyr; thus Mannaz was a tribal ancestor god. He, in turn, had three sons: Ingvio, whose name appeared in the Inguz rune, Irmio and Istio. These are also the names of the three main branches of the West Germanic tribes. He corresponds to Heimdallr who, according to the Rigsthula (one of the poems of the Eddas), is the progenitor of the three classes. This poem describes how Heimdal sets off on a journey to Midgard and is the guest of three married couples in turn. He spends the night with each couple, sleeping between the man and the woman. Later all three women give birth to his sons. The Mannaz rune signifies cooperation between people sharing the same environment for the benefit of the whole of the tribe or community. Mannaz is a double Wunjo. We already discussed that Wunjo is the rune of the clan.

Mannaz expresses the development of man's intellectual powers, and of his awareness as co-creator of Nature. Initially primitive and subject to the environment, man (via money) now unfortunately has too much control over Nature and is destroying his environment. However, it seems that the Mother has had enough! Shame about the plants and animals though.

Mannaz is the rune of man, the human condition, social structures and responsibilities. Like Wunjo this rune refers to people as a collective, be it a family a village or a magical fraternity, not the individual per se. Mannaz is the rune of the rational mind in contradistinction to the previous rune being more in tune with the instinctual side. Intelligence is the greatest power of humans, being composed of thinking and remembering, the functions of Hugin and Muninn Odin's ravens. Mannaz is known as the rune of the perfected intellect melding reason and intuition, with reason being the senior partner. There is a connection with Mimir, the giant Odin consulted with at the Well named after him, the Well of Mimir, or the well of memories. In divination this rune refers to people in general. The sort of people referred to and their relationship with the querant can be deduced from the other runes appearing with it in a reading. Inverted Mannaz sometimes indicates an enemy or at least indicates opposition offered by people, not necessarily men. In readings Mannaz may suggest legal affairs and matters of mutual cooperation, especially when it occurs in conjunction with Ehwaz. Together with Raido, it will often denote assistance from, or counsel given by people.

Magically, this rune can used to attract support from one's peer-group in a dispute. Combined in a bind-rune with Ansuz, it can be employed to win an intellectual argument or pass an examination. Both runes strengthen the mind when necessity arises. An even stronger sigil for this purpose can be constructed by adding Raido and Ehwaz to the bind-rune.

Spiritual aspects: intelligent awareness, open-mindedness
Emotional and Psychological aspect: awareness of kinship

with all humans as part of the greater whole.
Negative aspects: intellectual arrogance, bigotry

Germanic name: LAGUZ ↑
Anglo-Saxon name: LAGU
Old Norse name: LOGR
Phonetic value: L
Traditional meaning: lake or water.

Most likely the sea, or a large lake. Some of the Germanic
tribes lived near the sea and some more inland. The goddess I
associate with this rune is Nerthus. A goddess not much heard
of in the Norse mythos but who in Germany and Holland,
according to the Roman historian Tacitus, was once held in
a similar regard as Isis. This rune, therefore is predominantly
feminine. The goddess Nerthus is probably the oldest goddess
known from Germanic sources and she was worshipped on an
island in a lake, possibly in Frisia. She was supposed to bestow
blessings wherever she visited. Once a year, everyone laid down
their weapons while her wagon was ritually driven around the
mainland, a remnant of this was observed in medieval Holland,
where a decorated ship was used in processions. Laguz was also
probably connected with the custom of ship-burial and Odin
as the Ferry man.

In Holland we know of a native goddess named Nehelennia.
Statues dating from the first century has been found in
Walcheren, one of the islands of the Dutch province of Zealand.
She is portrayed with a dog and a basket of apples, and she was
sometimes described as holding a horn of plenty. She may be
a localised form of Nerthus. Laguz represents the waters of life
without which we would not be, but it also represents death,
as many of our people, especially those from the Lowlands, lost
their lives at sea.

Laguz is nevertheless one of the runes of healing, and initiation.
In may traditions water is used to 'initiate' a newborn into the
clan. In modern magical work water plays a major role.

Laguz, according to some German writers, also means love. There is some truth in this interpretation, for if Laguz symbolises the waters of life, it may represent the forces of mutual attraction called "love." In fact the shape of Laguz is that of a half-Ehwaz, as has been already explained, and Ehwaz is the rune of partnerships and marriages. The meaning of the Laguz rune as we have seen is on first glance, water, sea or lake. This is the rune of 'the deep', the hidden ebb and flow of the currents of life, as water is the first (well, second after oxygen) necessity of all life, be it plant, animal or human. To pay a visit to Loch Ness, with or without monster can confer an appreciation of the meaning of the Laguz Rune. Laguz is a very occult rune, deep undercurrents of the collective unconscious seeping and shaping wyrd. Laguz has a sympathetic resonance with Perthro, as both are interconnected in the Well of Wyrd, metaphorically and psychologically. As is well known from dream experiences, water is representative of the emotions. As the emotions often precede a deed they are first and foremost the medium of shaping Wyrd. To master rather than control or suppress emotions is the objective of the aware person into order to shape or alter wyrd. To dive into the deep of once own mind, accessing all negative or hindering emotional baggage, without fear of failure and to use the obtained knowledge constructively to rewrite the script (so to speak) will endow the individual to transmute these into helpful lessons for personal and spiritual growth. The Laguz rune being representative of water therefore also can be perceived of symbolising the 'etheric' or 'astral' forces. Water will assume the shape of any container it is poured into, so too will astral substance take shape of whatever the imagination and visualisation will dictate. The Laguz rune therefore can be used for all magical workings involving astral plane work.

Spiritual aspect: emphatic, sympathetic.
Emotional and Psychological aspect: the deep mind and the collective unconscious
Negative aspects: winging manipulative, lack of moral fibre.

Germanic name: INGUZ ◇
Anglo-Saxon name: ING
Old Norse name: none
Phonetic value: NG
Traditional meaning: the god Yngvi - Frey

The name of this rune is the oldest name of the Scandinavian fertility god Yngvi or Frey. Nothing is known of him from Germanic source material but as his name appears in the common Germanic Futhark, which incidentally is the oldest recorded rune row, he most certainly was known. His name has been preserved by the Anglo-Saxons in their rune poem. Frey was especially important among the Danes, he was a god of peace and plenty. Out of all the gods probably the nicest one for humans to follow. (Having said that, I stick with Odin though!) The Inguz rune is associated with agriculture, very much so. Frey is not just a god of male and animal fertility, he is also the god of vegetation and food, a far more civilised character than 'Old Pan' for example. However, he definitely is also the god of male sexuality, and that includes male homosexuality. Although frowned upon in the Viking Age, it wasn't actually illegal and priests of Frey in Sweden engaged in cross dressing, dancing and were involved in Seith, a magical discipline usually preserved for women. But all this does not exclude the notion that Frey is one of the gods of procreation. If we take the Anglo-Saxon form of Inguz and duplicate it a few times: we can see that it has a remarkable similarity to a double helix, the shape of a D.N.A. chain. Inguz is the carrier of genetic material and confers upon the individual inherited characteristics of his or her ancestors.

Inguz is closely related to Kenaz and Jera and can be seen as a progression of these runes. We can view Kenaz as either the male or female half of a polarity, depending on the gender of the person who is working with it. With Jera we see two similar shapes circling around each other. In the shape of the Inguz rune we find both halves joined and integrated, symbolising completion, totality and fulfillment. The Inguz rune is closely connected to the earth and the land, as is the next rune Othala. Inguz especially is the rune for 'seed', gestation and germination. Frey, as we have seen, is the god of vegetation and food, and he is in that respect the Lord of the good earth, the nourishing earth. Yes the earth is female and often seen correctly as the mother of all what lives on her, but she cannot create alone. The Inguz rune represents the male aspect of nature in this respect. But that's only one half of the story, the shape of the Inguz rune is reminiscent of the female genitalia. This rune then contains elements of both genders, perhaps this is the reason that in modern Asatru the god Frey is very often the patron god of homosexual man, and men who have integrated their female side. To complicate matters even more Inguz furthermore is associated with children, both female and male. As Berkana is the rune of pregnancy and Perthro the rune of birth, Inguz is the rune of the child. In Anglo-Saxon and Frisian, patronymic names were formed by adding -ing after the father's forename. This is the usual significance of the "-ing" in names like Bunting and Hadding which are found in England and Holland. The old Swedish royal family was called the Ynglingar. They were supposedly descended from Yngvi-Frey Like Gebo, Inguz is perfectly equilateral in shape, it looks the same from all angles, there is a deep magical significance in these runes combined as we perceive when we examine the next rune.

> Spiritual aspect: complete integration of the four Selves; physical, emotional, mental and spiritual.
> Emotional and Psychological aspect: the inner child
> Negative aspects: frivolity, immaturity

Germanic name: OTHALA ᛟ
Anglo-Saxon name: ETHEL
old Norse name: none
Phonetic value: O
Traditional meaning: inherited land

Inherited land or family estate are the meanings associated with this, the seventh rune of the third aettir, and quite correctly. But there is more to this when we examine the literally meaning of the word Othila. The oldest literal meaning is "noble." In is directly related to the Anglo-Saxon word 'atheling', meaning "prince" or "noble."

The god who is most obviously related to this rune is Odin. The shape of Othala is a combination of Inguz and Gebo, and Othala can be interpreted as the "gift of Ing," which probably explains why Othala has been related to the concept of inheritance by some modern commentators. By comparing the concept of inheritance to the Inguz rune, the previous rune in the futhark, it can be established that it is genetic material that is being inherited. Genetic material contains the Hamingja of the ancestral stream into which one is born. This was even more so in case one was of noble birth. In the Germanic tribal system the king was believed to be descended from the gods, usually from Odin but sometimes from Frey or Tyr. Furthermore, the king was seen as the bearer of the Hamingja of his people. It was customary that when a king ran out of luck and could no longer guarantee prosperity and the power of fertility for the land, he was sacrificed. And the land as well as the Hamingja or 'luck' of the country or tribe was inherited by the successor. The son became the next king if he was fit for the office; if not, another member of the royal family would succeed to the throne. Therefore the choice of the king's wife was considered especially important in light of the belief that the offspring would be the next to bear the Hamingja of the people and would inherit the king's "luck" and vital abilities, such as courage and wisdom.

Esoterically then the concept of inheritance can be established as the genetic material that is being inherited. This genetic material physically speaking has all the genetic components, either active or dormant of the whole of the human race,(and possibly more than that). I just used the word 'race', this is one of the runes which has been defiled in the past, to, justify racism. The rune Othala does represents the inheritance of the ancestral lands by a specific people, and that applies equally to all specific people, be they Germanic, Celtic, Aborigine or Native American. Each people has a mystical bond with the land, but I would like to expand on this by now largely redundant definition and state that in its broader concept the Othala rune refers to the inheritance by the whole of humanity of the planet itself and the inherent responsibilities thereof. It is no longer a matter of natives inheriting their lands, be it Germanic, Celtic natives, native Americans, we all collectively have inherited the planet itself, this most precious heritage we have to pass on to our descendants.

Spiritually, the Othala rune can be used to invoke Odin in his aspect of wanderer and teacher. Odin as we have seen with Ansuz and Wunjo, has three main aspects ascribed to him, mythological his brothers. These are Odin, Vili and Ve; these names correspond to his aspects of warrior, shaman and wanderer, respectively. Besides being one of Odin's names, "Ve" also means "sacred enclosure," which was that part of the hof or temple where only the officiating gothar had access. It was the place where the ritual regalia was kept. The Othala rune has been associated with an enclave by other rune-workers, and an enclave invokes images of safety and protection. It is mythological comparable to the walls of Asgard, which were built as a protection against the giants.

Spiritual aspect: Connectedness with all Earth dwellers.
Emotional and Psychological aspect: Love for one's home and security thereof.
Negative aspect: Racialism.

Germanic name: DAGAZ ᛞ
Anglo-Saxon name: DAEG
Old Norse name: none

Phonetic value: D
Traditional and literally meaning: Day

The associations with this rune are primarily with either the dawning of the new day, with the mid-point of the day, when the Sun is at its zenith, at noon and with setting sun at the end of the day. The two tides of rising and setting sun are most powerful for working magick, the other realms are more accessible.

And of course the 'Witching hour' at midnight is another transition point of the sun, namely its absence is a favorable time for magick works. As a spiritual discipline one may compose a short invocation to one's Deity and perform this at these four aforementioned times. It is remarkable however, that the Germanic people did not have a rune for 'night' but only for 'day'. They counted their days by nights. In English this is even preserved in the expression "a fortnight (fourteen nights) from now..." Likewise there is a rune for 'sun' but not for 'moon'.

Dagaz is assigned to the controlling power of Loki and, to a certain extent, Heimdal, Loki's counterpart. The 'hidden' invisible transforming power of Dagaz is indisputable, Odin. Dagaz is a hidden doorway to the other worlds: yes, other planets and universes and all.

The shape of the rune resembles a lemniscate, the symbol of infinity. It also resembles a Moebius strip, a symbol of timelessness and unlimited possibilities, as such Dagaz takes us beyond time and space. This is, in my opinion, correctly assigned as the last rune of the futhark completing a cycle—which, in actual fact, is a spiral pathway of evolution, and continues on the next level. Dagaz is above and beyond all levels of being. It is both being and non-being and stands for the supreme mystery of existence. Dagaz operates between light and darkness, mediating in both directions but partaking of neither. Dagaz synthesises, transmutes and dissolves all opposing polarities. This rune can transform consciousness, and it is the most appropriate rune for the purpose of initiation. This is the last rune in the futhark in my opinion, however it is fair to say that others have chosen

Othala as the last rune. Historically both are possible according to archaeological finds. The reasons that I allocate this rune as the last one, are metaphysical in nature. Dagaz as the last rune when laid out in a circle will appear opposite Jera, as Jera represents the winter solstice when the darkness is greatest, Dagaz represent the summer solstice when the light is strongest.

Thus, like Jera, Dagaz is a rune of change. Jera, as mentioned previously, is a rune of gentle change, whereas Dagaz is the rune of cataclysmic change. Whenever energy reaches a saturation point, it is forcefully converted into its opposite nature. Whatever is fully positive will turn into a negative. At this point Dagaz fuses with Fehu and initiates a new cycle. Dagaz is a rune with many layers of kenning. It represents the end of an era and the beginning of the next cycle. It acts as a catalyst initiating change without changing its own nature. Dagaz is the rune of transcendence. Most of the runes partake of Fire and Ice. Some are fire. Some are ice. Some are both or neither. Dagaz transcends this duality. Dagaz goes beyond duality to synthesis, unity and transcension to another reality. Dagaz is the rune which can confer cosmic consciousness   Dagaz can be used to combine other runes in a synergistic bind rune, to braid and weave runes for a specific working into a framework of Dagaz will fuse their energies and combine their powers to be used for an act of magical intent.

On a very mundane level, Dagaz can be used to hide things from view. An experiment with this could be conducted as follows, select an object such as a photograph of someone or a vase, place this in a fairly prominent position in your room (however not too obvious). Trace a Dagaz rune over the object, with the intent to make it invisible. See whether anyone coming in notices it. Likewise with practice, one can shield oneself in Dagaz and become invisible to other people. What happens is that they don't consciously register you. It is extremely helpful for jumping queues. It has saved me no amount of time and bother over and over again.

Spiritual aspect: total enlightenment

Emotional and Psychological aspect: positive focus in life.

Negative aspect: destructive impulsiveness

COMMON GERMANIC FUTHARK

Rune	Name	Traditional meaning	Letter	Meaning
f	FEHU	Cattle	F	Money, wealth, hamingja
u	URUZ	Aurochs	U	Health, strength, endurance
T	THURISAZ	giant	Th	Conflicts, annoyance, strife
a	ANSUZ	god	A	Communication, inspiration, ancestors
r	RAIDO	riding	R	Control, direction, right or wrong, advice
k	KENAZ	torch	K	Knowledge, learning and teaching
g	GEBO	gift	G	to give, to receive, to share or agree.
w	WUNJO	joy	W	Joy, perfected Will
h	HAGALAZ	hail	H	Disruption, unresolved matters, the past
n	NAUTHIZ	need	N	Necessity, friction, warnings
i	ISA	ice	I	Restrictions, delay
j	JERA	year/harvest	J/Y	Harvest, year, seasonal/karmic returns; the Earth
I	EIHWAZ	yew	EI	Yew, Tree of Life/Death, Yggdrasil
p	PERTHRO	fruit(?)	P	Womb, space, Well of Urd/Mimir
z	ALGIZ	protection	Z	Protection, defence, wardings
S	SOWULO	sun	S	Life-force, clarity, understanding
t	TEIWAZ	Tyr	T	Spiritual Warrior, Justices, Thing
b	BERKAND	birch	B	Great Mother, nurturing, protecting children
e	EHWAZ	horse	E	The Body's various vehicles for travel and movement
m	MANNAZ	human	M	Intelligence, reason, Law
l	LAGUZ	water	L	Feeling, intuition, empathy
N	INGUZ	Yngvi/Frey	Ng	Fruitfulness, peace and plenty
O	OTHALA	inherited land	O	Genetic, cultural and spiritual inheritance.
D	DAGAZ	day	D	Day, dawn, enlightenment

tHREE

Divination with Runes

Divination is an art, in which some practitioners are more naturally gifted than others; but most people, nevertheless, have sufficient innate ability to become reasonably proficient – be it with Tarot, Astrology or the runes. Where the runes differ from more familiar, widespread divinatory systems is that they are, on face value only twenty-four angular signs. They are more sparse in appearance than Tarot decks for example, which might seem, at first sight, a disadvantage. However, as we shall see, these simple forms have a remarkable capacity to stimulate the clairvoyant imagination.

Competent rune-workers disdain the use of the so-called "blank rune" – a spurious modern innovation, without any traditional meaning. After all, all there are no such things as blank letters of the alphabet or blank Tarot cards! This bogus practice probably started when some foolish novice bought a set of runes from a supplier whose habit was to make one or two blank slips of wood as spares with each rune-set, so that if a rune were lost or damaged, it could easily be replaced just by carving it on the blank. The ignorant buyer assumed the blank pieces had some special significance of their own – and this idea was soon incorporated into a badly researched book on the runes. Whatever the case may be, there is no doubt that the twenty-four rune system of the Elder futhark is complete in itself and needs no modern, New Age additions. The numerology works well: there are twenty-four hours in a day, twelve months in a year and eight recognised festivals in modern Pagan tradition, Nordic as well as Celtic. The blank rune muddles up and invalidates this rich system of correspondences.

To be able to divine with runes one has to develop a high degree of intuition and invest a measure of trust in one's own ability. To be able to perceive accurate data relevant to someone having a reading, one must make a sympathetic link with the runes and even more important, one must imprint the subconscious mind with each one of them. This has to be done on at least two levels. Firstly, one must read everything one can about runes, but with an critical mind, asking questions all long the way, and referring always back to one's own understanding. Second, to infuse the runic symbols into the deep mind, use them for art! Play with them, make pictures; write your name in runes and see whether the runes of your name spell out something you recognise as part of your personality. Combine the runes of your name into an aesthetically pleasing sigil and use this as a signature to mark your personal property, such as books. For those who are right handed, draw runes with the left and visa versa; this will link in with the intuitive part of the brain, the right hemisphere. Simple exercises like these, enjoyable in themselves, will give intuitive understanding of the meaning of runes.

A daily meditation on one rune in futhark order, with notes taken of any thought coming up during the meditation will prove most helpful; it will lay the basis in twenty-four days for a practical working knowledge of the runes. After this, do the same exercise again but at random. Just pull a rune out of the pouch and observe over the next twenty-four hours any events that take place, no matter how apparently insignificant, which are in keeping with the meaning of the rune. A silly but accurate example would be if the rune is Raido inverted, and the next day you miss the bus! Observe and record simple happenings like this, for future reference. If you are sceptically minded, it might occur to you that something is bound to happen, each day, that would fit any of the runes, so the connections you observe are just 'coincidence'. This is partly true, but it misses the point. By routinely observing 'coincidences', you stimulate your clairvoyant instincts – to the point where, increasingly, you just know that something specific is going to happen as soon as you look at the rune. And when it does happen, just as you expect, you will know this is beyond coincidence.

Concerning the interaction between the client and the reader on a subliminal level during divination, I have some new probabilities to evaluate based on my practice doing readings and my close observation of other readers. I think that on a subliminal level, in trance or clairvoyantly, the interactions between the two parties are a lot more involved and complicated than hitherto suspected. I have observed people setting so much store on the revelations of readers that they actually create the reality as 'predicted' to them. It really made me examine my own position, and the responsibility. Suppose that once a rapport has been established between the two parties, for example by act of money being passed from one the other before the reading, a unified field is formed in the web of wyrd between A and B. B, being the reader, has access to A's data by interacting subliminally with A's subconscious. How much is projection and reflection from A to B? And would this create a reality which then is picked up and 'predicted' or advised? Perhaps a third party should be invoked before each reading to 'supervise' the operation, like the triune A plus B and whatever deity is appropriate: Odin, Norns, Frigga, or Freya to give a few examples. When working with runes for divination it is always a good idea to pay respect to Odin, with a simple acknowledgement or thank you.

When you feel you're ready, try to ask a simple question and pull one or more runes to see whether the meaning of these runes can be interpreted as an answer. On your birthday, draw a circle with twelve segments. Pull one rune in each one, note them down and see whether they forecast something in the next twelve months. After a year you will recognise when looking back that certain runes did mean certain things. You can also do all of these little exercises for your friends and family – just to get familiarity with the runes, their meanings and how to interpret them, in the light of specific issues. Runes are excellent for answering questions. They give a direct accurate answer if the question is properly phrased, including simple yes/no questions.

When you feel ready, you should make your own set of runes, preferably from the wood of a fruit-bearing tree. Avoid commercially produced runes, which are sometimes supplied with books. If you have difficulty crafting your own runes, ask a trusted friend or kinsman/woman to make them for you, in exchange for an appropriate gift. Runes are magical tools; therefore it is wise to consider whose energy they will carry as well as you own. Once you have made your runes, you have to infuse them with personal might and make them your own. Sleep with them in a pouch carried under your clothing around you neck or waist. After nine days and nights, construct a simple dedication ceremony in which you ask for the blessing of the gods or goddesses of your choice. Especially recommended are: Odin, Freyja and the Norns.

However, having asked for the help and blessing of the god/desses in this way, if you should ever make money out of a reading it is strongly recommended that you give nine per cent of your earnings to a charity of choice, preferably a environmental one — trees or animals for instance.

Runic divination is a kind of mediumship, in which one has to mediate between the querant and the runes and interpret messages from spiritual levels, of which the physical rune-stones or rune-staves are focal points or doorways. Divination is, in origin, a sacred act of asking counsel from the gods, as the word 'divine' from which divination is derived clearly suggests. Long-term forecasts, however, cannot be as detailed. They may contain contradictions as well, since before they are realised the individual concerned may make certain decisions or carry out various actions, which will have changed the pattern of the web. There is no such thing as a fixed future; there is only the web, which appears to us in time-bound and space-bound conditions as past, present and future. In reality, the web is a complex whole and, when understood as such, can be utilised to exercise a degree of control over one's circumstances. The so-called 'past' is actually part of the present; it is that part of one's wyrd, which actively creates and shapes current circumstances. As such, the past is not a clear-cut, separate realm of non-existence; it has not ceased its influence on the so-called present. By means of divination, it is possible to gain information concerning a specific aspect of the web, for example the past. Because the web is an integral whole, any interference with one part of the web causes reverberations in other parts of it.

Once the information needed has been acquired, magic can be applied. Magic is the act of interfering in one part of the web through operation of the will and thereby deliberately altering wyrd. The power of magic and divination in the Northern Tradition is most strongly associated with Odin. These powers in terms of past, present, and hypothetical future time are identified with the three goddesses or Norns: Urd, Verdandi and Skuld who weave the web of fate or Orlog. Divination has a close relationship with magic. After all, there is little point in investigating future probabilities if not with the intent to magically manipulate these probabilities in one's favour. Definite predictions can certainly occur as part of a rune reading, but usually happen spontaneously and not necessarily under the conscious control of the reader. If you view divination solely as a means of prediction, you will find that it is not reliable.

This is only to be expected. Matters which are hidden in the future and predestined are usually unavoidable and form part of a person's own wyrd – they are consequences of one's own past actions, one's own being and inner nature. But the more the individual is conscious of the hidden side of Nature and of his own self, the more control he has over his circumstances, in which case it is all the less likely that undesirable "predictions" will come true.

PROFESSIONAL ETHICS

Two simple ethical principles should be followed when doing readings for other people, commercially or otherwise. Firstly, no matter how bad the reading looks, never predict a disaster, merely point out dangers and suggest ways of minimising them. Second, don't assume expertise you don't possess; instead, give accurate advice on whom to turn to for specific needs. For example, I've picked up more than once on serious medical conditions in a client. In cases like this, never speculate – just tell them in the strongest possible terms to have a check up as soon as possible. When doing readings on a regular basis in your own area, familiarise yourself with support agencies such as: local Citizen's Advice Bureaus, social services, rape crisis centres, STD and family planning clinics, Samaritans, etc. so that when you encounter a particular problem, you can offer the client a sympathetic hearing but refer them on! Third, respect your client's confidentiality. To be a decent psychic consultant, one has to develop a working ethic similar to a priest(ess) or social worker. If you decide to 'go professional' in this field, it will be very helpful to take a course in co-counseling. Most local authorities run affordable courses in this field. Many people coming for readings are lonely and troubled, perhaps going through one of the familiar life-crises: adolescence, parenthood, menopause, or retirement. Rather than dazzling displays of your psychic powers, such people really want a sympathetic ear and some common sense advice. Sometimes, however, one may come across a serious problem case, such as people with evident psychiatric needs. I once had a young lad coming up to me in Camden Psychic centre who insisted he was 'The Beast' and was looking for confirmation by begging every psychic there to find the number 666 tattooed on his body! If you take up reading professionally, you will meet someone like this before too long! However, most people are relatively normal – and common sense, a sense of humour and objectivity will go a long way.

EXERCISES FOR BEGINNERS

Before you start, make sure you are clear of mind. Take your time and be sure are well rested and not distracted by noises, phone calls and such. Make sure you know why you are doing the reading you are doing. Avoid people who just want you to predict their future. Wyrd is flexible; there is no such thing as a fixed future. Make sure the energy between you and the querant is sympathetic. If you feel uncomfortable reading for any particular person, say so. Get a firm and clear statement of intent that you querant is open minded and willing for you to do the reading now. Take a moment to consider any possible expectations either you or your querant may have and clarify these as needed. Make sure that you and your querant both are willing for the duration to believe in and open yourself up to divine insight and the "magic" of the Runes. Proceed with Spiritual preparation and Invocation. Take a moment to fully feel the presence of your personal spiritual guide, call on Odin, Freyja or the Norns or any god or goddess of your choice. Ask for help and guidance from Deity as you understand it and formally set your personal interest aside. Finally imagine a connection of energy, for example a circle of light around you and the querant.

If you are a complete newcomer to psychic readings, I can suggest ways to gain some practice and experience firstly, in developing the ability to 'see' and deduce certain facts from the layout of the runes, and secondly, in presenting the information to the client. You may be highly intuitive in picking up things but not very skilled in communicating them in a manner that is helpful and comprehensible. Buy a daily newspaper or a woman's magazine and turn to the problem-page. Pick a problem letter and, *without* reading the agony-aunt's comments first, do a reading on it, as if the letter-writer were your client. Copy the letter out and lay it in the middle of your cloth. Focus on it and lay the runes out face down. Turn the runes over and see whether the runes correspond at all with the issues written about, then imagine you have to speak to a client and advise her or him. How would you put the information in words? How would you present it? Only after you have done this, read what the agony aunt has said. It will be mundane, of course, but usually reasonable, common-sense advice – and professional agony aunts (or uncles) have to be able to write clearly and comprehensibly.

To develop some real clairvoyance, get a friend to select a problem-letter from a paper or magazine and write it out. Have her seal this in an envelope and lay it in front of you as you do a reading. Try to see whether the runes make sense. At first, look at the runes and try to 'guess' what the possible issues are in the letter, and write your impressions down. Then open the envelope and compare your impressions with the actual letter. Give yourself marks like four out of ten or six out of ten. Do this once a week and keep a diary or journal of the results. You will soon find that you're getting accurate readings.   Allow yourself to spiritually ask for and receive inspiration. Consciously look for information and interpretations beyond what you "think" the runes mean or have known them to mean in the past. Pay attention to "little" details such as runes falling out of the bag, slips of the tongue and other unexpected indications of Spirit.

CONSULTING THE RUNES

Runes can be consulted by two main methods: casting and laying of the runes. The most ancient method is described by the Roman historian Tacitus in chapter 10 of his 'Germania' (ca98 CE). A branch was cut from a nut-bearing tree and cut into slips, on which 'signs' were written; these slips then were thrown randomly onto a white cloth. After this the priest, or if the reading was private, the head of the household, would utter a silent prayer to the gods. Facing northwards, he would take up three slips, one after the other and interpret their meanings. The signs used in Tacitus' time have not survived, but they were almost certainly early versions of the runes. Similar casting techniques are mentioned in Icelandic sagas, written down some twelve centuries later; the technique, therefore, stood the test of time! Casting the runes then, is basically the technique of throwing them from a cup or a pouch onto a white cloth, on which may be embroidered a design outlining specific areas with special meanings – for example, a circle divided into twelve or eight segments.

The advantage of casting the runes is that it allows the powers that be – or the forces of chaos, if you prefer – to determine the cast, which can then be interpreted as a whole. Alternatively, one may prefer, as I do, a more controlled method whereby the runes are laid face down and selected by the querant. This can be done by moving the hand slowly and deliberately over the runes, asking oneself if they feel hot or cold. Selecting them on this basis, one at a time, the querant lays the runes out on the cloth, face down, according to a pre-determined pattern. (I will discuss various layouts later on.) The reader then turns up the runes one after another and interprets each one in the light of its position. When using a structured layout, the interpretation can be enhanced by cross-referencing each rune with those in related positions such as the one opposite. This can suggest deeper insights and subtleties which contribute to a thorough consultation. The advantage of this method is that the subconscious of the querant can guide the hand while selecting relevant runes, and thus will communicate specific issues to be discussed.

Before we look at practical examples of reading the runes, one final aspect of the theory of runic divination must be discussed: the distinction between invertible and non-invertible runes. An invertible rune is a rune which, when placed upside-down, looks different from its positive or upright version. Thus, for example, Fehu: ᚠ is normally shown in its upright form, but inverted it looks like: ᚼ Fehu is therefore an invertible rune. However, Eihwaz, when placed upside-down, looks the same. It is therefore a noninvertible rune. Only fifteen out of the twenty four runes can be inverted, while nine runes of special power cannot. Inversion of runes negates the meaning of those runes in their positive upright positions and consequently turns their meanings opposite. A very simple example would be Fehu. At its most basic level, it would normally mean money. Inverted Fehu, however, would mean lack of money. Now clearly, this reversal of meaning is only possible when the runes actually can be turned upside-down.

Sometimes it can be difficult to know whether the light - meaning or the dark-meaning side of the rune is shown in any given reading. For example, Uruz inverted may be interpreted as a weakness. Uruz inverted therefore takes on the meaning of the opposite of Uruz upright. Thurisaz, however, as we have seen, is quite aggressive in its normal upright mode. Inverted it become more defensive rather that aggressive and may therefore take on a more positive function. Different spreads may be used to define what is helpful and what is an obstacle. It may also be good to phrase your question so as to leave no doubt for instance, "What kind of problems will I face in my new job?" In a more general reading, certain runes, which I refer to below as "companion runes", may help to brighten the aspects of others. If the rune in the primary place is companioned well, its better meanings may be taken, but if it is companioned with one or more of the darker or more difficult runes, it may represent an upcoming challenge. Some of the runic meanings or divinatory indications overlap; when these runes appear together it strengthens whatever aspects they have in common.

BASIC GUIDELINES IN RUNIC MEANINGS FOR DIVINATION

The basic meanings of the runes applied in divination are listed below. These are only generalisations and guidelines. Do not apply them mechanically – always use your own intuition and judgment first. These meanings are very simple and on the most basic level serve as the background for a straightforward reading in which questions of a practical nature can be answered.

Fehu: Financial strength and prosperity of the client in the present and near future; inverted or in combination with Nauthiz, Isa or Hagalaz: possible indebtedness.

> In questions of finances: money or trouble with money, quarrels over money.
> In questions of romance: sexual attraction or jealousy.
> In questions of health: life-force or fever.
> In questions of the workplace: energy, financial improvement or trouble with co-workers, burnout.

> Companion runes: Gebo, Jera, Berkana, Inguz

Uruz: Health matters, positive or otherwise depending on other runes selected; with Thurisaz or Teiwaz, possible surgery. Inverted, usually a lack of physical stamina. Natural aspect: rain, water. Vital strength, cleansing, healing. Badly aspected, it can show uncontrolled power.

> In questions of finance: vigour or an uncontrolled investment.
> In questions of romance: welcome strength and energy or coming on too strongly.
> In questions of health: good health, healing; always a good rune.
> In questions of the workplace: enthusiasm or a bull in the china shop.

> Companion runes: Perthro, Laguz, Sowulo

Thurisaz: Conflicts and complexities of an aggressive nature, disputes, psychological problems. "Thorn". Natural aspect: thunderstorms "Thurse", an elemental giant. Lightning, breakthrough, disruption, aggressive male sexuality, battering down barriers. The "thorn" can also be the thorn of awakening. Used in warding and in works related to magical and/or bodily combat.

In questions of finance: disruption, conflict.

In questions of romance: strong male energy or badly disruptive male energy.

In questions of health: for a male, the genitals; for a female, harm or pain to the feminine parts.

In questions of the workplace: a strong attack, trouble, strife.

Companion runes: Uruz, Sowulo, Teiwaz

Ansuz: Communications and transmissions; points things back to sources in the past. Also authority, elders, possible promotions in job related readings. Natural aspect: wind. "God": Odin.

In questions of finance: answers, open paths of communication, advice or bad advice.

In questions of romance: communication.

In questions of health: relates to the lungs, air-passages, mouth, and throat.

In questions of the workplace: communication, inspiration.

Companion runes: Raido, Kenaz, Wunjo, Mannaz

Raido: What is right or not right; what move to make; decision making. Travelling, journeys, moving house or job. Natural aspect: the sun's journey "Riding". Timing; ritual activity, especially dance and processions; proportion, legal matters. A journey; used magically to ensure that a journey goes well and safely.

In questions of finance: regularity, consistency.

In questions of romance: the relationship is proceeding on the proper path.

In questions of health: a gradual, natural process of healing or a slow decline.

In questions of the workplace: regular progression; may also show an upcoming business trip or change of job.

Companion runes: Jera, Sowulo, Eihwaz, Ehwaz

Kenaz: Opening up of new ways; opportunities; information, solutions, creativity, studies. Natural aspect: burning, decay. "Torch", also "boil or sore". Craftsmanship, dissolution and recreation; the raw power of fehu controlled. Its darker side is dissolution without reformulation.

> In questions of finance: control.
> In questions of romance: sexual energies sublimated or directed towards a goal.
> In questions of health: infection, inflammation.
> In questions of the workplace: creativity, skill.

Companion runes: Uruz, Teiwaz, Inguz

Gebo: All matters of an exchanging nature; contracts; personal relationships. Gifts, favours and obligations. Agreements of a reciprocal nature. Exchange of power or money, a wedding, a relationship between equals, a two-way flow of communication, mutual oath-taking, mutual loyalty. Also the relationships of sacrifice and blessing between humans and the god/desses.

> In questions of finance: a good partnership.
> In questions of romance: a good relationship.
> In questions of health: balance, taking care of one's body properly.
> In questions of the workplace: a good working partnership.

Companion runes: Ehwaz, Mannaz, Othala

Wunjo: Gain; accomplishments; that which is wished for; success. When inverted or badly aspected with other runes, caution. Natural aspect: warmth "Joy". Happiness, calm, the well-integrated ego; friendship and clan- or oath-fellowship; badly aspected, blindness to danger or trouble.

> In questions of finance: security or a false sense of security.
> In questions of romance: contentment or blindness to deception.
> In questions of health: good health, the immune system.

In questions of the workplace: a sense of satisfaction.

Companion runes: Uruz, Gebo, Sowulo, Othala

Hagalaz: The uncontrolled forces in the unconscious originating from the past, of a disruptive nature but causing changes for the long term better. Short term disappointments and possible loss. Natural aspect: hail. In divination, this rune most often shows a major disruption which, if one stands firm, can resolve into a better situation.

In questions of finance: a sudden disruptive shock.
In questions of romance: a new person or event
  intruding.
In questions of health: a sudden difficulty or attack.
In questions of the workplace: surprise, disruption, a major
  change.

Companion runes: Uruz, Raido, Eihwaz, Sowulo

Nauthiz: Restrictive forces; fears; anxieties; feelings of guilt, sometimes poverty.

Also often a warning and therefore a possible help. Forewarned is forearmed. Natural aspect: fire "Need". The need-fire (fire kindled by friction); a strengthening trial, achievement won through effort, cleansing. The ability to "write around" Wyrd.

In questions of finance: unusual care or effort needed to
  improve things; hard work will make you better off than
  before.
In questions of romance: trouble ahead, effort is needed to
  make the relationship work, but it is worth it. In
  questions of health: a strain which ultimately will make
  you stronger.
In questions of the workplace: a serious challenge.

Companion runes: Kenaz, Eihwaz, Dagaz

Isa: Blocks; stultified conditions; grievances; anything that the individual is not prepared to let go of; obstacles, delays, inactivity, frustrations.
Natural aspect: "Ice". Immobility, absence of energy, solidification, blockage or lack of change; also stability, peace, and the cohesiveness of the ego. The antithesis of Fehu in all ways.

In questions of finance: frozen cash-flow.
In questions of romance: an icy reception, loss of feelings.
In questions of health: low vital force, but also
   consolidation, an end to pain, the halting of a difficulty.
In questions of the workplace: nothing happening, no gain.

Companion runes: Wunjo, Othala

Jera: Hopes and expectations; turning points; gradual changes for the better; results of earlier actions, pay offs and improvements. Natural aspect: the turning of the year, the growing of plants. "Year" or "harvest". Cyclic progression, natural development, rewards coming in the course of time; also used for fruitfulness and fulfilment. The turning of Wyrd.

In questions of finance: investments paying off, plans
   coming to maturity.
In questions of romance: a relationship coming to fullness.
In questions of health: regular improvement, fertility.
In questions of the workplace: fulfilment, improvement.

Companion runes: Fehu, Berkana, Inguz, Dagaz
Eihwaz: The driving forces and motivations; sense of purpose. Challenges and tests. Endurance, courage and daring. Natural aspect: yew-tree. "Yew". The World-Tree. Death, initiation, sudden transformation. The soul surviving the body's death; the reclaiming and awakening of ancestral memories and power in oneself, inner strength. Also used in dealing with the dead (a risky practise at best). Odin's self-sacrifice on the World-Tree.

In questions of finance: a major transformation.

In questions of romance: deep feelings expressed.

In questions of health: the skeletal system, revival of strength; also, hereditary factors coming out.

In questions of the workplace: a major crisis.

Companion runes: Wunjo, Sowulo, Mannaz, Othala

Perthro: The deepest creative part of the unconscious; the hidden realm of higher material that is waiting to come to fruition and birth; hidden talents; occult or psychic abilities. Unexpected positive changes. Perthro; Natural aspect: fruit, a well, a womb.

"Lot-cup". The Well of Wyrd, divination, the workings of Wyrd becoming manifest; also birth and the first setting of wyrd which shapes all that comes after it.

In questions of finance: an apparently random event which cannot be avoided, with good or ill effects depending on the rest of the reading.

In questions of romance: discovery of whether or not a relationship is meant to be.

In questions of health: the female parts, genetic inheritance from the mother's side.

In questions of the workplace: the fulfilment of whatever conditions currently hold.

Companion runes: Kenaz, Berkana, Inguz

Algiz: Influences which will protect you; religious aspirations. Higher spiritual awareness, assistance from higher powers or in the form of spiritual guidance by a teacher or mentor. Natural aspect: elk, swan. Warding, might summoned down from the worlds above, knowledge and inspiration coming from the god/desses.

In questions of finance: security, good yield.

In questions of romance: a spiritual relationship.

In questions of health: good health, protection from disease and difficulties.

In questions of the workplace: security, ambition.

Companion runes: Gebo, Sowulo, Dagaz

Sowulo: energy, power, enthusiasm, positive thoughts, success and accomplishments. Good physical health. Natural aspect: the sun "Sun". The will, success, rulership, holiness. Safe passage through difficulties. Badly aspected, insensitivity and arrogance.

In questions of finance: good fortune, perseverance.
In questions of romance: things are likely to go well.
In questions of health: endurance, a positive attitude.
In questions of the workplace: ambition, success.

Companion runes: Wunjo, Teiwaz

Teiwaz: Personal strengths, taking initiatives; honor and justice; leadership and authority. Competitiveness in business or sport. Physical activities. "Týr". Victory, bravery, stability, honour; this rune was carved upon weapons. Badly aspected, legalism and oppression of the spirit by regulations.

In questions of finance: legalistic matters will be of great importance, either helping or hindering as the rest of the reading shows.

In questions of romance: lack of flexibility, conventionality.

In questions of health: strength and order; if a problem exists, standard medical treatments will be either highly recommended or should be supplemented with alternative forms of healing, depending on how the rune is aspected.

In questions of the workplace: rules and regulations.

Companion runes: Raido, Wunjo, Sowulo

Berkana: Fertility; birth; rebirth; growth; maternity; family life; feminine Mysteries. New beginnings and opportunities, healing and rejuvenation. Natural aspect: a birch-tree, a peat bog "Birch". Birth and bringing into being; concealment, nurturing; gestation and the womb; protection, fruitfulness; secrecy. Also death, as the goddess Hel ("The Concealer").

In questions of finance: an investment which will prove to be fruitful.

In questions of romance: strong female energy; if badly aspected, a smothering or unfriendly woman.

In questions of health: fertility, female parts, hormonal balance.

In questions of the workplace: a dominant female figure or secrets being kept.

Companion runes: Kenaz, Perthro, Jera, Inguz

Eihwaz: Adaptability; joint efforts; cooperation; sexuality. Natural aspect: a horse "Horse". Relationships (especially superior/inferior relationships), wedding or partnership, control; also travel, communication of holy wisdom, finding the means to make achievement possible. Odin's steed Sleipnir. Recklessness, betrayal by false friends when inverted, broken trust.

In questions of finance: a partnership, general success.

In questions of romance: a solid relationship.

In questions of health: how it is aspected shows the overall physical condition.

In questions of the workplace: partnership; shows how you are working with superiors, inferiors, or team members.

Companion runes: Raido, Gebo, Inguz

Mannaz: People at large; attitude towards others; other people's attitudes towards you; legal matters; friends and enemies; intellectual pursuits, favorable for exams when paired with

Ansuz. Natural aspect: the human mind "Human being". Intelligence, memory, the realising of our godly inheritance, the conscious mind, the reaching of full mental and spiritual potential. Used for mental skills and wisdom.

In questions of finance: a decision to be made which will require much thought.
In questions of romance: a close soul-mate or a purely intellectual relationship.
In questions of health: indicates mental health and the brain.
In questions of the workplace: intellectual effort will be required.

Companion runes: Ansuz, Kenaz, Sowulo

Laguz: Emotions; stability; imagination; psychic matters; affections. Natural aspect: water (flowing water for weal, stagnant water for woe) "Water" (also called laukaz, "leek"). Life-force, concealment and bringing forth; cleansing, particularly against poison. The subconscious; secrets. Badly aspected, something destructive in hiding or about to come forth.

In questions of finance: a steady flow
In questions of romance: strong attraction or destructive secrets being kept
In questions of health: life-force, the bloodstream; badly aspected, shows a need for cleansing
In questions of the workplace: a steady stream of work, something about to be revealed

Companion runes: Uruz, Ansuz, Algiz, Sowulo

Inguz: Integration; gestation; expectations; progeny. Natural aspect: a seed "Ing", or Freyr. Self-sacrifice, the planting of a seed, fruitfulness. In the form ◊, the castrated male; in the form ♦, the whole male.

In questions of finance: good fortune, the start of a fruitful enterprise.

In questions of romance: a promising relationship that may call for some self-sacrifice to make it work.

In questions of health: the male parts, hormonal balance, sexuality in general.

In questions of the workplace: the start of fruitful endeavours.

Companion runes: Kenaz, Gebo, Jera, Berkana

Othala: Home country; spiritual heritage; house, land, properties; established fundamental values, conservative. Natural aspect: "Udal (inherited) lands". Family matters, ancestral possessions or powers, inheritance, the kin-soul. The border between the human world and the "wild space" outside.

In questions of finance: an inheritance, a solid investment, matters affecting the home

In questions of romance: a stable, long-term relationship.

In questions of health: inherited conditions, good or bad according to how the rune is aspected.

In questions of the workplace: a stable, long-term job.

Companion runes: Ansuz, Gebo, Wunjo, Eihwaz, Mannaz

Dagaz: Position between light and darkness; traversing between the worlds. Cosmic consciousness; change from one thing into its opposite; new beginnings. With negative runes possible violent destruction. Natural aspect: dawn, "Day" (or, more properly, "dawn"). A sudden burst of illumination, transcendent spiritual unity, fulfillment. In all matters, this rune shows fulfillment, inspiration, and the sudden beginning of better things, it activates and brings out the best in all other runes.

EXAMPLE READINGS AND SPREADS

1. A THREE RUNE SPREAD: URD, VERDANDI AND SKULD

The simplest and often the most effective reading is a three rune spread. There are various ways of doing this depending on what type of runes you have made. Round or square runes are best for casting, rectangular ones are more suitable for a spread.

THREE RUNE SPREAD: URUZ, SOWULO AND WUNJO

The question is job related. M is in a dead-end but relatively safe job, the wages aren't terrific but there is at least some job security. However, she has been told by a friend that there is a better paid job available in New Zealand, with prospects of valuable training and the possibility of promotion. The question is therefore "Shall I go for it?"

The Runes are: Uruz, Sowulo, Wunjo

This is very clear and direct – it says: *Go for it! Big time!*

Uruz in the first place means: *This is an opportunity not to be missed; it is earned from your past actions. You deserve it.* The first place is associated with *Urd*, 'that which has been'.

Sowulo in the second place reaffirms this, as the second place is associated with *Verdandi* or the becoming-present: it indicates success, enjoyment, and lots of sunshine.

Wunjo in the third place, the place of Skuld, 'That which must be'; reinforces all of the above, though its meaning of joy and perfection. Based on these runes I would say this person cannot go wrong in taking the offer, it is unlikely that such an opportunity will come along twice!

Now, this is a very simple Runic divination, but nevertheless, it may lead to a complete change of life for M.

R has a different problem. He's due to undergo minor, but necessary surgery and is rather afraid. He is actually thinking of refusing, as he has a phobia about anaesthetics, due to an unpleasant experience as a child.

The rune-worker does not deal directly with medical problems (unless of course he or she is qualified in that field and doing readings is a secondary job or hobby). The objective is here, on the contrary, to get to the bottom of R's fears, to increase his confidence so that he can face the operation calmly.

CHREE RUNE SPREAD: ISA, LAGUZ INVERCED, GEBO

Isa here represent the trauma in his early childhood, when, it appears, he had a tonsillectomy and felt as if he was choking. Laguz inverted tells me that he has serious unresolved emotional problems with this issue and really should receive some psychological counselling as his fears are deeply seated within the subconscious. Gebo, however, as the outcome is clearly stating that the operation itself holds no danger at all; his problem is totally subjective and should be dealt with accordingly. It would help if R cut down on the smoking a few weeks before the anaesthetic!

2.   A NINE-FOLD SPREAD:
THE NINE WORLDS; YGGDRASIL

For the purposes of this spread, I have correlated the nine worlds of Northern Mythology to the nine-fold division of the soul and the nine-fold division of the psycho/somatic complex as stated by Carl Gustav Jung, probably the most spiritual advanced psychologist of this century. This system is more complicated and requires some study of the Norse Myths as well getting familiar with the basic theories of Jung. However, doing a reading in this format can deliver a wealth of useful data, on all levels of being. Indeed, a whole reading can be devoted to a particular level, for instance, the level of the soul or the level of the personality, or spiritual levels of a higher order; thus, a very thorough consultation might involve three readings of nine runes each.   All of these will overlap at some point, but the approach is generally useful.   One can narrow things down and fine tune a reading by means of follow-up readings – looking at the previous reading information from a different perspective. For example, if a psychological reading gives certain data which suggest a need for further enquiry one can then use the same nine-fold cloth but this time from the perspective of the soul's point of view.   The third type of reading is purely for those people who accept the reality of gods and goddesses, never mind in what form; so a consultation with the god or goddess directly can be obtained through the mediation of the runes and the reader, if at all sympathetic toward the idea of 'the Gods'. I will offer my opinion as to which god/dess goes with which square, but from here those people who want to take things further should seek a religious commitment to these gods, before proceeding.

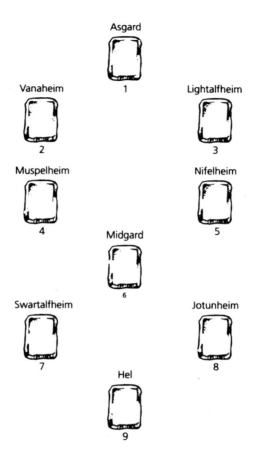

CHE NINE WORLDS SPREAD

Nine Worlds: the significance of each position in the spread

1 *Asgard*, the realm of the Gods, ruled by Odin. In a reading, any rune placed here denotes higher spiritual ambitions, Higher Self, Inspiration, matters of honour and sacrifice. The realm of pure spirit. This realm can be accessed through the rune Gebo. Psychological attribute: Individuation.

2 *Vanaheim*, the realm of feelings; ruled by the Vanir, the gods of fertility, pleasure, peace and prosperity, all matters of eroticism and relationships. The element of Water. This realm can be accessed through the rune Inguz. Psychological faculty: Feeling. Gods to call on for access to this very pleasant realm are *Freyr* and *Freyja*. Psychological attribute: feeling.

3 *Lightalfheim*, realm of the light Alfar, spirits of Air, all mental activities, art, creativity, plants trees. Ruled over by Freyr. Access rune: Sowulo. Psychological faculty: thinking.

These are the three Upper Worlds, which can be accessed through pathwork and guided meditation using the given rune as a doorway; also power can be invoked from that realm through the associated rune.

The next three areas are the three Middle worlds. The first two, Nifelheim and Muspelheim, generated all of the other worlds.

4 *Muspelheim*, the realm of fire. Together with Nifelheim, a source of eternal creation through the interactions of their respective elements: Fire and Frost. Muspelheim is inhabited by the sons of Surt, who at the Ragnarok will destroy Midgard. It is a very hostile world; no travelling attempts have ever been recorded. For those whom want to tempt wyrd, Dagaz is the ruling rune. For the purpose of divination, however, we focus upon the creative aspect of Muspelheim and allocate this word the psychological faculty of intuition.

5 *Nifelheim*, the realm of fog and mists. Like Muspelheim, its polar opposite, it is a very hostile world. It is ruled by an all devouring serpent, Niddhog, the eater of the evil dead and drinker of blood. Nifelheim is the realm of the shadow side of the unconscious; again a no-go area. The rune belonging with it is, not surprisingly, Nauthiz. Psychological attribute: shadow.

6 *Midgard*, the realm of humanity, the inhabited world, civilisation. Its guardian and protector is *Thor*. The rune

connected with it is Jera. In psychological terms, this world represents the ordinary waking consciousness, the Ego or Lower self, personality.

Now we come to the Under Worlds.

7 *Swartalfheim*, the realm of the Inner Earth. It is the domain of the dwarves who deal with precious metals and stones, working smith craft, taking base materials and transforming them into higher materials by means similar to alchemy. Their ruler is *Modsognir* and the rune that will grant access is Eihwaz. The psychological faculty attributed is: sensation.

8 *Jotunheim,* realm of the frost giants, raw disruptive male power. For this reason I have assigned to this realm the Jungian psychological concept of the *Animus*, the male part of the self. This realm is tricky and confusing, it contains the realm of *Utgard Loki*, a King of Giants who cheats Thor by trickery and sorcery. The access rune is Isa. In this realm, nothing is as it appears, however those well versed in shamanic travelling should be able to go there.

9 *Hel,* the realm of Hella, the goddess of death, counterpart of Odin in this respect. Odin takes the heroic dead which are shared with Freyja. Those dead who in life were devoted to a specific God or Goddess may hope to share their realm with this deity after death. All others go to Hella's ream, a recycling plant for reincarnation. This realm can be travelled to quite safely provided certain precautions are taken. A pathworking to give to access this realm will be provided in the next chapter on runes and magic. The realm of Hella can be accessed to communicate with ancestors and obtain knowledge from the dead. The access-rune is Hagalaz. Psychological aspect: Anima, female part of the self.

You will have noticed that Asgard, Midgard, and Hel are connected linearly. I am very pleased with the result of this diagram as it is actually an improvement on the earlier one given in *Leaves of Yggdrasil* and repeated in *Northern Mysteries and Magic*. Thank you Odin!

All the access runes provided are non-invertible, which means they are completely neutral in term of positive or negative. They function purely from a impersonal perspective as keys to power – indeed, they might be called the nine 'songs of power'. Each one can mediate energy and information from their respective realm.

EXAMPLE OF A NINE WORLDS SPREAD:

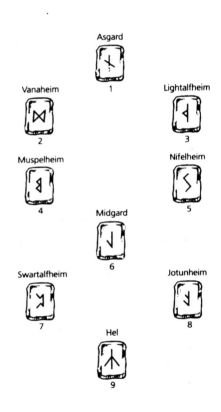

Asgard rune:	Nauthiz	
Vanaheim rune:	Dagaz	
Lightalfheim rune:	Thurisaz	*inverted*
Muspelheim rune:	Berkana	*inverted*
Nifelheim rune:	Sowulo	
Midgard rune:	Laguz	*inverted*
Swartalfheim rune:	Perthro	*inverted*
Jotunheim rune:	Ansuz	*inverted*
Hel rune:	Algiz	*inverted*

For interpretation, one views the runes in connection with the three levels in turn. Examining the runes in the three Upperwolds we see that only Dagaz is in a favourable position. Nauthiz in Asgard is not favourable and neither is Thurisaz inverted in Lightalfheim. When interpreting, one has to develop the art of sensing which runes are compatible with which world and how they may affect each other. Nauthiz in Asgard means in this reading that the person is not yet or is not at this moment on-line with the forces and intelligences of Asgard. Lightalfheim, too, is inaccessible as there is a negative Thurisaz blocking any influences for this realm. Dagaz, the rune of Dawn and enlightenment in Vanaheim, the word of Vanir, states clearly that from this realm help can be sought and inspiration obtained – perhaps in the form of artistic endeavours. More could be said about this but space is limited.

Turning to the middle-worlds, we find Berkana inverted in Muspelheim. Obviously, in the context of this particular reading, this realm is hostile to creative forces represented by Berkana inverted. Nifelheim contains Sowulo, indicating that this realm can be of help. Since Nifelheim, amongst other things, represents the darkest, unacknowledged aspects of the personal subconscious as well as being a foggy place, Sowulo provides a redeeming factor. The sun can drive the fog away and shine her light on the amassed negative garbage in the subconscious. From this, practical knowledge can be obtained and unpleasant 'forgotten' traumatic memories can be accessed, processed, assimilated and transmuted to positive power, through meditation on the relationship between the Sun and the Shadow. Midgard has Laguz inverted; this suggests feelings of loss and disappointment in a worldly matter, a let down by a loved one, being 'ripped off' or deceived; perhaps tears. This says you have to get in touch with your feelings, perhaps sort things out in a confrontation before you can let go and move on.

Moving on to the three Underworld runes, we come to Perthro inverted in Swartalfheim. Swartalfheim is the world of hidden treasure and Perthro is the rune of hidden mysteries and unexpected wind falls... but since it is inverted, there is nothing to expect from this combination. No access. In Jotunheim we find a situation of conflicting perspectives. Ansuz, the rune of Odin representing, amongst other things, communication, appears inverted. Aesir and Giants are opposing forces as we know, and this combination tells me that a row is about to ensue and the advise given is, be very careful with any form of communication, keep your mouth shut for the moment and don't put anything in writing; there is a clear danger of abuse. Finally we come to the inverted Algiz in Hella's realm. While its upright form indicates the tree of life, inverted is the symbol of the tree of death. This combination is not entirely negative however. The advice offered is seek the counsel of your elders who went before, ask for protection, information is available, call on your ancestors and watch you dream space as it is likely that they will respond through this.

This is quite a comprehensive reading, and my interpretation is condensed by necessity into a few paragraphs. It is given as an example to experiment with. The only way to become proficient with this as well as any other type of reading is practice and hard work. Always keep records and give yourself marks. Review the record at least once a year to assess your progress. The interpretations I have given for the above reading are not fixed hard and fast. Play about with it and the sooner you can dispense with this interpretation offered by me and develop your own, the better you will become at reading the runes.

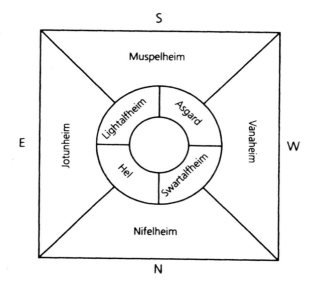

A Traditional Casting Cloth

This cloth can be used to cast the runes, and I suggest the following method. First, spread the cloth over a table. Then hold all the runes in a vessel: a cup, a beaker, a horn, etc. Invoke the three Norns saying 'Urd, Verdandi, Skuld', and throw the runes onto the cloth. Discount any runes which fall outside the cloth and view the others where they have fallen, interpreting them against the backdrop of the space they each occupy. For example, Dagaz falling in Asgard suggests a sudden spiritual awakening.

3. A YEARLY FORECAST.
DATE 1/1/1999

For a Rune casting cloth I can recommend the traditional design, and the attributes allocated to the nine worlds can be applied:

Example of a Yearly Forecast

The dominating rune, placed in the centre of the spread to represent the querant, is Isa. The monthly runes are as follows:

January	Berkana
February	Eihwaz
March	Jera
April	Gebo
May	Ehwaz
June	Teiwaz *inverted*
July	Laguz *inverted*
August	Dagaz
September	Fehu *inverted*
October	Inguz
November	Thurisaz *inverted*
December	Raido.

Isa in the centre means that the person concerned is stuck in a situation, at the time of the reading, and has been for about three months prior to this, with limited opportunity to move in a desired direction. The basic issue may be job related, but it may also affect other areas of life since the personal web of wyrd is an holistic concept.

January starts very well with a new opportunity, possibly leading to a new job, enterprise or any new project. Eihwaz in February adds some tension to this and a effort has to be made to prove oneself. In March a positive result arising from something originating a year beforehand may come to the fore, possible leading to the settling or signing of an agreement, with Gebo in April. This may involve finances; for instance, it could indicate the recovering of moneys owed. In May one may meet another person and invest hopes in a possible relationship; this, however, will not come to anything — with Teiwaz inverted in June, the person concerned is not what he or she pretended to be initially. As a consequence, there will be some sadness and regret in July with Laguz inverted. August, with Dagaz, is the optimum time for a holiday. There is the danger of overspending on this with the result that in September, with Fehu inverted, there is a financial restraint or loss. In October, however, this should restore itself and something which started in April with the signing of a deal will pay off and come to fruition. November is the month where one has to watch one's health and energies and attempt to avoid any conflicts and postpone possible disputes to December when, with Raido, one stands a better chance of obtaining a positive outcome.

This is a example made up with no real querant involved. It is given to show how one can read the runes in connection with the following runes, in a sequence, according to the yearly cycle. One can use the same reading to pick out issues which will influence the coming year as a whole, without necessarily including the months. To do this one looks at runes opposite and at right angles or in triangular formations with other runes, as is done with planets in Astrology.

Like this: Jera opposite Fehu inverted squared 90° by Teiwaz inverted which is opposite Raido, which therefore also square Jera and Fehu. This suggests a possible complicated legal situation, with a mixed result or an undecided result for that year, i.e., litigation, which could go on for some time, and involving financial costs. This is one example, of how to connect runes in a pattern and draw a conclusion from this. Those of you familiar with Astrology have an obvious advantage, but anyone can learn this if they put the effort in.

Let's look at the same runes divorced from the yearly cycle and instead let's use the twelve segments as the twelve houses derived from Astrology. Those who have their own horoscope can overlay this with runes as an exercise and try to do a reading from this. The object of giving readings is assisting people in their life's situations to the best of one's ability, anything that can add to this can be used. In fact, the twelve-house system is already made very popular in combination with rune reading. Those who want to limit themselves to Norse symbolism exclusively can always re-name the houses after the twelve palaces of Asgard (see table in appendix of this book).

The runes are read anti clockwise from the first house to the twelfth house. For those totally unfamiliar with the attributes of the houses I will give short definition of these.

First house: the self; basic personality, how others perceive you, psychological motivation, ambitions, all beginnings, self-image and self expression.

Second house: security, money, natural talents, material values, personal possessions.

Third house: early education, peers, siblings and relatives, short journeys, communication ability.

Fourth house: childhood environment., influence of the mother, physiological foundations, domestic issues.

Fifth house: luck, pleasure, children, love, creativity.

Sixth house: health, employment, service.

Seventh house: the other, spouse, partnerships, relationships. Opponents

Eighth house: Transformation, sex, death, regeneration.

Ninth house: spiritual ambitions, philosophy, religion, higher education, long journeys.

Tenth house: career, public image, social contributions,

authority, influence of the father.

Eleventh house: ideals, friendships, groups, social values and aspirations. Hopes and wishes.

Twelfth house: Endings, enemies' constraints, wyrd, past lives. Subconscious, and sacrifices.

Let's attempt to interpret the previous runes on this model. Isa remains also in the centre, representing in this context, the situation as it is.

EXAMPLE OF AN ASTROLOGICAL FORECAST

First House: Berkana. Pleasant personal appearance and demeanour, this person is well liked and appreciated, by her peers, a gentle disposition, possible problem with assertiveness. Must develop persistence in any project.

Second house: Raido. A well-defined sense of fairness and what is right, a balanced view concerning finances and possessions. Possibly in the position to offer advice to other in the forenamed subject.

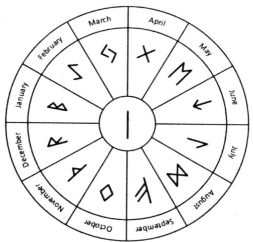

Example of an Astrological Forecast

Third house: Thurisaz (inverted). Not enough self-preservation. Must learn to say No. Tendency to avoid confrontation at the expense of self.

Fourth house: Inguz. Stable home background, perhaps too much protected by the mother; must be more independent. Tend to rely too much on other people's opinion.

Fifth house: Fehu inverted. Tends to be to optimistic financially and has to avoid gambling or taking risks with money. Also must develop more self-esteem based on own observation, not other people's views.

Sixth house: Dagaz. A worry is to be resolved; a possible fear relating to health is unfounded, possibly a sudden improvement in employment opportunity.

Seventh house: Laguz. Possibly emotionally too dependent on partner; however, very loving in nature.

Eight house: Teiwaz (inverted) A happening which may result in having to face death in one form or another, possible an elder in the family. A confrontation with death which forces one to contemplate one's own mortality. This does not mean this person dies themselves.

Ninth house: Ehwaz. Idealistic approach and a positive outlook in life; an optimistic personal philosophy. This would be a good time to seek out like-minded people and/or explore alternative religions.

Tenth house: Gebo. An accomplishment in career, a promotion and possible an acceptance of a contract leading to perhaps a position in management.

Eleventh house: Jera. All good things are returned; also the not-so-good things albeit fewer in number. The returns are through other people and social acquaintances as well as personal friends.

Twelfth house: Eihwaz. An unusual experience leading to insight into a past event, possible ancestral knowledge, the emergence of past life awareness. A possible personal sacrifice.

The above reading are very general and are offered to show how one may interpret the runes, by synthesising the meaning of the runes with the meaning of the houses. In the first reading involving these runes, Laguz was interpreted as being inverted, in the second Laguz is interpreted as upright, this is because if one views the wheel of the year the runes move sunwise, and in the second reading with the houses the runes move anti clockwise.

4. THE CROSS SPREAD.

This spread is very useful where the querant needs help making either-or decisions, but it can also answer more complicated questions and solve problems.

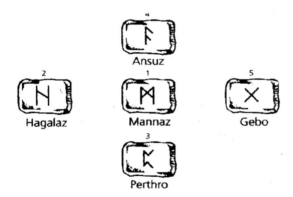

A Cross Spread

Here I have a simple spread to answer questions by looking at the questions from five different angles. I have allocated runes to each square, and each rune conveys the interpretative meaning of the square.

One is Mannaz: the person who asks the question. The rune turning up here states where this person is coming from; it may relate to the subject matter of the question and in rare events may actually convey that it isn't the right question to ask or the right time to ask the question.

Two is Hagalaz: everything related to this question which arises from the immediate past, what gave rise to the need to ask the question. What went before.

Three is Perthro: gives information about a unseen or hidden aspect related to the question, it may point out something one had overlooked or was not aware of. What is behind this, or what is hidden from sight.

Four is Ansuz: gives a higher spiritual dimension in the light of the question and may show what spiritual forces are involved or may offer advice from the god/desses. What is above.

Five is Gebo: this should give the answer, interpreted in the light of the other four runes. Together they should give an accurate answer or advice regarding a question or problem.

Conclusion, or what's next.

The method is the same as with the Nine Worlds spread. Select runes unseen on each of the squares, turn them over one by one, interpret each one in turn and than connect them and draw a conclusion from this. Of course one can instead of one rune draw two or three for each square; this may give more elaborate information but is also more difficult to interpret, however as with all the above techniques, experimentation will generate experience.

FOUR

The use of runes for personal empowerment, self-development and relationship counselling.

In this chapter we are looking at how Runes can help you to improve your personal life, relationships, and circumstances through understanding and growth.

Self analysis through the runes

Runes can be employed periodically as a means of psychological exploration, to assess one's own strengths and weaknesses. For this purpose, I will extend the meaning of the runes from divination into psychotherapeutic applications, both for individuals and to offer a technique to explore the dynamics in a relationship. This goes further than the traditional methods of divination, but in this book I attempt to strike a balance between tradition and innovation.

The meanings of the runes given for this exercise are derived from their general meanings as given in chapter 2 but more specifically adapted to the purpose of psychology. The runes, projected at a more complicated level of depth psychology, will not change their essential nature. The runes may change their "octave" but not their "ground-tone". For example, Fehu, in the mundane sense, means wealth and money prospects, whereas the same rune in a psychological reading means creative energies, in particular the level of self-confidence and self esteem available to the querant which allow him to improve his life's circumstances. The more one meditates on and experiments with these creative variations within the runes, the more inner resonance one will develop; thereby a working relationship with the runes will be established.

The basic meanings of the runes of the first aett to be applied in this type of consultation are listed below.

Fehu: Personal strength, attitude and expectations regarding money. Self-esteem and confidence.

Uruz: Instinctual drives versus personal discipline, raw force, physical health endurance and assertiveness.

Thurisaz: The power of self-assertion and non-destructive confrontation with others. The individual's will and the opposing willpower from the environment.

Ansuz: Communication skills, the ability to adequately express oneself. The ability to listen to others, leadership qualities. Higher sources of communications, from either within or outside the self.

Raido: The ability to realise right from wrong and the power to base your decisions on this. The amount of control that the individual can exert in his or her life.

Kenaz: The ability to recognise a situation for what it is, to be objective and apply common sense. The individual's knowledge and capacities.

Gebo: The ability to compromise, to give and receive in equal measure. The individual's attitude towards giving and receiving.

Wunjo: The ability to relax and enjoy oneself, sense of humour. The individual's potential for enjoyment and his or her chances of being successful.

Hagalaz: How the past may be exerting its influence on the present, and how to use this for change.

Nauthiz: The need to recognise one's limitations and to either accept these or transform these.

Isa: Areas one prefers not to look into, possible suppressed memories, which have to be assimilated. Blockages, frustrations and hindrances.

Jera: What one may expect to occur as a result from previous actions. The time factor in the individual's development.

Eihwaz: What one fears, either in objective terms or subjectively, and the ability to achieve one's desires; tensions between fear and desires.

Perthro: That which is hidden and ready to emerge. The ability of introspection and self-awareness.

Algiz: Matters of self-protection and self-preservation. The ability to defend oneself and protect one's own.

Sowulo: The source within, which guides the individual on his or her path.

Teiwaz: The ability of objective judgement and fairness; the potential for handling a conflict correctly; warrior attitudes and courage.

Berkana: Creative abilities, nurturing and birth-processes in the self.

Ehwaz: Emotional responses, adaptability and subjectivity.

Mannaz: The mind and intellectual objectivity.

Laguz: The intuitive faculties and the potential for sympathetic feelings.

Inguz: The potential for individuation and integration.

Othala: The relatedness to heritage and kin, and awareness of origins.

Dagaz: The ultimate aim of transformation of the personal consciousness into whatever one envisages being the greater whole.

The beginner may find it helpful if an example of a reading using this technique is given in detail. We use here a 'futhark spread', in which the querant selects all of the runes, one at a time. These are laid out in Futhark order, in three rows of eight, so that their positions each correspond to one of the runes of the futhark. The following is a description of a profile reading for a middle-aged professional female who anticipated some trouble in her personal and professional life. As the reading below clearly shows, most of the perceived trouble was in her own mind in the form of fear, anxiety and some depression, without any actual indication of major disasters about to happen. It is in such circumstances that a psychological profile reading can be very helpful as it mostly points out subjective realities.

Fehu pairs with: *Gebo* – A very positive combination, meaning that she may expect a monetary gift, possibly as a reward.

Uruz pairs with: *Isa* – Must watch her health and examine her diet.

Thurisaz pairs with: *Jera* – Possible repercussions of an argument or a falling out last year.

Ansuz pairs with: *Hagalaz* – Negative feedback possible: malicious gossip and severe criticism by an elder person.

Raido pairs with: *Dagaz* – A decisive move will have a positive change as result.

Kenaz pairs with: *Eihwaz* – understanding has to precede the taking of any risk.

Gebo pairs with: *Kenaz* inv. – Cannot see how well off she really is; must look more on the bright side.

Wunjo pairs with: *Wunjo inverted* – A lack of enjoyment or pessimistic outlook, which she can change herself by being aware that she creates her own reality. By changing her negative expectation, she may alter her wyrd in this respect.

Hagalaz pairs with: *Teiwaz* – She must take decisive action to break with certain aspects of her emotional past and move on, possibly in a complete different direction.

Nauthiz pairs with: *Sowulo* – Again the need to seek the sun, i.e., to cultivate more a positive outlook and approach.

Isa pairs with: *Ansuz inverted* – This is a major block in her psyche; perhaps outside advice is required.

Jera pairs with: *Othala inverted* – Excessive anxiety that she may lose her home in a year's time, but she can take action right now to prevent this.

Eihwaz pairs with: *Mannaz* – She must maintain her integrity within a small group of people and not be persuaded by majority opinion if this is not in her own best interest.

Perthro pairs with: *Perthro* inverted This area is closed off for now.

Algiz pairs with: *Inguz* – She can expect protection and nurturing within her family and probably do well to forge closer bonds with some of them.

Sowulo pairs with: *Uruz* – She must balance fire (yang) and ice (yin) to counteract the above effect of Uruz paired with Isa. The Chinese concepts of Yin and Yang correspond very neatly to 'fire' and ice'. Spend more time outdoors.

Teiwaz pairs with: *Laguz* – A balance has to be struck between justice and mercy. She must be less rigid with implementing the rules and be more understanding of other people's failings.

Berkana pairs with: *Fehu inverted* – A creative project may have to be shelved due to lack of funds.

Ehwaz pairs with: *Raido inverted* - She must not follow advice given by people around her but follow her own instincts for now.

Mannaz pairs with: *Thurisaz* - She must not give in to pressure from other people attempting to persuade her in a certain direction by decisively cutting ties where and when needed.

Laguz pairs with: *Ehwaz* – She has to use her intuition to distinguish between true and false friends, and consider her own interest a bit more.

Inguz pairs with: *Berkana inverted* – She has to consolidate what she's got and for the moment be content with this, any premature attempt to expand will meet with failure.

Othala pairs with *Algiz*: – She must ward her home, and secure her position within it.

Dagaz pairs with: *Nauthiz* – for the moment, lay off any heavy occult work.

So, that is that. As this reading is not specifically bound to any given time it seems that some of it was already known to the client, and was applicable in the immediate past. Some of it seems to give caution in the near future. This is the only weak point of this type of reading; it is very difficult to pin it down time wise. It seems to give a wide-angle snapshot of the 'now' – with the immediate past and present included! The art here is to intuitively feel which is which. The lady concerned contacted me today, ten days after the reading, and we went over it again. Since then she has noticed things happening in her surroundings which where clearly indicated in the reading. Looking at it with hindsight proved most interesting: virtually everything felt into place and very specifically so. Perhaps it is a good idea to do this type of reading once a month and then read back every week and review it. It seems from the above reading and the subsequent events in the lady's life that some of the information provided by the reading pertained to the immediate past – perhaps even from a few weeks previously – and some of it materialised virtually the same week! That's quite exceptional, but it does prove the validity of this type of reading.

PARTNERSHIP READING

The next technique I developed was for the use of exploring areas within a relationship and creating opportunities to discuss, iron-out, prevent or deal with any possible problems, which may occur at any time in any relationship. To utilise this reading technique, however, the relationship must be basically strong. Both partners must be totally honest and trusting with each other, as areas of vulnerability could be exposed and therefore exploited. It works in a similar way to synastry astrology, although it is easier and faster. Each rune is interpreted to refer to a specific area of a relationship between people in keeping with the rune's basic meaning. For this technique to work, both partners must be present and each must have a set of runes. A cloth is made portraying all the twenty-four runes and arranged either in a circle or in a straight line.

The following is an example. The first rune on the cloth is Fehu, which represents the financial prospects in the relationship and the attitude of both partners towards finance. Both partners select a rune unseen and place it on each side of Fehu. The rune-reader turns the runes over and interprets them in the light of the Fehu rune. If, for example, the first partner draws Gebo and the second Nauthiz, it could be the case that the second partner regards the first one as having too generous or irresponsible an attitude towards money and security within the relationship, whereas the first partner may well take the view that the other is too restrictive with finance and too afraid to take risks. Clearly this could give rise to an area of possible conflict. By being aware of potential problems, they can work together to reach a compromise in the matter.

Below are listed the meanings of the runes adapted to the framework of personal relationships.

Fehu: The financial prospect in this relationship and the attitudes of the partners towards money, sharing wealth.

Uruz: Endurability and strength of the relationship.

Thurisaz: General conflict area; battle of wills, outside disruptive influences.

Ansuz: Communication within and about the relationship

Raido: The rights of each individual within the relationship.

Kenaz: Learning and teaching from and with each other.

Gebo: Giving and taking.

Wunjo: Enjoyment and fun, shared sense of humour.

Hagalaz: Unconscious influences operating within the relationship; past experiences in other relationships, including projections from parents.

Nauthiz: Areas of restriction; possessiveness; mutual needs.

Isa: Privacy; those areas in the relationship that are not shared.

Jera: Long or short-term influences upon each other.

Perthro: Hidden aspects within the relationship, secrets.

Eihwaz: Ideals; expectations, and disappointments thereof.

Algiz: Caring and protecting.

Sowulo: Maintaining one's individuality within the relationship.

Teiwaz: Combined strength; authority; who is the boss?

Berkana: Fertility; children; parental projection of the self onto the children.

Ehwaz: Sexuality; any joint efforts; adaptability, tolerance.

Mannaz: Intellectual compatibility and mutual understanding; shared ideas

Laguz: Empathy, affection, emotionality, loving.

Inguz: Family matters, parents and children

Othala: Home life, property, "in-laws", social life

Dagaz: Contributions to each other's well being, growth and change.

To experiment with this reading I reiterate, one must be willing to face unpleasant truths without, however, using the reading to cast blame.

This type of reading can also be adapted to a business partnership. In that case, just adapt the given meaning of the runes above for a personal relationship to less emotional interpretations.

ChAKRA REAOIŊGS FOR SELF-CEACHIŊGS

Chakras are vortexes in the energy body which connect with the physical. This reading is useful to asses the energies working or nor working properly through the chakras, and their consequences in every day life and spiritual potential. Blockages or any unwanted cords can be deduced by means of the Runes and subsequently dealt with by Runic magick.

I would use this method in a situation where I am so f**ked that I don't even know where to start or how I feel or what to do — those frustrating times in life where you know you must do something but you don't even know where to begin.

Crown, spiritual connection, knowledge and
understanding.
The spiritual lesson in the situation.

Third eye. Intuitiveness. What you see when you step
outside the situation and connect to your primal instinct

Throat. Communication and self expression. How you
manifest your attitude about the situation to others.

Heart. Compassion. How your decisions effect those around you and how what they feel influences your choices.

Solar plexus. Wisdom. Intellectually what you know. The facts.

Sacral, emotion and feeling. How you feel about this. Your gut feeling.

Root. The foundation of the problem or situation.

This divination method is read from the bottom up. You pour the runes out of your bag and lay them face down on the table. Then they are swirled to mix them and the runes are chosen one at a time and laid from the bottom up in eight rows of three (see above).

The root is the basic problem or situation. Sometimes the reason we can't figure out what to do is because we don't even recognise what the real problem is.

Sacral is how you feel about it, or how your emotions are effecting the mix.

Solar plexus is the basic facts, regardless of how you feel about them.

Heart is how your decision or actions will affect others and therefore the influence that has on you. Decisions involving other people should always take into account their feelings as well as your own. If the decision is yours alone to make, this would be how the advice and feelings of others is influencing your decision. Sometimes we don't even realise we're doing something because someone else wants us to, instead of us wanting it ourselves. You must come to terms with their influence and choose if you wish to let it make or break your decision.

Throat is how you are communicating this situation to other people. Your attitude sometimes may mislead others as to how you really feel— sometimes we don't even realise we are sending the wrong message to others in this way and it needs to be pointed out to us.

Third eye is what you see when you step outside the situation emotionally and intellectually and just view it from an instinctual intuitive perspective. Too often we disconnect from that primal part of ourselves when it could truly help us.

Crown is, of course, the spiritual lesson in it all. Everything has some spiritual lesson to it if we look close enough, even if it is only that we need to be more receptive to our needs or the needs of others.

The final three runes, I chose not to make "the outcome" because I don't feel that any situation important enough to consult the runes on can be summed up in a nutshell that way. Instead I chose to make this a next step, or launching point for solving or rectifying the situation. A point to start from to begin making things happen, especially since this spread would be used when you just don't know where to begin to get things straightened out.

Five

RUNES AND MAGIC

Runes always were and still are used for magical workings. In popular culture, this magical aspect of the runes has long been seen as dark and sinister – as, for instance, in M.R. James' classic horror story, which was made into the film 'Night of the Demon'. Central to the story is a traditional method of planting a curse using runes, employed by a magician whose character was supposedly modelled on Aleister Crowley! The truth is, however, that the runes are more commonly used for constructive magical purposes, such as healing and helping. Sometimes a more aggressive working is needful; I myself only have done a destructive working once, back in 1983, when my life was being threatened. It worked, quite dramatically, but succeeded also in terrifying me – and everyone else living in the house at the time. Generally speaking, aggressive workings are best avoided unless absolutely necessary, both on ethical and common sense grounds; negativity has a habit of bouncing back!

To work magic with runes one has to work with the gods, whatever one perceives them to be. Their existence, be it subjective or objective in your interpretation, has to be assumed as a working paradigm. It is really of no consequence whether these gods and goddesses live in your own head, in the collective unconscious or on Mars, for that matter. What is of consequence is that when working magic with the runes one is aware that one is accessing forces behind the runes; namely, what we call the 'gods' or 'archetypes' if you prefer.

So let's introduce the main gods and their associated runes in order of prominence and relevance to magic. For an more in-depth description of these gods and others I refer the reader to my earlier book, *Northern Mysteries and Magic.*

The Gods and Goddesses and the Runes

Since there are so many different forms of several of our deities' names, and since this is a simple introduction, for the ease of the newcomer I have chosen to use the Anglicised Norse forms which are best known to most people.

The High Gods

Odin*:* Originally a god of death, whose range later came to encompass magic (especially runic magic), battle (giving victory by choosing who should die), and poetry. In modern terms he is already being identified as the god of the Internet.

Odin won the runes by hanging on a tree for nine days and nights, wounded with his own spear. He gave up one of his eyes for a drink from the Well of Mimir ("Memory"). To his chosen ones, he gives victory, inspiration, magic, madness, and death when he sees fit. He is seen as especially a god of wisdom, a patron of poets, thinkers, and singers and magicians.

Odin usually appears as a grey-bearded man, tall and thin, with a blue-black cloak and an eye-patch or wide-brimmed hat tilted to hide his missing eye. His weapon is the casting spear Gungnir, with which he dooms his chosen ones to die in battle. He has two wolves, Geri and Freki (both names mean "the Greedy"); two ravens, Huginn ("the Thoughtful") and Muninn ("the Mindful")and a grey, eight-legged horse called Sleipnir ("Slipper"). He is the husband of Frigga and the father of many gods and human heroes. As the leader of the Wild Hunt, he also brings fruitfulness to the fields. Odin is a god of foresight, careful weaving of plots, and long-term agendas.

Associated Runes: Ansuz, Gebo, Wunjo, Eihwaz, Othala, Dagaz

Thor: "Thunder", son of Odin and Earth. The most popular god of the Viking Age, and often known now as "god of the common man". Thor appears as a big, muscular man with red hair and beard and huge fiery eyes. He drives a wagon which is drawn by two goats, Tanngrísnir (Teeth-Barer or Teeth-Gnasher) and Tanngnjóstr (Tooth-Gritter). His weapon is his hammer; Mjolnir and his belt of strength. Thor is often underestimated in intelligence, by people who only have a superficial understanding of the Norse Myths. This is an erroneous misconception one of Thor's names is the 'Deep Minded'.

Associated Runes: Thurisaz, Uruz, Raido, Sowulo.

Freya: Freya is probably the best-known and best-loved of the goddesses today. Her title simply means "Lady"; her original name is not known. Freya is the "wild woman" among the deities of the North: free with her sexual favours (though furious when an attempt is made to marry her off against her will); mistress of Odin and several other gods and men; skilled in the form of ecstatic, consciousness-altering, and sometimes malicious magic called Seithr; and chooser of half the slain on the battlefield (Odin gets the other half).

Freya's chief attribute is the necklace called Brisingamen, which she bought from four dwarves at the price of four nights of her love.   This goddess drives a wagon drawn by two cats, perhaps large forest-cats such as lynxes, and is seen today as the patron goddesses of cats and those who keep them.

Associated Runes: Fehu, Perthro, Hagalaz, Berkana, Laguz

Frey: "Frey" is a title simply meaning "Lord"; his original name was apparently some form of Yngvi/Ing. Together with Thor, Frey was one of the best-loved gods of the Viking Age. His holy animal was the boar.

Frey was called for frith (fruitful peace) at home, and for good weather and gentle rains. He was, and is, often thought of as a giver of riches, whose blessing is called on for fruitfulness and growth in all fields of endeavour.

Associated Runes: Jera, Ehwaz, Inguz

Frigga: Wife of Odin, Frigga is the patron goddess of the home and of the mysteries of the married woman. She is seen as Odin's match (and sometimes his better) in wisdom; she shares his high-seat, from which they look out over the worlds together.

Frigga is especially concerned with keeping social order. She is called on for blessings when women are giving birth and for help in matters of traditional women's crafts (spinning, weaving, cooking, sewing) and the magics worked thereby. Frigga can also be called on by mothers who want to protect their children.

Frigga is the mother of Balder, and is often thought of as still mourning for him. She is a seeress, who knows all fates, though she seldom speaks of them Old Norse Frigg, Anglo-Saxon Frige, Old High German and modern Dutch Frija, Wagnerian Fricka.

Associated Runes: Perthro, Bekana

Tyr: His name simply means "god"; at one time, he may have been the Germanic equivalent of Zeus or Jupiter, the "Sky-Father" of the Indo-Europeans. In Old Norse, Tyr appears only in the myth in which he gives up his hand so that the gods can bind the Wolf Fenrir. However, there are suggesting strongly that he may originally have been a god of justice.

Associated Runes: Teiwaz, Raido, Gebo.

Heimdall: Watcher at the gates of Asgard, he can hear the grass growing on the ground and the wool on a sheep's back, and needs no sleep. He came to Midgard in order to father the three tribes of humans - thralls, freemen, and rulers - and to teach runes and lore to the last. Heimdall is sometimes seen as a rather aloof god and lacking in humour; however, he is a great teacher, and an especially good god to call on for those who work in subjects calling for cool intellect rather than the furious inspiration given by Odin.

Associated Runes: Algiz, Mannaz, Dagaz

OTHER GOD/DESSES AND WIGHTS

Austri, Sudri, Vestri, Nordri: The four dwarves who hold up the four corners of the sky (Ymir's skull): East, South, West, and North. Sometimes also thought to be the four who forged Freya's necklace Brisingamen. Their names in Old Norse – Austri, Suðri, Vestri, and Norðri – mean East, South, West and North respectively. Runes to use on the quarters to invoke the powers are: East: Gebo, South: Dagaz , West: Inguz, North: Isa.

Balder: Son of Odin and Frigga. After his death was foretold, Frigga persuaded everything in the Nine Worlds to swear not to harm him, but neglected the mistletoe, which she thought was too small and weak to harm him. Making a game of his invulnerability, the gods cast weapons at him; meanwhile, Loki made an arrow of mistletoe and put it in the hand of Balder's blind brother Hod, aiming it for him. Old Norse: Baldr, Old English: Bealdor.

Associated Rune: Sowulo

Loki: A giant brought among the Aesir by Odin, who swore blood-brotherhood with him, Loki wavers between a culture-hero/Trickster and a destroyer. He is responsible for getting the gods most of their goods, but only after he has led them to the edge of destruction. Loki also brings a surprising amount of humour into the Norse tales. As well as being the father of the Wolf Fenrir, the Midgard Serpent, and, allegedly, Hel, he is also the mother of Odin's eight-legged horse Sleipnir, and cross-dresses in the typically feminine falcon-hides of Frigga and Freyja when he needs to fly between the worlds.

Associated Rune: Dagaz

Eir: Goddess of healing, patroness of health-care workers, called on against sickness or injury. She is one of the goddesses on the mountain called Lyfia ("to heal through magic"), and gives both physical and psychic means of healing; shamanic healing, especially, falls into her realm.

Associated Runes: Uruz, Jera, Laguz

Hel: Ruler of the kingdom of death. The Prose Edda describes her as half-black, half-white (she is sometimes seen as half-rotting, half alive) and of grim and unmistakable appearance. There is no evidence for the worship of the goddess Hel in elder times, but there are several folk who work with her today. Also called Hella.

Associated Runes: Hagalaz, Ehwaz.

Holda: A goddess known through German folklore, her name means "the Gracious One". I think she is a friendlier form of Hella; however she has also much in common with Frigga, being the patroness of spinners and the keeper of social order, especially enforcing taboos about working on holy days. She is also said to be the keeper of the souls of anabaptised (or sometimes simply young) children, and women who want to bear children ask for them at her well. Holda also appears at times as the leader of the Wild Hunt. According to one tale, it was she who taught humans how to plant and process flax. When it snows, Holda is supposed to be shaking out her feather-bed.

Associated Runes: Hagalaz, Perthro, Berkana

Idunna: the goddess who keeps the apples of youth, by which the gods stay ever-young. Apples are one of the oldest and holiest symbols of life and rebirth among the Germanic folk, appearing as grave-gifts from the Bronze Age onward.

Associated Runes: Jera, Berkana, Inguz.

Land-wights: The beings who dwell in rocks, springs, and so forth. They are shy and easily driven away (especially by noise or strife); when they have fled, the land will not prosper. The land-wights are friendly towards humans who treat them well. Gifts of food and drink were often left by their dwelling places.

Associated Runes: Inguz, Jera.

Mimir: A giant, perhaps the brother of Odin's etin-mother Bestla. Keeper of the Well of Mimir, in which all wisdom lies – the spring where Odin gave up his eye to drink. Odin learned the runes from Mimir's head.

Associated Runes: Kenaz, Ansuz.

Nerthus: The "Mother Earth" worshipped by the North Sea Germans, according to the Roman historian Tacitus. Her worship included the springtime procession of a wagon which bore her image about the countryside, finally arriving at a holy island within a lake.

Associated Rune: Laguz.

Norns: The three Norns, Urd (Wyrd), Verdandi, and Skuld, are etin-maidens who guard the Well of Urd from which the World-Tree springs. They reach into the Well's waters (the past) and sprinkle the Tree to shape the present and future. They are also said to do their shaping by cutting runes and/or by spinning and weaving.

Associated Runes: Hagalaz, Nauthiz, Isa, Jera, Perthro

Skadi: Daughter of the giant Thjazi, who came among the Aesir in full armour to take revenge for her father. As part of her weregild, she demanded a husband; she had wanted Balder, but, being forced to choose among the gods by their feet alone, ended up with Njord. His sea-home was as unpleasant to her as her mountain-home was to him, and so they parted. She later bore a son to Odin. Skadi is a goddess of skiing, hunting, revenge, protection of the clan, and those women who follow the path of the "Maiden Warrior".

Associated Runes: Isa, Eihwaz, Teiwaz

Valkyries: "Choosers of the Slain", these maidens were originally seen as frightful battle-spirits accompanying Odin in his work of marking men for death in war. The most famous of the Valkyries, known chiefly through Wagner's Ring Cycle, is Brunnhilde, demoted from her position for defending a hero against Odin's will and punished by being forced to fall in love with Siegfried the Dragon-Slayer.

Associated Rune: Algiz

SOUL-LORE

Before we can explore magic from a specific Northern perspective we need to look at the metaphysical thoughts and concepts held by our ancestors. To work effective magic it is necessary to develop the various parts of the magical self. Our people had a fairly complex set of beliefs regarding the several parts which went to make up a human being. These parts are the physical body and the psychic part of the self usually called soul. The soul itself was thought to comprise various components which we list below.

Fylgja – English equivalent fetch – a sort of Guardian spirit, usually embodied in an animal or human shape which reflects your inner nature. Wolves and bears were particularly common among the Northern folk, as the plethora of Germanic wolf/bear names suggest. The fetch is related to the shamanic "totem animal" or "power animal" or personal Norn. In other occult traditions such as Cabala and Thelema, the fylgja would be the Holy Guardian Angel, sometimes called the Higher Self. It attaches itself to the individual from birth, or more correctly from conception, and remains with the individual during its lifetime. It may then seek attachment to another member of the clan or family of the person. It is a being which mediates between Gods and humans. It is said that if you see your own fetch, you're to die shortly.

Hamingja - close to the concept of *mana*, but also has a tenuous link to Wyrd and the Old Norse concept of luck. Hamingja is related to the deeds of ancestors and one's own behaviour in terms of the concept of honour. The hamingja is the personal magical power source and the shape shifting force, the shape-shifting material being the hamr and the intent or will being part of the hugr. Another definition is 'luck' or 'fortune'. Like the fylgja, the hamingja is also seen as being a protective spiritual being and indeed differentiation between hamingja and fylgja is not always easy to determine.

Hamr - a subtle, plastic, image-forming material surrounding the physical body, roughly comparable with etheric material or ectoplasm. It can be used by the Hugr for shape-shifting and sending and can operate outside the physical body. It is used to leave the body during sleep or shamanic trance and can assume animal form. The idea of course is to train oneself magically to do this at will.

Hugr - the personality aspect of the soul, which is built up during a specific incarnation in a specific time and place and relates to that life alone. The term hugr is related to the name of one of Odin's raven, Huginn, which may mean "the thoughtful" or "the bold". Mythology has it that Odin each day sent out his two ravens Huginn and Muninn so that they might return with information for him. Likewise it is possible with considerable effort to send one's hugr out to gain or retrieve knowledge from other realities or to perform workings. The hugr represent qualities which are now called thought, wish, desire and temperament. There is a strong connection between the hugr and the fylgja. In a sending or a shape-shift the matter is taken from the hamr and shaped at will, the will being an aspect of the hugr.

Minne - the several forms of memory – personal memory, ancestral memory, racial or folk memory. It twins with the hugr. Much of Norse magic is concerned with the hugr, which is comparable with the Christian notion of the soul, but differs from it in an important respect. The soul is regarded as being a part of the godhead, whereas the hugr relates to the mental rather than spiritual life of the individual and manifests as personality, feelings, thoughts, desires etc. Manipulation of the hugr is the real basis of magic, as will be demonstrated shortly. Being able to control the hugr not only places you firmly in control of yourself but enables you to safeguard others. Instances are known of the uncontrolled hugr proving strong enough to affect animals and even other humans adversely.

Nose itches? Ears ringing? Toe throbbing? Someone is thinking about you. Hiccough or tickling sensation? Same thing. What you're experiencing is the hugr of the person who's thinking of you impinging upon your physical body because they've not learned how to control it properly. Or, if they have, then the action must be regarded as deliberate and you must safeguard yourself when dealing with them.

MAGIC - TYPES AND DEFINITIONS

Magic has always been difficult to define, and virtually every school of magic defines it differently. The main problem is a compulsion to attempt a single definition which is all-purpose; this can only work for a single set of circumstances, not as a generality. We offer three main definitions of magic in the Northern tradition because the definition varies according to circumstances.

1) Practical magic - advancing materially - Magic is the art of manipulating the law of probabilities in your favour.
2) Personal magic - advancing psychologically - Magic is the art of knowing oneself through working with the mythology and inculcating the folk-soul and its precepts.
3) Devotional magic - advancing spiritually - Magic is the art of transmuting oneself into a being capable of conversing and interacting with the deities.

In the Northern Tradition two distinct branches of the magical art were in use: Galdr and Seidhr.

Galdr is the magic of charms, a mixture of poetry and sorcery. It has an implied acceptability for the peoples amongst whom it was found, certainly in pre-Christian times, though that acceptability doesn't imply a universal welcome, as is the case with magic almost universally. Galdr magic is more akin to the modern practice of ceremonial magic.

Seidhr however, is both a complementary practice and an exact opposite. Whilst Galdr is acceptable, Seidhr is regarded with total anathema, despite its impeccable mythological origins. It positively reeks of antisocial and perverse behaviour. The reason is found in the Ynglinga Saga, when Odin's powers as the archetypal Norse magician are described. With his various galdr accomplishments listed the author, Snorri Sturluson, adds that Odin also understood the practises in which the most potent magic was to be found, and used them himself. Its purposes appear mostly dark grey by modern standards, if not positively black, and include foreknowledge of an individual's destiny, the imposition of death, illness, misfortune and bad luck, as well as the befuddlement of mind and bodily strength. In many respects Seidhr is closely allied to shamanism, and both make use of trance states. I think the most compatible modern term to Seidhr would be 'sorcery' in its darker meaning. The prophetic aspect of seidhr, appears to be form of mediumship, similar to spiritualism: the beings are called into the stead and entertained by the recital of a specific chant before they will reveal their knowledge of what shall come to pass.

The difference in appreciation between those two traditional forms of Norse magic is a bit like the Victorian respectable Golden Dawn magician and the village witch! Another distinction is that Galdr magic is used under the control of the will of the magician, with a specific intent.

The magical form known to our forebears as galdr is generally seen as the most characteristic form of Germanic magic, encompassing as it does both magical songs and the might of the runes. In contrast to seidhr and spae-craft, both of which are chiefly trance-oriented and involve a direct contact of the worker's soul with the Otherworld to affect the thoughts and spirits of others (seidhr) or receive prophetic wisdom (spae-craft), galdr is focused and directed within the Middle-Garth by the conscious mind of the worker. Galdr is, by definition, a vocal magic: the word means literally "song", and the connection between the noun *galdr* and the verb *gala*, "to sing" (or: to croak or crow like a bird), was as obvious to our Norse forebears as to any modern etymologist. It may be observed that vocalisation is an important part of most forms of magic, and that seidhr is also characterised by certain chants.

Closely related to seidhr is another magical, more spiritual practice namely; Spae-Craft. In the earlier days of our folk, the most honoured female leaders of the tribes were the spae-women who advised the war-chiefs concerning their battles with Rome. The most notable of these women was the Veleda, who fore-saw the victory of the Batavi and gave advice for the tribe to rise against the Romans in 69 C.E. Of her, Tacitus says, 'She was a maiden from the tribe of the Bructeri who possessed great powers, according to the old custom of the Germanic peoples to regard many women as seeresses, and in an extended superstition to consider them even to be goddesses' (*Histories* IV, 61).

Spae-craft is still practiced today, it is a more communal form of magic in which whole communities participate, whereas both Seidhr and Galdr are practised in closed groups or solitary.

I can recommend two resources of Oracular Seith, in the USA the Teachers are first and foremost Diana Paxton and her Hrafnar Group and in the UK Jenny Blain. These two sources are the ONLY ones I personally can vouch for as to their integrity. Anyone else who 'drops my name' is suspect of manipulation and exploitation. Sorry, I don't mean to be egotistical but the fact is for right or wrong I am famous and therefore open to namedropping'. I am only just informed that a former member of the UK Rune Gild who was been asked to leave in the early nineties, by myself and second by Edred Thorsson for exactly the kind of exploitative behaviour is using my name to ingratiate himself with women who wish to learn Seith, by claiming that he was trained by me or with me, whatever, the person in question attended just two meetings where Seith was practiced.

What to be on the look out for: Sadly predators are now latching themselves onto this most sacred of spiritual Nordic disciplines. They are secretly soliciting women for so called Seith training. I myself have almost been the victim of such a predator in the early nineties and it was only due to some intervention of members of my group that I escaped. People from my group where approached by this man ostensibly to give 'Freya Aswynn' a Big surprise, they were instructed NOT to tell me about it but just have me blindfolded at a certain pick up point in the Midlands where I would be taken to a secret location for 'something special', one can speculate about the nature of this surprise.

In the early nineties there was a problem with the someone advertising himself as Seith Teacher for Woman ONLY. Unfortunately this individual had attended two meeting in my own group and based on that started soliciting woman for 'Seith'! Yeah Right!

Just in case any females are being approached in a similar cloak and dagger fashion, Please BEWARE. I Freya Aswynn give due warning.

In my opinion Seith is taught by woman to woman and secondary to gay men. I have never seen a heterosexual man perform Seith, it would be extremely rare! But some could be bisexual and I wouldn't necessarily recognise this, and they wouldn't necessarily say so.

And this brings us to the theory that there may be a genetic requirement.

For example someone who is half Nordic half African could be taught Seith or any Norse trance technique, however for example someone half Chinese and half Slavic or Iberian Celt would have more difficulty. They may very well be an excellent trance medium, however what would come through isn't necessarily Norse! For people to be able to 'connect' so to speak there has to be 'signature', now I have had both Black and White women in the High seat in New York, and it's very likely that the Black female had some white ancestry, so it's NOT a matter of racial purity per se. Black people have their own excellent system of trance and they have in fact be very helpful in sharing their know how about possession, another Hot Potato open to the same abuse as Seith; Spiritualism and Drawing Down in Wicca, not that any of these system are abusive in themselves, but can be exploited due to the lack of traditional Occult Training where One is taught to discriminate and recognise fraudsters. This kind of training is becoming slowly available for the Northern Mysteries derived and adapted from the excellent training traditions of the Western Mysteries.

SO HOW DO YOU RECOGNISE A FRAUD?

Those who are for real find the Seith experience too sensitive for public consumption and guard it. Being all too difficult to explain to begin with, they would never think to 'use' their relationship with the gods; spirits etc. as a tool with which to impress others or in any way advance themselves. Let alone veiled threats to woman who say NO to an invitation. This includes bragging about the Deity, their Forces, frequency of visits and/or the parading about of teachings given at these occasions. There is much study with long years of seeking and 'trying' the Spirit/s as the soul's evolution ensues. This isn't an 'I read a book, agreed with the material and adopt something as a truth overnight' kind of thing; or 'I did a weekend workshop and now I am a shaman' type of thing. Be warned also that even one of these people in a large magically alert group real or virtual can destroy any opportunity for trustful learning. At their worst these kind tend to draw a crowd round about them through the 'secret' sharing of their scintillating, deeply irreverent tales. Be aware of those who like to focus in on only certain aspects of their God, not the whole package. Also take notice of those who 'work' the quiet crowd fringes, drawing the vulnerable to them.

RUNIC COMBINATIONS FOR SPECIFIC MAGICAL PURPOSE

Runes can be combined together in a sigil or symbol to synthesise their energies and with intent, for this so-called bind runes are part of the magical practice. First, as always, one must develop a well defined understanding of the meaning of runes as applied to result magic. If you have read the foregoing chapters that should not be a problem. Second, one has to phrase exactly a statement of intent, outlining the objective of the working. It would be superb if that was done in a poetic format, especially if a repetitive chant would be composed, i.e., a verbal spell to be recited over the bind rune whilst it is constructed. Once, while living in a communal house, I was placed in a position where it became necessary to 'persuade' someone to leave. I constructed a runic sigil incorporating the persons name, and a Raido rune to get him moving and whilst beating the drum, simply chanted: "Eni, mini, miny mo; Roger the lodger has to go!" He left six o'clock the next morning. A simple repetitive spell, in rhyme, is all it takes. One might chant: "Tip, tap, tob, I will get this job", just before an interview having combined the runes Eihwaz, Ansuz, Mannaz and Wunjo in a sigel drawn either on a rune stave carried in a pouch around the throat (the throat chakra relates to communications) or else painted below one's collar.

EXAMPLE CHARM FOR TRAVELLING

Safely I ride, safely I go,
Safe fare over road and foam,
Rightly beneath me all roads flow,
And whole come again to home.

Examples:

Bind rune for job: 𐌼 or ᛏᛘ

Suggested Runes: Eihwaz, Wunjo, Kenaz

Example spell in verse:

> *The wheat grows fair in fields all,*
> *The worker earns his/her pay,*
> *The crafter hears keen tools' call,*
> *I'll find fit work this day.*

For travel: ᛗᚱ or ᚱᛏ
Suggested Runes: Raido, Eihwaz, (laguz)

> *Gladly ride I roads aright,*
> *In faring fear no ill,*
> *By Sun's warmth and Moon's soft light,*
> *I fare, and find my fill.*

For healing: ᚼ
Suggested Runes: Uruz, Jera, Sowulo, Laguz

> *By healing honey, herbs of strength,*
> *By flood-tide's flowing roll,*
> *By depths of sea and heaven's length,*
> *Be body here made whole!*

For wealth: ᛉ
Suggested Runes: Fehu, Gebo, Kenaz

> *Fire of gold and field's grain,*
> *The gifts of good all glow.*
> *Upon my work shall weal rain,*
> *Wealth to my winning flow.*

For reconciliation or restoration of peace after a bad argument with someone you rather not lose permanently: ᛗᛏ
Suggested Runes: Gebo, Wunjo, Ehwaz

No strife nor storm shall struggle on,
Our troubled times are past.
Bring friendship's ale forth from store,
And love return at last.

For losing excess weight backing up a sensible diet. Suggested
Runes: ᛉ
Uruz, Raido, Nauthiz

Lithe and slender, slim and light,
As birch at break of day,
My body shape with shining might,
And drop all dross away.

Bind runes and charms can be constructed for any purpose,
one only has to think runically, translating the objective in
runic format.

RUNIC SIGILS

A sigil is an artificial symbol, representing anything from a
person or other living being to a situation or magical intention.
Sigil workings come in three phases:

1.   Creating the intention.
2.   Formulating the sigil.
3.   Activating the sigil.

Creating the intention. In all likelihood the intention already
exists, and might be anything from simple healing to an
adjunct to more complex magical matters. As with formulating
the purpose of a ritual the intention requires to be basic and
clearly accessible. It is symbolic of the will of the magician. A
sigil created to do ?... something is likely to be both harder to
control and less accurate of intent than is the case with a sigil
created to *do* a specific something. Be as exact as possible. The
idea is to think out as clear as possible a sentence representing
your intent, and then distort it in a sigel beyond any recognition

but as an image, the deep mind doesn't do words, it assimilates and interprets images.

Formulating the sigil. The best known, and most efficient, means of creating a sigil is to write a short statement of intent. This adds additional potency to the working and sends a clear message to your unconscious mind about the importance of achieving the goal.

For example: "I want five hundred pounds." This statement will, via its letters, form the building bricks which enable the sigil to be constructed. The script employed runes. Strangeness wonderfully affects the mind, and an unfamiliar script, so long as it is understood, increases potency. Begin by writing the statement down in runes: I want five hundred pounds.

I ᛈᚨᚾ☐   ᚠᛁᛟᛗ ᚺᚾᛏᚾᚲᛖᛗᚾ ᛈᛟᚾᛏᚾ

Next eliminate any letters which appear more than once:

I ᛈᚨᚾ☐   ᚠᛗᚺᛟᚾᛋᚲ ᚾ ᛈ

These remaining letters are now subjected to a personal design process to achieve the shape of the sigil.

During this process the mind of the magician is focussing unconsciously on the purpose for which the sigil is being created and imbuing it with that basic desire.

Once you have a functional symbol it can be empowered by embellishment to improve its appearance and appeal. Using these Runes this basic sigil can be created:

Activating the Sigil

Once created the sigil is now ready to be activated or charged, setting it about the business for which it was created.

The sigil draws upon the internal power of the operator by being set within the unconscious mind until such time as it is no longer of service, having completed the purpose for which it was intended. In order for this to be achieved the so-called 'psychic sensor' has to be tricked or bypassed. In other words once you've dropped the sigel in the well of the deep mind you must forget it. This is the most difficult part.

I suggest that you make the sigil as complicated as possible, so it will be difficult to remember what it looks like; create the sigel on paper, wood or painted it on a drum. Charge it with intent, use the runes which constitute the bindrune and create 'chaotic word' a total nonsensical sequence of letter, but either harmoniously or very raw and powerful. Chant this repeatedly whilst visualising energy going into the sigel. Store the energy and lock it in. Bury it, hide it, for nine days, this will enhance its power, providing you don't think about it. Then destroy it, unseen, this should set the energy contained as sigel free to do it's work.

Technology of Runic Magic.

We now turn to the Rúnatál section of the poem called "Hávamál" ("The Words of the High One") in the Elder Edda, which is spoken from the point of view of Odin himself. This poem is full of magical implications as we shall see when we look at the commentary.

RÚNATÁL

I wot that I hung on the windy tree
    nights all nine,
wounded by spear, given to Odin,
    given, self to myself,
on that tree of which no man knows
    where it rises from roots.

They comforted me not with loaf nor with horn,
    I pried below me,
took up the runes, screaming I took them,
    and fell back from there afterwards.

Fimbul-songs nine took I from the famous son
    of Bölthoru, Bestla's father,
and a drink I got of the dear mead,
    sprinkled from Óðroerir.

I took to waxing fruitful and becoming wise,
    and grew and held myself well,
Word led me from word to another word,
work led me from work to another work.

You should find runes and read the staves,
    mickle great staves,
    mickle strong staves,
    which Fimbul-Thulr stained
    and the Ginn-Reginn readied,
    and Hroptr risted.

Odin among Æsir, for the alfs, Dáinn,
    Dvalinn for the dwarves
    Ásvidhr for the etins,
    I risted some myself.

Know you, how to rist, know you, how to read?
know you, how to colour, know you, how to test?

know you, how to ask, know you, how to bless?
know you, how to send, know you, how to sacrifice?

Better 'tis unoffered than to be over-offered,
*   a gift ever looks for a gift;*
better 'tis unsent than over-sacrificed,
So Thundr wrote before the history of folks,
there he rose up when he came again.

I know a song no folk-ruler's queen knows,
*   nor human kinsman;*
"Help" it hight, because it shall help,
against sicknesses and hurts and sorrows full-wrought.

That ken I secondly which men's sons need,
*   they who wish to live as healers.*

That ken I third, if great need is on me,
*   chains against my foe,*
edges I deafen of shots against me,
*   that those weapons cannot bite.*

That ken I fourth, if are set upon me
*   bonds upon my limbs,*
so I chant that I can go,
fetters spring from my feet,
and chains spring from hands.

That ken I fifth, if I see the shot on the way,
*   throwing-spear in the folk's throng,*
it cannot fly so fast that I cannot stop it,
*   if I should sight it.*

That ken I sixth, if some thane would scathe me,
*   with a young root's wood,*
I will be hale but who says he hates me,
*   the harm eats him rather than me.*

That ken I seventh if I see high fire,
   in the hall around my seat-kin,
It does not burn so broadly that I cannot protect from it,
   I know how to sing that galdr.

That ken I eighth which is for all,
   a needful thing to take.
Where hate waxes among warriors' sons,
   I am able to quell it.

That ken I ninth if there is need for me
   to save my ship on the flood,
I can still the wind upon the waves,
   and make all the sea to sleep.

That ken I tenth if I see garth-riders
   playing aloft
I can work it that the wild ones fare (home),
   their shapes home,
   their souls home.

That ken I eleventh if I shall into battle
   lead old friends,
I sing under shield-rim that they fare with power,
   hale to the battle,
   hale from the battle,
   they come hale, where-ever.

That ken I twelfth if I see up on a tree
   a dangling hanged-man,
so I rist and in runes I colour,
   that so the man walks
   and speaks with me.

That ken I thirteenth if I shall cast water
   upon a young thane,
he shall not fall although he comes in battle,
   he shall not sink before swords.

That ken I fourteenth if I shall, before the folk,
  speak of the gods,
of Æsir and elves I can tell all,
  few unwise ones are so able.

That ken I fifteenth which Thjódhrørin sang,
  dwarf, before Delling's door.
Might sang he to Æsir power to the elves,
  understanding to Odin.

That ken I sixteenth if of a young maid I will
  have all heart and pleasure,
I turn the soul of the white-armed woman
  and wend to bed with her.

That ken I seventeenth that she will not scorn me
  for any other man -
this song remember you, Loddfáfnir,
  long will you lack it,
  but it will be good to you, if you get it,
  useful, if you take it,
  needful, if you receive it.

That ken I eighteenth which I will not make known,
  not to maid nor man's wife -
it is wholly better which one alone knows -
  that follows to the end of the song -
except for her who lies in my arms
  or be my sister.

Now are Hár's sayings spoken in the hall of Hár,
  all-needful for men's sons,
  un-needful for etins' sons.
Hail the one who spoke! Hail the one who knows!
  useful to the one who takes,
  help, to the one who heeds.

Translation from Old Norse by Dr Stephan Grundy PhD

RÚNATÁL
(COMMENTARY)

The version of this portion of the "Hávamál" given above clearly divides into at least two distinct sections, both of which are highly relevant for any understanding of Northern magic and runecraft. Our main concern here is with the list of charms, numbered 1 to 18, and the attitudes they reveal towards magical practices and techniques.

Most important magical teaching is to be found in the eighth verse.

> *Know you, how to rist, know you, how to read?*
> *know you, how to colour, know you, how to test?*
> *know you, how to ask, know you, how to bless?*
> *know you, how to send, know you, how to sacrifice?*

We've already remarked that eight is one of the principal magical numbers of runic magic, as the runes themselves are divided into aettir. We examine first, eight separate techniques which must be mastered by the rune-magician.

The first three, cutting, reading and staining. In order to cut runes you have to be able to read them, otherwise what you cut is gibberish.

The third technique, staining, was practised on many items, including standing stones, with traces of pigment enabling the original colouring to be reconstructed to red. Red is the colour of magic in many cultures, including our own. The Old Norse word *taufr* (talisman) and modern German *Zauber* (magic) come from a root meaning "to make red" (a development which may actually have come from the practice of staining runes). Thus the reddening of the staves, either with actual blood or the red paint symbolising it, is the specific action which loads them with power and gives them life, together with the chanting of the rune-name, which is the stave's breath as the red stain is its blood.

"Cutting" for magical purposes involved a concentration on the characters being cut to liberate their powers in the mind of the runemaster. It would have been a ritual in itself, with the runes being chanted to accompany the action: as the runes are scored in the earthly wood or stone, they are also cut in the World-Tree's roots, within the Well of Wyrd.

"Reading" involved knowing the runes and their correspondences, to ensure that what was being cut, be it for secular or magical purposes, was appropriate. The runic names, of course, were the most important of these. There is also a close correspondence to the modern concept of "reading", i.e. consulting for divination.

"Testing" referred to both the runes being cut and the person doing the cutting. The concept of magical initiation is as valid for a runemaster as for anyone else, and the knowledge imparted orally and in practical demonstration would only have been communicated to the worthy, who would have to "prove" their worth. One must have the ability to test someone before accepting them as apprentice. For this of course one must oneself be thoroughly versed in rune craft.

"Asking" can refer to the wisdom to ask the right questions of the runes, questions that can be answered within the scope of rune-casting and can also mean to "invoke". The Old Norse word used here, *bidja*, can also, in some contexts, mean to direct (cognate to English "bid"). In both cases, the skill required is that of understanding the runes, and the gods themselves and the function they are meant to fulfil; it is no use asking, or bidding, them against their nature.

"Blessing" has to do with the consecration of the symbols used, each of which, whilst a component, made up a magical whole. The word used here, *blóta*, can (and usually does) mean to sacrifice to the gods, or worship through sacrifice, but it also signifies hallowing or dedicating something to a specific end.

"Sending" refers to the actual activation of the completed rune charm, the making-it-do-its-job part of the process. Like any other form of spell, a rune charm needs to be directed, and whilst this could be assumed in the purposeful way in which the charm is created, the extra reinforcement of the completed charm ensures an inescapable certainty equivalent to a statement of intent.

"Sacrificing" refers to the process by which the strength of the run ester is poured into the runes: the more you give, the stronger your results are. The runes can be made to be work by types of physical "sacrificing", as well; for instance, by carving and staining them lightly, then scraping them into a drink or other magical substance meant to be empowered by them.

The eighteen "Hávamál" charms form the longest extant list in any recorded sequence. Because of its range the "Hávamál" list is the best to work with at this stage. If we examine each in turn we will discover its use and, in many cases, some hints as to its application.

1] is an uplifter, bringing cheer back during periods of illness or depression.
Runes applied to activate this charm are: Fehu, Inguz and Wunjo.
2] is a charm for leech craft, and complements the first. Its use will enable the user to become skilled in the healing arts. Runes: Uruz, Sowulo, Laguz.

3] This is a battle charm, preventing adversaries from harming the possessor in conflict. It would most likely have been worn as a talisman, possibly in such a way, and of such material, to make its presence obvious, thus providing a psychological advantage to the wearer. Runes: Thurisaz, Isa, and Nauthiz.

4] is a charm which, because it relates to freeing oneself from bonds, does not actually require cutting. The runes could be employed in other ways, as sound vibrations and mental tuning devices, in addition to being cut, stained, etc. Odin uses the verb "galdra", specifically implying that this enchantment is meant to be sung. Runes: Ansuz, Fehu, Eihwaz.

5] turns a flying missile from its harmful course and sends it harmlessly to earth, assuming it's spotted in time. This is a protective devise, to neutralise malicious magic and for psychic self defence. Runes: Kenaz, Isa, Raido, Jera, and Dagaz

6] This is a counter-magic spell. Using this charm will invoke the much-vaunted Law of Return, by which a thwarted spell rebounds upon its caster. Runes: Hagalaz, Jera, Raido, Nauthiz, Thurisaz.

7] A fire-extinguisher, we may interpret this as a spell to intervene in a situation where a cool head is needed, for example to avid a violent confrontation. Runes: Isa, Laguz, Nauthiz, Gebo, Wunjo.

8] Strife between folk was one of the main hazards of Germanic life. The custom of feud, only somewhat mitigated by the processes of Thing (the judgement assembly) and the paying of weregild, brought many great heroes to grief. The Germanic people were (and still are) a bunch of proud, touchy individualists. This charm was, and is, indeed a very needful one to know. Runes: Gebo, Mannaz, Wunjo.

9] For a seafaring people such as the Vikings, this spell, which kept ships afloat and brought them to safe harbour, was another important one. It provides an instance of weather magic in a specific environment, and as the North Sea was the main adversary of any voyage from the Northern lands was an essential for seafarers who wanted to see home again. Runes: Gebo, Raidio, Ehwaz, Laguz.

10] The translation here is ambiguous, with some scholars citing ghosts and others citing witches. Both of them were known to ride on top of house-roofs at night. Northern myth and magic played a formative part in the creation of the Medieval witch-myth, and broomstick-riders spring readily to mind. There is also an implication of shape-shifting into bird form, as both the hides and the souls of the garth-riders have to be sent home. Runes: Ehwaz, Dagaz, Laguz.

11] Another spell for protection in battle. The fact that the spell is chanted does not necessarily mean it wasn't written first. It is chanted behind the shield, and could well form part of the back of the shield. Tacitus mentions that Germanic warriors preparing for battle had a special cry, called the "baritus", from which they could read how the battle would turn, and comments that, "the object they specially seek is a certain volume of hoarseness, a crashing roar, their shields being brought up to their lips, that the voice may swell to a fuller and deeper note by means of the echo". Runes: Ansuz, Sowulo, Algiz

12] This is real necromancy. Hanging was the primary method of sacrifice to Odin, as well as a means of execution. Here we have runes both written and stained to bring the victim back to life. Communicating with the dead. Runes: Hagalaz, Kenaz, Eihwaz, Ansuz.

13] The act of pouring water would have been the "sending" part of this charm. This process, magically recreating the name-giving day in which the nine-day-old infant received name, doom, and gifts, allows the setting of an örløg ("fate") against harm in battle upon a young warrior. This can be repeated in a ritual where one accepts a new magical name, for example as part of an initiation ceremony. Runes: Perthro, Laguz, Berkana, Othala.

14] This charm tells us of the close ties between Germanic religion and Germanic magic. To be able to work in the worlds of the astral means to know the mighty ones who dwell there, when to call on them, how to befriend them, and how to read the messages they send us. Both the Gods and the elves, and other beings are important for the rune magician to deal with and understand. Runes: Eihwaz, Mannaz, Ansuz.

15] This is an unusual charm in that our narrator (Odin) acknowledges it came from another. Thjodrerir is a dwarf who doesn't appear elsewhere in the Elder Edda. A literal translation of Delling is "shining"; his door is the dawn light. Dwarves traditionally turned to stone at dawn; but in addition to Thjodrerir, the charm is also of benefit to the Gods and elves and Odin, and presumably human beings as well. Runes: Raido, Kenaz, Sowulo

16] This charm is written from a male point of view, but can work for either sex. This is making the desired one do the right thing by your inclinations, and shows the role sexual magic had to play in Northern society. This is the charm for getting your lover. Runes: For a male: Kenaz, Inguz, Nauthiz. For Female: Thurisaz, Kenaz, Nauthiz.

17] And this is the charm for keeping your lover once you've used 16 to get her (or him). Again, singing doesn't negate the possibility that the eight stages described above are implicit in creating the charm. It simply means that its oral recitation activated the charm.
Runes: Gebo, Ehwaz, Wunjo.

18] This is the hardest of the eighteen to interpret or comment upon. The only members of the opposite sex to whom it may be told are blood siblings or committed mates. This implies it has a personal power which remains personal and powerful whilst its secrecy is maintained. Its purpose is probably to promote a personal union, either with the beloved or with other aspects of oneself. This is a personal combination of runes which is a secret for each individual to discover.

Any act of active magic is best done in a sacred and secure space, assuring protection from all negative possible influences both from within as well as from outside.

For this purpose, a simple format has been developed and has now been in use for a number of years, proving its effectiveness.

RUNIC CIRCLE

All your rituals should begin with the casting of a runic circle, both to ward yourself and to draw up and concentrate the might of the runes for your use. Regular casting of the circle has several effects. Firstly, when you start magical studies, you start giving off more energy and, at the same time, become more sensitive – you shine through the Inner Planes as an unwarded beacon, which can draw all sorts of nastiness and unwanted attention. The regular casting of a runic circle protects you from such things. Also, as you practise it, you will become more and more used to the sounds, shapes, and magical feelings of each of the runes.

Begin by standing with feet together and hands at your sides, facing north. Raise your wand, ritual knife, or your strong hand and trace the stave-shape of fehu, chanting "Fehu, fehu, fehu" as you see a bright beam of red might springing from you and shaping the stave to burn in the air before you. Make this a big, bold gesture, with the bottom of the stave at groin-level and the top at head-level - you are putting your whole strength into it, so don't hold back or be cowardly about it. If circumstances do not permit you to chant at the top of your lungs, intensity must do in place of volume.

Turn to the northeast with your arm still held straight out, seeing a line of red might forming an arc from north to northeast, trace Uruz and chant it thrice. Continue through the first aett thus, until you are surrounded by a burning circle marked at the eight directions with runes: Thurisaz eastward, Ansuz south-east, Raidho south, Kenaz south-west, Gebo west, Wunjo north-west.

Finish the circle at the north. Not losing sight of the fehu rune burning in front of you, send your energy through its heart to trace a Hagalaz rune behind it as you chant "Hagalaz" thrice. Do the second aett in the same manner, as a second circle outside the first. When you reach the north again, still aware of the fehu and Hagalaz burning before you, send your energy out one more step to shape Tiwaz as you chant its name thrice, and continue with the third aett.

When you have closed off the third circle, stand for a moment, seeing the threefold ring of runes about you. Then trace the sign of the Hammer seeing a huge Hammer-sign of blazing lightning spring forth outside the Fehu-Hagalaz-Tiwaz triad as you chant, "Hammer ward me, where-ever I go!" Repeat the Hammer-signing and the chant in each of the eight directions.

When you have reached the North again, reach over your head and trace a deosil swastika in glowing white light (or, if you have not yet come to emotional terms with the misuse of this mighty sign by the Nazis, another Hammer-sign), chanting, "Hammer shall always hallow my ways!" See the sign spinning faster and faster, funnelling a flood of brilliant whiteness down into your red circle. The crown of your head should tingle with this energy.

Below you, trace a widdershins swastika (or Hammer) in glowing black light, chanting, "Hammer shall always hallow my ways!" See the sign spinning faster and faster, funnelling a flood of brilliant blackness up into your red circle. The soles of your feet should tingle with this might.

Lift head and hands so that you stand in the shape of an Algiz-rune and cry, "Above me Ásgarðr's awesome might!"

Spread your feet shoulder-width apart and cry, "Roots below in Hella's realm! Bring your feet together and lower your arms until you stand with them stretched out in the shape of a cross. Chant, "Rune-might ringed around me, rune-might roar within me. Rune-might wards me, works my will where-ever I shall go!" Stand still for a moment, eyes closed if needful, seeing the whole ring as a burning triple wheel whose spokes are the eight runic triads and Hammers, whose axle is you yourself between the whirling swastikas. The rune ring is now set and will move everywhere with you.

There is a simpler form of this ritual, chant and trace with your magical tool or your index finger, the runes in a complete circle around you, starting with Fehu in the North and ending with Dagaz. Then make the Hammer sign on the four quarters and invoke the protection of Thor. For example. "Hammer in the North hallow and protect this space, hammer in the East etc etc." When working in the Northern tradition always start North, if you use an harrow or altar place this in the North.

The regular practice of either of these rituals will put you in close touch with the runes and give great protection in every given situation.

Once the circle has been set up one is totally safe to work magic, healing and/or divination.

RUNES FOR ACTIVATING THE CHAKRAS.

From virtually every mystery tradition, we learn about psychic energy centres or vortexes in the physical/emotional/and mental bodies, interconnecting these.

There are many different theories as to how many of these centres there actually are. The traditional view promoted by Theosophy in the last century was that there were seven Chakras. Some other esoteric traditions postulate that there are as much as much as twelve, some of them are not located on the Earth plane. We are going to work with eight; the seven well known ones and in addition the one below the feet, the tap-root chakra, which for the sake of not altering the conventional numbering of the seven chakras, I have designated the Zero chakra.

This exercise is experimental, therefore feel free to alter the runic order relating to the chakras in accordance with your insight.

Zero chakra: Located beneath ones feet extending as a taproot into the Inner earth. The Rune Hagalaz allows you to access this realm and draw energy form there.

First chakra: The base chakra associated sense of identity and security.   Isa is one of the runes which can be used to activate this chakra.

Second chakra: Traditionally associated with sexuality, pleasure and procreation, also in a creative sense. Nauthiz is most fitting to be used.

Third chakra: The chakra of power and will.   Eihwaz or Jera either will do, even better both as a bindrune, Eihwaz as the spine and Jera circling around this, somewhere around the waist.

Fourth chakra: The heart chakra. To activate this, Sowulo is a mighty force.

Fifth chakra: Throat chakra, the chakra of communications: Gebo as well as Ansuz are two of the best runes to use to activate this chakra and improve communication on all levels.

Sixth chakra: The Third eye. Inguz chanted and projected into this space will develop clairvoyant abilities, if present.

Seventh chakra: The crown. Dagaz, as the rune of enlightenment, will facilitate understanding and processing of information obtained from higher levels of consciousness.

There are other runes which are very good in connection with various charka's but I suggest you find these for yourself by experimentation, as long as you are in a secured space you will unlikely come to any harm. This is an exercise for which one may create a daily schedule and keep records of results obtained, changes of consciousness and/or dream space. Also these chakra exercises can improve one's health and personal stamina a great deal.

First, one must be able to *feel* these chakras: where they are situated and whether and in which direction they are spinning. Exercises for this can be found in many books for beginners; especially recommended is *Wheels of Life* by Anodea Judith, someone I have worked with, who is both knowledgeable and reliable.

Practice the chanting, singing and whispering of runes, see them taking shape whilst you doing this. Develop visualisation of runes; visualisation exercises can also be found in most esoteric books for beginners. Ones you master these two disciplines, chant and visualise the rune either from above one's head or from under one's feet, and draw this up to the chakra you're working with. Feel the rune settle here, and leave it for a while, note what you feel, both physically and emotionally. work steadily on this, the results will be interesting and worth while.

GUIDED MEDITATION TO HELLA'S REALM

From the Caballa we have become familiar with the concept of 'path' workings. This means literally, travelling on the interconnecting paths given in the traditional Cabbalistic teachings on their version of the Tree of life.

Well, we've got a Tree of life of our own, and within the Norse mythology is given an exact description of the nine different realms and where they are located. Why would that be, if not to suggest that in at least some of the shamanic traditions of the North these paths were also explored?

Incidentally, I must state categorically that the further or deeper one gets into a mystery system, whatever it may be, Cabbala, Seidhr or Voodoo, the more one realises the similarities suggesting that their is a universal principle behind them all and that they are localised versions of Mystery teachings of an Higher order. This for the benefit of people who have known me to work exclusively in my own ancestral and ethnic traditions. All ethnic traditions have similarities and possible similar origins in a different time span, at lest from a linear perspective, they can and will interface in the forthcoming Age of enlightenment.

PATHWORKING TO HELLA'S REALM

First of all this is not ever to be done alone. Always have a trusted companion with you and preferably a tape recorder running. The pathworking below is used to obtain communication with the shamanic underworld. i.e., the realm of the dead. The reason that one is warned not to undertake this alone is that the experience one undergoes can be extremely intense, and recollection may be very difficult after one returns. It is not actually dangerous in the physical sense unless one has an already existing health problem such as a weak heart, but psychologically it is dangerous to venture out in such uncharted territory without a guide.

Get as relaxed as possible, by whatever means you are used to, if not trained at all in this, join a meditation centre or a Yoga school. Once you are totally relaxed, visualise yourself in an environment in nature, a place if possible from memory where you really felt good.

Here is the text to be read out by you companion, while you follow the instructions.

"You stand outside with your eyes closed. The breeze rustles softly through your hair; you smell the green scents of grass and leaves, feel the steady might of the earth beneath your feet.

Now you open your eyes. You are standing in a place well known to you, a natural stead, far from cities and roadways, where you feel at home and at one with the earth.

After you have stood gathering your strength for a few moments, you begin to walk. The way leads you on towards a woodland - grass rising to bushes at the path's edges, bushes to small trees, small trees to greater trees. The sunlight shines through the branches arching above the way, dappled by leaf-shadows that shift with every breeze. As you go farther, the trees become older and thicker, the pathway darkening beneath the shade of their great limbs and heavy leaves.

Ahead of you lies a clearing. In the middle of the clearing rises the greatest tree of all - an ancient yew, red berries shining like drops of blood against its dark needles. It is so tall that you cannot see its crown, only the grey clouds streaming about it like tails of foam; its roots rise higher than your head, and sink deeper into the earth than you can guess. You walk about the tree to the north. There, beneath one of the roots, you see a dark cave. You step within, breathing in the musty scent of the yew's bark as you make your way downward through the blackness.

As you go, you begin to hear the clanging of metal - of hammer striking anvil, steel ringing from stone, iron clashing with gold. The sparks glimmer through the darkness all about you, flying from the anvils of the Swart-Alfs as they forge swords and helms, rings and spears and shield-bosses; the eyes of the dwarves glow red. You follow your path along, downward and to the east.

The wind from the east is cold and raw with the storms of spring, slapping icy against your face. Far off, you hear the thundering of the giants and rime-thurses casting boulders through the air, the sound of rocks cracking beneath their feet as they stride through the wastelands of Jotunheim; their shapes are shadows among the crags. But you are on the path that keeps you safe; you follow it along, downwards and to the south.

The wind from the south is dry and burning-hot, crackling with sparks from the darkness as it sears your face. The leaping flames light the southern way - the all-devouring flames of the Muspilli, that will eat the world at the end of time; the sparks from Muspellheim hiss about you on the hot desert wind, and you can hear the fiery laughter of Surt's sons. But you are on the path that keeps you safe; you follow it along, downwards and to the south.

The wind from the west is damp and cool, salty with the smell of the sea and rich with the scents of moist earth. From western Vanaheim, you hear the crashing of the waves against the shore and the soft evening lowing of cattle; glimmers of gold glint from the dark earth about you. But still you follow your path, down and to the north.

Now the shadows on your way grow deeper. From the darkness, you hear the running of waters - large streams and small, some winding slowly and some rushing in torrents from the fells. You must step over the lesser streams, wade through the greater. Some are hot and stink of sulphur; some freeze your feet with the cold of mountain ice, but all must be crossed.

Ahead of you, you hear the rushing of a much greater river, like a stormwind through trees. Beneath the sound of its frothing waters, you hear the clashing of weapons - sword on sword, spear on shield, steel clanging against rock, blade ringing against iron blade. The river runs sharp-edged with weapons, beating the lead-grey waters to froth in white rivulets about the harsh rocks. It is too deep and fierce to ford; you must follow the path on to where a huge boulder stands shadowy by the side of the steel bridge.

A shape rises beside the boulder - a great woman's shape, roughly wrought in craggy grey stone. This is Modhgudhr, the giant-maid who wards the bridge to Hel. She speaks to you in a deep harsh voice like rock grating on rock, asking who you are and what you are doing here; you must answer her.

When you have answered the giant-maid's challenge, you stride forth over the steel bridge. Cross boldly and it will be a wide and easy way; should your feet falter, it will be narrow and keen as the edge of a sword. Your tread rings from the steel as if a whole host rode with you - but a host of the dead would not make it ring so; it sounds only beneath the feet of the living.

On the other side, a huge black wall rises before you, its shadow falling cold and dark across your face. You stand at the northern march of Hel; the wall wards the way. You start walking towards the east, following the slow dark curve of the wall.

At the east is a huge gate, black oak bound with bars of black iron. As you gaze at it, it swings slowly open: the land of Hel, of night and mist, lies before you".

Here your companion asks you questions, "what do you see, is there a path, is there an animal, human, trees, flowers?" You keep voice contact with your companion at all times, do not allow yourself to drift off. When you feel you've had enough, signal to your companion by a before hand agreed signal with the hand or the head.

To return; your companion reads:

"The gates of Hel swing closed behind you. You follow the path back around to the north, crossing the ringing bridge. Moðguðr lets you pass freely; the path turns upwards and towards the west, over the many waters you crossed before. You leap the smaller and ford the greater, following the path about and up. The salty damp winds of Vanaheim blow against your face; upward and south, you pass again through the burning wind of Muspelheim; upward and east, through the raw harsh winds of Jotunheim. Now you hear the hammering of the dwarves again, steel and stone, and see the glowing fires of their forges. Above you is the cave-mouth where you came in, a ring of light beneath the black roots of the tree; you climb upward, stepping out from under the arching root and blinking against the brightness of day. Slowly you follow the path back through the woods - old trees giving way to younger trees, younger trees to saplings, saplings to bushes, bushes to grass - till you reach the stead where you began. You stand there for a few moments, breathing in the soft breath of the wind, feeling your feet firmly rooted in the earth, your soul firmly rooted within your body, your self whole and one within the Middle-Garth's Ring".

APPENDICES

PRONUNCIATION GUIDE

RUNE-NAMES:

The rune names given here are in Proto-Germanic. Pronunciation is relatively simple: there are only five vowel sounds, and the consonants, with the exceptions below, are as in Modern English.

a — as in "father"
e — *ay*, as in "day"
i — *ee*, as in "speed"
o — as in "home"
u — *oo*, as in moon
dh — a soft *th*, as in "leather"
g — always hard, as in "give"
h — may be heavily aspirated, almost as a *ch*
j —always pronounced as y
k — always a hard sound; no soft *c* exists
r — trilled
th — as in "thorn"
z — always buzzed, halfway between an *r* and a *z*

OLD NORSE

a — as in "law"
á — as in "father"
e — as i in "gin"
é — as ay in "day"
i— as in "is"
í — *ee* as in "speed"
o — as in "omit"

ó — as in "owe"

ø, ö — as in "not"

u — *oo,* as in soot"

a — *oo,* as in "droop"

y — *u,* as in French "*tu*"

ý — *u,* as in German "*Túr*"

ae — *e* as in "get"

au — *ou* as in "house"

ei — *ay* as in "day"

ey — as *ei*

dh — a soft *th,* as in "leather"

f — pronounced as English f initially, as English v in
   medial and final positions.

g —hard as in "give"

j — always as English y

ng —as in "sing"

r — trilled; r on the end of the word is not given an extra
   syllable.

s — always voiceless, as in "blast"

th — as in "thorn"

z — pronounced as *ts* (as in German)

ANDREW CLIFTON'S UNSCIENTIFIC RUNIC PRONOUNCIATION GUIDE

Note, in this pronunciation guide, the syllable which is normally
given the greatest stress is shown in bold.

RUNE NAME PRONUNCIATION NOTES

FEHU	FAY-HOO	*OO as in moon*
URUZ	OO-ROOZ	
THURISAZ	THUU-RI-SAHZZ	*UU as in murder*
ANSUZ	AHN-SOOZ	
RAIDHO	RYE-DOUGH	*an easy way to remember this one!*

KENAZ	KAY-NAHZZ	
GEBO	GAY-BOW	*'bow' as in 'bow and arrow'*
WUNJO and	WOON-YOH	*'yoh' rhymes with bow dough*
HAGALAZ	HAH-GAH-LAHZZ	
NAUTHIZ	NOW-THIZZ	*the th, is hard, as in 'thorn'*
ISA	EEE-SAH	
JERA	YEH-RAH	*'yeh' as in"heh'*
EIHWAZ	EYE-WAHZZ	
PERTHRO	PER-THROW	
ALGIZ	AHL-GEEZ	*g hard, as in golf… not 'Jeez'!!*
SOWULO	SO-WILL-OH	
TIWAZ	TEEOO-WAHZZ	*run the ee and oo sounds together*
BERKANO	BEAR-KAH-NOH	
EIHWAZ	EH-WAHZZ	*'eh' should rhyme with hay, nay, say*
MANNAZ	MAH-NAHZZ	
LAGUZ	LAH-GOOZ	
INGUZ	EENG-OOZ	*'ing' as in drinking*
OTHALA	OH-THAH-LAH	
DAGAZ	DAH-GAHZZ	

RECOMMENDED READING LIST:

Northern Mysteries and Magick: Freya Aswynn   Publisher Llewellyn

Futhark: Edred Thorsson Publisher Redwheel/Weiser

Runelore: Edred Thorsson Publisher Redwheel/Weiser

Teutonic Magic: Kveldulfr Gundarsson, Thoth Publications, Loughborough.

Teutonic Religion: Kveldulfr Gundarsson, Thoth Publications, Loughborough.

Wotan: The Road To Valhalla: Kveldulfr Gundarsson, Thoth Publications, Loughborough.

Taking up the Runes: Diana Paxson Publisher   Redwheel/ Weiser

Runic States: Kevin Steffen Publisher Eschaton.

Other titles available from Thoth Publications.

TEUTONIC MAGIC
by Kveldulf Gundarsson

Tales of Teutonic magic have thrilled the world for centuries. Now bringing together the dark stuff of sagas, Kveldulf Gundarsson reveals the personal magical path behind the legends and explains the practical techniques of the Northern Tradition. Gundarsson is a well known expert on the esoteric lore of the Teutonic people. This, his first book Teutonic Magic brings a lifetime's worth of expertise to the subject. Blending historical lore with practical experience of esoteric skills, Gundarsson presents the reader with a spiritual path walked by the ancient and dark age Germans.

He explains the magical writings of the Teutonics, including the uses of each of the runestaves, the laws of magic in the Northern tradition. He provides the key to unleash the awesome might of the Northern magical tradition.

This is not a dry academic book. Gundarsson's writing flows like the sagas themselves, covering subjects such as Norse deities and rituals. It describes the structure of the spiritual realms in which the Norse Magician would walk. It is no wonder that this book is considered the classic text book of anyone who would study Runes or the Northern Tradition.

ISBN 978-1-870450-22-5

PRIESTESS: THE LIFE AND MAGIC OF DION FORTUNE
by Alan Richardson

Dion Fortune was the pen-name of Violet Firth, one of the most luminous and striking personalities of the twentieth century, Womanhood's answer to Aleister Crowley, and quite possibly the Shakti of the Age.

This new, revised, expanded and beautifully-written edition tells the full story of a woman who hid behind a veil of secrecy and who became a cult figure in the years after hew death in 1946. A brilliant writer and pioneer psychologist her whole life was devoted to living out an eternal Myth in a story that can be told in terms of Virgins and Dragons, Moons and Oceans, and the spirit of the land itself.

As a powerful psychic and medium, obsessed with the study and practise of Magic, and a high-grade initiate within the Hermetic Order of the Golden Dawn, her career was never entirely in this world, and her companions not always human. In her own eyes at least she was a Priestess, a channel for the Great Goddess, an exponent of the time-lost Mysteries of Women long before the present generations of feminists and goddess-worshippers where ever born.

From her birth in Llandudno, through her years in the drowned lands of Somerset, Alan Richardson unfolds the luminous and very moving patterns of her life: her early career as a psychoanalyst, her nervous breakdown, her time as a Land Girl and her developing psychism; her memories of past lives on Atlantis and relationships with Inner Plane beings who have an evolutionary interest in our world; her romance with a man she believed to be non-human and her fraught marriage to a doctor whom everyone knew as Merlin; the foundation of her own group devoted to bringing through the Western Mysteries at a time when few people knew that there was such a thing; her occult battles against the Nazis and fellow magicians – and the start of her long, hard, and always stormy journey into the Otherworld, toward the heart of the Goddess that she saw as sleeping within the Earth itself, and who needed awkening...

ISBN 978-1-870450-11-9